INSECT PHYSIOLOGY AND ECOLOGY

Edited by **Vonnie D.C. Shields**

Insect Physiology and Ecology

http://dx.doi.org/10.5772/67619
Edited by Vonnie D.C. Shields

Contributors

Samir Touaylia, Vincenzo Fogliano, Daylan Tzompa Sosa, Womeni Hilaire, Tiencheu Bernard, Alvaro Castañeda Vildozola, Mario Alfonso Urias López, Luis Martin Hernandez Fuentes, Carlos Rosales, Miroslav Kostic, Sladjan Stankovic, Harlan Ronald Storti Rodrigues, Ronald Rodrigues Guimarães Júnior, Ronald Rodrigues Guimaraes, Tania Rosas-Pérez, Arturo Vera-Ponce De León, Shamayim T. Ramírez-Puebla, Reiner Rincón-Rosales, Julio Martínez-Romero, Michael F. Dunn, Esperanza Martínez-Romero, Monica Rosenblueth, Sneha Vissa, Richard Hofstetter, Ramón Rebolledo

CBS, Edition **2018**

Published by InTech

Janeza Trdine 9, 51000 Rijeka, Croatia

Publishing Process Manager Maja Bozicevic

Technical Editor SPi Global

Cover Designer InTech Design Team

Additional hard copies can be obtained from orders@intechopen.com

Insect Physiology and Ecology, Edited by Vonnie D.C. Shields
p. cm.
Print ISBN 978-953-51-3033-8
Online ISBN 978-953-51-3034-5

Contents

Preface

This book contains chapters focusing on the theme of insect physiology and ecology. Chapter contributions fall under the following headings: I. "Morphology, Ecological Importance, and Distribution," II. "Symbiotic Relationships," III. "Resistance and Defense Mechanisms," and IV. "The Edible Insect: Used as Food Sources." In the first section, the authors discuss the morphology, ecology, and distribution of water beetles, dung beetles, weevils, and tabanids. In the second section, the authors review insects, such as scale insects, bark beetles, and ambrosia beetles and their microorganismal interactions with bacteria, fungi, or mites. In the third section, the authors present insecticide detoxification and insect defense mechanisms against infections. In the last section, the chapters cover insects as sustainable food. This book targets a wide audience of general biologists, as well as entomologists, ecologists, zoologists, virologists, and epidemiologists, including both teachers and students in gaining a better appreciation of this rapidly growing field.

In Chapter 1, Section 1, "Morphological, Ecological Importance, and Distribution," "Thoughts on Water Beetles in a Mediterranean Environment," Samir provides data collected from a survey on water beetles in freshwater ecosystems in the northern part of Tunisia. The author provided a list of 123 species as a result of monthly collecting surveys over the course of a year.

In Chapter 2, "Dung Beetles of Chile, with Emphasis in La Araucania Region," Ranz et al. examine dung beetles or scarabs, as they provide a role in the decomposition of manure from livestock, as well as the recycling of soil nutrients, stimulation of plant development, spreading of seeds for reforestation, biological control of parasite load in manure, soil aeration, help in pollination, and maintenance of ecological balance. The authors provide details with respect to the diversity and distribution of these species.

In Chapter 3, "Weevil Borers in Tropical Fruit Crops: Importance, Biology and Management," Hernández-Fuentes et al. examine weevils as an economically important group of coleopteran insects. The authors consider the main species of weevils, such as the big avocado seed weevil, small avocado seed weevil, avocado branch weevil borer, avocado weevil, avocado stem weevil, guava weevil, coffee berry borer, and diurnal weevil, with respect to fruits, such as avocado, coffee, guava, and anonas. The authors provide a description, distribution, list of host plants and biology, damage, and control methods for each of these species.

In Chapter 4, "Tabanids in South America," Guimarães et al. provide information on the taxonomy, morphological data, distribution, and bionomy of tabanid species in South America. These insects are the largest blood sucking Diptera and are known for their painful sting. They are vectors of several helminths, viruses, bacteria, and protozoa that can affect humans, as well as wild and domestic animals.

In Chapter 5, Section 2 "Symbiotic Relationships," "The Symbiome of *Llaveia* Cochineals (Hemiptera: Coccoidea: Monophlebidae) Includes a Gammaproteobacterial Co-symbiont *Sodalis* TME1 and the Known *Candidatus* Walczuchella monophlebidarum," Rosas-Pérez et al., review beneficial associations of scale insects with symbiotic microbes.

In Chapter 6, "The Role of Mites in Bark and Ambrosia Beetle-Fungal Interactions," Vissa and Hofstetter explore the complex interactions that insects share with mites and fungi. They provide a detailed review of specific beetle-fungal and mite-fungal associations, mutualistic and antagonistic effects of these fungal relations, and ecological and evolutionary consequences of beetle-fungal-mite relationships within the host complex.

In Chapter 7, Section 3 "Resistance and Defense Mechanisms," "Role of Carboxylesterases (ALiE) Regarding Resistance to Insecticides: Case Study of Colorado Potato Beetle (*Leptinotarsa decemlineata* Say)," Stankovic, S. and Kostic address details on the mechanisms by which the Colorado potato beetle has been successful in developing resistance against new insecticides by degrading or detoxifying toxins more quickly than susceptible insects or by using higher levels or more efficient forms of specific enzymes, resulting in a decrease in the duration and intensity of exposure and the likelihood of lowering the probability of a lethal outcome.

In Chapter 8, "Cellular and Molecular Mechanisms of Insect Immunity," Rosales assesses challenges that insects face when they are exposed to microorganisms (i.e., bacteria, viruses, and fungi), as well as parasites on a daily basis. The author discusses innate immunity, including both cellular (i.e., involving phagocytosis of bacteria and encapsulation of parasites mediated by hemocytes) and humoral or systemic responses (i.e., secretion of soluble antimicrobial peptides into the hemolymph). These authors include a description of specific receptors for sensing infection and the signaling pathways that activate genes for the production of antimicrobial peptides, as well as recent advances in insect antiviral immune responses.

In Chapter 9, Section 3 "The Edible Insect Used as Food Sources," "Potential of Insect-Derived Ingredients for Food Applications," Tzompa-Sosa and Fogliano study how insects can be used as sustainable and efficient protein and lipid sources, with respect to their nutritional value for human consumption. The authors include information with respect to insect processing, protein and lipid extraction, perspectives of food applications of insect-derived ingredients, and information on the chemical and physical characteristics of insect proteins and oils that will be useful in assessing possibilities of their use in different food applications.

In the last Chapter, Chapter 10 "Entomophagy: Insect as Food," Bernard and Macaire discuss the benefits of edible insects as rich sources of protein, unsaturated fatty acids, minerals, vitamins, and carbohydrates, for humans, as well as for providing feed to animals. The authors comment, in addition, how edible insects have valuable medicinal, commercial, and ecological importance and can potentially replace conventional animal sources.

I wish to thank InTech Open Access Publisher for initiating this book project and inviting me to serve as editor. I would like to recognize the Publishing Process Manager, Maja Bozicevic, assigned to the task of publishing this book and for guiding me through the process. I would also like to acknowledge all the authors for their hard work in submitting and editing their contributions. Lastly, I wish to express a special thanks to my husband, Dr. Thomas Heinbockel, Professor and Director of Graduate Studies, Department of Anatomy, Howard University, College of Medicine, and our son, Torben Heinbockel, for their patience and understanding in the last year when I was working on this book project.

Dr. Vonnie D.C. Shields
Associate Dean, Jess and Mildred Fisher College of Science and Mathematics
Professor, Biological Sciences Department
Towson University, Towson, Maryland, USA

Section 1

Morphology, Ecological Importance, and Distribution

Chapter 1

Thoughts on Water Beetles in a Mediterranean Environment

Touaylia Samir

Additional information is available at the end of the chapter

http://dx.doi.org/10.5772/66639

Abstract

The chapter provides data from a survey carried out on water beetles in various freshwater ecosystems in Tunisia as a Mediterranean country of considerable diversity. Studies dealing with these insects are fragmentary not only in comparison with the European fauna but also in comparison to other zoogeographical areas. A compiled checklist of beetle species collected from Tunisia is given with an insight on new recorded species. Diversity, altitudinal distribution, and geographical pattern of water beetles in Northern Tunisia are discussed with regard to other Mediterranean areas. They include various chorotypes related to the history of the Mediterranean basin. Several species are threatened and require conservation. According to the criteria of the IUCN, several water beetle species can be included in the list of threatened species.

Keywords: water beetles, diversity, phenology, biogeography

1. Introduction

Water beetles are holometabolous insects characterized by a strongly sclerotized body with the forewings hardened into elytra [1]. They occur in a wide variety of habitats, living in virtually every kind of fresh- and brackish-water habitat, from the smallest ponds to lagoons and wetlands and from streams to irrigation ditches and reservoirs [2]. They exhibit high species richness in the Mediterranean area and are primarily found in the ecotone between land and inland waters [3]. They are of great ecological interest as bioindicators of the quality of limnic ecosystems, the type of water, and habitats in danger [4]. Tunisia, a Mediterranean country, has important water beetles diversity. Studies dealing with these insects are fragmentary not only in comparison with the European fauna but also in comparison to other zoogeographical areas [5–16]. Synonymy of species is established by following the "*Catalogue of Palearctic*

Coleoptera, vol. 2" edited by Löbl and Smetana [17]. Abellán et al. [2] developed a system for ranging species according to their conservation priority or vulnerability on local, national, and global scales. IUCN [18] also gave rigorous criteria and categories that should classify the species according to extinction opportunities for a given period. Of particular interest, Northern Tunisia has endemic biota under increasing anthropogenic threats, where several species require conservation [19].

The fauna of North Africa could be considered as originating from the passage of Euroasian species to the African continent as a result of plate tectonics, leading to the connection of the two continents. The Mediterranean and the Atlantic were later connected (in the Pliocene), thereby isolating the two continents [20]. Water beetles of Tunisia include various chorotypes related to the history of the Mediterranean basin of which Tunisia is a part. The northern part of Tunisia includes two mountain ranges: the Tell (Kroumir and Mogods Mountains) and the Dorsale (Châambi Range reaching the Cap Bon Peninsula) [21]. Tunisia has a humid to Saharan climate. The humid area is limited to the Kroumir Mountains [22]. Annual rainfall decreases from the north to the south, with most of the rainfall in winter [23]. Water resources are unevenly distributed within the country; the northern part, covering an area of only 17% of the territory, has 60% of the total water resources [24]. This highly influences the water permanence and, therefore, the phenology, composition, and distribution of the water beetles' communities. We analyze the faunistic, chorological, and phenological aspects of the aquatic coleopteran species in the study area.

As the species distribution is determined by a set of ecological and historical filters acting on several spatial and temporal scales [25], we analyze the assemblage of aquatic beetles in response to environmental variables characterizing the explored streams.

Understanding patterns in biological diversity along major geographical gradients is an important topic in ecology. Garrido Gonzalez et al. [26] reported that the species distribution of water beetles was greatly influenced by altitude affecting the characteristics of aquatic settings. We tested if such finding is similar for the coleopteran fauna in Northern Tunisia.

2. Knowledge status of water beetles

Previous to the exploration of Northern Tunisia during the last years, very little was known of the water beetle fauna of Tunisia. So far as can be ascertained, there are only a few published notes on their biogeography, and the other a compiled checklist of 214 species collected by [5–16, 27–34]. Compilation of studies focusing on taxonomy of water beetles of Tunisia indicated a checklist of 236 species taking into account the eventual synonymous. The result of our recent researches is a list of 123 species, including all that I have found on the Northern Tunisia (**Figure 1** and **Table 1**). It is mainly the result of monthly collecting surveys over the course of a year (May 2005–April 2006), supplemented by some species collected by friends while working at other groups of insects. We considered only the water beetles' families sampled in the recent survey. About 1420 species in about 40 genera belong

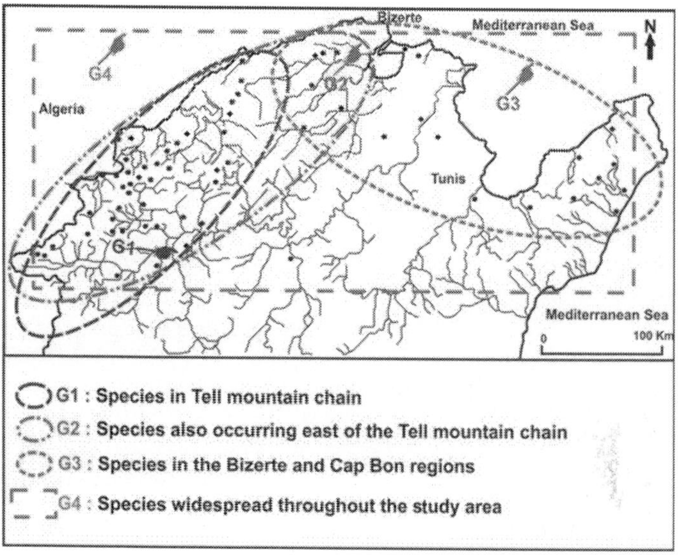

Figure 1. Map showing sampling sites and distribution of water beetles in the study area.

to Hydraenidae Mulsant, 1844 and are encountered on all continents, and some inhabit even the Subantarctic Islands, where only a few insects are able to cope with the hostile climatic conditions [1]. The Hydraenids of Tunisia comprise 57 species. Recently, we sampled 24 species including four recorded for the first time in Tunisia: *Hydraena atrata, Hydraena scabrosa, Ochthebius mauretanicus,* and *Ochthebius mediterraneus. Limnebius irmelae* is also an endemic species and thus require protection with regard to the anthropogenic threat on its habitat. By the North African scale, *H. atrata* shows demographic and geographic scarcity [19, 35]). Several species (*Hydraena rivularis, Hydraena numidica, Hydraena pici, Limnebius nitifarus, Limnebius theryi,* and *O. mauretanicus*) have their distribution area restricted to the extreme northern part between Algeria and Tunisia, and they could be considered as threatened species (reduction of their numbers and their biogeography). According to the criteria of the IUCN [17] for threatened species, we suggest the inclusion of *Hydraena kroumiriana* in the national red list of protected species [19].

Elmids occur on all the continents with about 1330 species in 146 genera. Members of this family are generally living in lotic habitats, and very few species are encountered on lake shores or in ponds, whereas Dryopids are represented by about 300 species in 33 genera occurring in all biogeographical regions, except for the Australian continent. Larvae are generally riparian or terrestrial; adults of about 75% of the species are regarded as aquatic (lotic and lentic habitats), and the remaining ones are riparian or terrestrial (humicolous and arboricolous) [1]. Studies on Elmidae and Dryopidae [13, 15, 31] reported 18 species. We collected nine species (**Table 1**), four of them are new recorded from Tunisia; *Oulimnius troglodytes, Potamophilus acu-*

Species	PH	CH	D	I (L;E)	"G"	Species	PH	CH	D	I (L;E)	"G"
*Hydraena testacea**[2,3]	F	EM	G1	0.0804	2.75	*Helophorus algiricus**[1,2,3]	F	NA	G2	0.0407	2.423
Hydraena atrata[1]	S	EM	G1	0.0042	1	*Helophorus asturiensis*[1,2,3]	F	WM	G4	0.0078	1.85
Hydraena rivularis[1,2,3]	P	NA	G4	0.0047	2.086	***Helophorus cincticollis***[1,3]	S	WM	G4	0.0154	2.342
*Hydraena leprieuri**[1,2,3]	P	v	G4	0.0545	2.432	***Helophorus paraminutus***[1]	F	P	G2	0.0173	1
Hydraena numidica[1,3]	P	NA	G4	0.0381	2.689	*Aulonogyrus striatus*[1,2,3]	P	TEM	G4	0.0104	1.811
Hydraena kroumiriana[3]	S	En	G1	0.0382	3	*Gyrinus urinator*[1,2]	P	TEM	G4	0.0223	1.352
Hydraena scabrosa[1]	S	NA	G4	0.0084	1	*Gyrinus dejeani*[1,2,3]	F	CEM	G4	0.0013	2.006
Hydraena pici[1]	S	NA	G1	0.0042	1	*Haliplus mucronatus*[1,2]	P	CEM	G4	0.0333	1.295
Limnebius furcatus[1,2]	S	EM	G4	0.0086	1.576	*Haliplus guttatus*[1]	P	M	G4	0.0218	1
Limnebius irmelae[1]	P	En	G4	0.0173	1	*Haliplus andalusicus*[1]	S	WM	G3	0.0042	1
Limnebius nitifarus[1]	F	NA	G1	0.0173	1	*Haliplus lineaticollis*[1,2,3]	P	SC	G4	0.0011	1.948
*Limnebius pilicauda**[1,2,3]	P	WM	G4	0.0598	2.568	*Peltodytes caesus*[1]	F	P	G2	0.0312	1
*Limnebius theryi**[1,3]	F	NA	G4	0.0564	2.606	*Peltodytes rotundatus**[1]	P	TEM	G4	0.0846	1
Limnebius perparvulus[1]	F	M	G2	0.0128	1	*Hygrobia hermanni*[1]	P	TEM	G3	0.0173	1
*Aulacochthebius exaratus**[1]	P	AEM	G4	0.0409	1	*Noterus laevis**[1,2]	P	M	G4	0.0416	1.266
Ochthebius aeneus[1]	S	EM	G4	0.0128	1	*Laccophilus hyalinus*[1,2,3]	P	M	G4	0.0057	1.91
Ochthebius dilatatus[1,2,3]	P	EM	G4	0.0264	2.128	*Laccophilus minutus**[1,2]	P	EMA	G4	0.0749	1.376
*Ochthebius viridescens**[1,2]	P	TEM	G4	0.0508	1.242	*Hyphydrus aubei*[1]	P	EM	G4	0.0360	1
Ochthebius punctatus[1]	P	EM	G4	0.0312	1	*Yola bicarinata*[1,2,3]	P	EM	G4	0.0172	1.719
Ochthebius difficilis[1,2]	F	TEM	G4	0.0192	1.376	*Bidessus minutissimus*[1,2,3]	P	EM	G4	0.0043	2.048
Ochthebius meredinicus[1]	F	WM	G4	0.0128	1	*Hydroglyphus geminus*[1,2,3]	P	EMA	G4	0.0017	2.082
Ochthebius mauretanicus[1]	F	NA	G3	0.0084	1	*Hydroglyphus signatellus*[1]	S	SC	G3	0.0042	1

Species	PH	CH	D	I (L;E)	"G"	Species	PH	CH	D	I (L;E)	"G"
Ochthebius praetermissus[1]	F	NA	G3	0.0128	1	***Hydroglyphus major***[1]	S	AM	G1	0.0042	1
Ochthebius mediterraneus[1]	S	EM	G2	0.0128	1	*Hydroporus feryi*[*1,2,3]	F	NA	G4	0.1152	2.698
Hydrophilus pistaceus[1]	F	WM	G3	0.0128	1	*Hydroporus obsoletus*[*2,3]	S	TEM	G1	0.1197	2.818
Laccobius revelierei[1,3]	F	AM	G2	0.0205	2.076	*Hydroporus pubescens*[*1,2,3]	P	CEM	G4	0.0458	2.441
Laccobius atrocephalus[1,2,3]	P	AM	G4	0.0076	1.849	*Hydroporus memnonius*[1]	S	EM	G1	0.0042	1
Laccobius orientalis[*1,2]	P	IM	G4	0.0471	1.364	*Hydroporus analis*[*2,3]	F	M	G1	0.0804	2.75
Laccobius atratus[1,2,3]	P	EM	G4	0.0387	1.744	*Hydroporus tessellatus*[*1,2,3]	P	CEM	G4	0.0665	2.704
Laccobius pommayi[1]	P	NA	G3	0.0218	1	*Stictonectes escheri*[1]	F	WM	G4	0.0173	1
Laccobius bipunctatus[1,2]	S	CEM	G2	0.0095	1.731	*Stictonectes optacus*[1,3]	S	WM	G4	0.0205	2.781
Berosus affinis[*1,2,3]	P	TEM	G4	0.0530	1.578	*Stictonectes samai*[*1,3]	P	NA	G4	0.1382	2.94
Berosus spinosus[1]	F	P	G3	0.0128	1	*Stictotarsus procerus* [1]	S	NA	G3	0.0084	1
Berosus signaticollis[1]	S	P	G3	0.0042	1	*Agabus africanus*[*1,2,3]	P	En	G1	0.0548	2.652
Anacaena lutescens[1,2,3]	P	H	G4	0.0266	1.912	*Agabus didymus*[*1,2,3]	P	TEM	G4	0.0422	2.384
Anacaena globulus[1,2,3]	P	EM	G4	0.0088	2.162	*Agabus nebulosus**	F	CEM	G4	0.0637	2.719
Anacaena bipustulata[1,2,3]	P	EM	G4	0.0079	2.148	*Agabus bipustulatus*	P	SC	G4	0.0152	2.462
Helochares lividus[*1,2]	P	TEM	G4	0.0693	1.446	*Agabus biguttatus*[*3]	S	EMA	G4	0.1741	3
Enochrus nigritus[1]	P	CEM	G4	0.0264	1	*Ilybius bedeli**	S	NA	G1	0.0532	2.884
Enochrus affinis[1]	S	P	G1	0.0042	1	*Dytiscus circumflexus*	S	CEM	G2	0.0128	1
Enochrus fuscipennis[1]	F	EM	G4	0.0173	1	*Rhithrodytes numidicus**	P	NAF	G1	0.0804	2.75
Paracymus scutellaris[1]	S	EM	G4	0.0084	1	*Graptodytes fractus*[*1,2,3]	F	M	G4	0.1152	2.698
Hydrobius fuscipes[1]	S	H	G2	0.0128	1	*Graptodytes ignotus*[*1,2,3]	S	M	G4	0.0487	2.673
Hydrobius convexus[1]	S	WM	G1	0.004	1	*Graptodytes flavipes*[1,2,3]	P	TEM	G4	0.0234	2.185
Hydrochus flavipennis[1,2,3]	P	EM	G4	0.003	2.056	*Graptodytes pietrii*[1,3]	F	WM	G4	0.0329	2.342

Species	PH	CH	D	I (L;E)	"G"	Species	PH	CH	D	I (L;E)	"G"
Hydrochus smaragdineus[1,2,3]	S	WM	G2	0.001	2.17	*Graptodytes varius**[1,2,3]	P	EM	G2	0.0609	2.56
Hydrochus grandicollis[1,2]	P	M	G4	0.009	1.521	*Hygrotus lagári*[1]	P	WM	G4	0.0312	1
Coelostoma hispanicum[1,2,3]	P	WM	G4	0.007	1.85	*Deronectes perrinae**[1,3]	F	NA	G1	0.0925	2.921
Hydrocyphon sp. [1,2]	F	–	G2	0.008	1.644	*Deronectes fairmairei*[1,2,3]	S	WM	G4	0.0065	2.311
Elodes sp. [1]	S	–	G3	0.004	1	*Melodema coriacea*[2]	S	TEM	G1	0.0314	2
Cyphon sp. [3]	S	–	G1	0.038	3	*Rhantus suturalis*[1]	S	SC	G3	0.0042	1
Dryops peyerimhoffi[1,2]	P	NA	G4	0.032	1.42	*Liopterus atriceps*[1]	S	WM	G2	0.0173	1
*Dryops sulcipennis**[1]	F	TEM	G4	0.045	1	*Colymbetes fuscus*[1]	P	P	G3	0.0084	1
Dryops algiricus[1]	S	TEM	G4	0.012	1	*Colymbetes schildknechti*[1]	S	WM	G3	0.0084	1
Pomatinus substriatus[1]	S	TEM	G2	0.008	1	*Hydrovatus cuspidatus*[1]	S	SC	G3	0.0084	1
Oulimnius rivularis[1,2,3]	P	EM	G4	0.010	2.095	*Hydrovatus clypealis*[1]	S	EM	G3	0.0042	1
Oulimnius troglodytes[1,3]	F	EM	G4	0.015	2.462	*Hydaticus leander*[1]	P	ACM	G4	0.0173	1
Limnius intermedius[1]	P	EM	G3	0.012	1	*Nebrioporus clarkii*[1,2]	P	M	G1	0.0137	1.844
Esolus filum[1,2]	S	NA	G1	0.013	1.844	*Nebrioporus cerisyi*[1]	P	CEM	G4	0.0084	1
Potamophilus acuminatus[1]	S	P	G3	0.004	1	*Nebrioporus* sp. [1,2]	S	–	G2	0.0137	1.844
Helophorus maritimus[1,3]	F	M	G4	0.020	2.076	*Cybister tripunctatus* ssp. *africanus*[1]	F	ACM	G3	0.0084	1
Helophorus alternans[1,2]	F	EM	G4	0.016	1.404						

asterisk (*) indicates species whose reciprocal information species-factor exceeds value 0.04. PH, phenology; P, permanent; F, frequent; S, seasonal; CH, chorology; En, endemic; NA, North African; TEM, Turano-Europeo-Mediterranean; ACM, Afrotropico-Centralasiatic-Mediterranean; EM, Europeo-Mediterranean; EMA, Europeo-Mediterraneo-Asiatic; CEM, Centralasiatic-Europeo-Mediterranean; IM, Indo-Mediterranean; AM, Afrotropico-Mediterranean; WM, West-Mediterranean; M, Mediterranean; P, Palearctic; H, Holarctic; SC, Subcosmopolitan; D, distribution in the study area; G1, occurring in the Kroumir and Mogods mountains; G2, also occurring east of the Tell mountain chain; G3, found only in Bizerte and Cap Bon regions; G4, widespread throughout the study area. Recently recorded species for Tunisia are indicated in bold. Altitudinal distribution; lowland (1), midland (2), highland (3).

Table. 1 Values of mutual information I (L; E) and the barycenter "G" for every species.

minatus, *Dryops peyerimhoffi*, and *Pomatinus substriatus* [36]. A particular attention was given to the Maghrebin endemic *Esolus filum* since the citations in Algeria are old and the occurrence of the species in North Africa refers to the recent catch conducted in the Moroccan Rif [37]. Hydrochidae is a monogeneric family with about 180 species occurring on all continents. All species are truly aquatic, living in well-vegetated stagnant water, and/or at the edges of very slowly flowing water [1]. Boumaïza [15] and Hansen [38] recorded two *Hydrochus* species from Tunisia. We sampled also *Hydrochus smaragdineus* (**Table 1**) for the first time [39].

Helophoridae is a monogeneric family with about 185 species, more or less confined to the Holarctic Realm [1]. Most species seem to prefer standing shallow water with plenty of organic debris, such as edges of small-to-medium sized water bodies [40]. Thirteen species of *Helophorus* are known in Tunisia [12, 38]. The checklist of Helophorids of Tunisia was bettered by new records of three species (**Table 1**): *Helophorus milleri*, *Helophorus paraminutus*, and *Helophorus cincticollis* [41]. Hydrophilidae consists of about 2652 species in 174 genera occurring on all continents, among them about 70% are aquatic [1]. In total, 21 *Hydrophilid* species were recently found. The most interesting ones from the zoogeographical point of view are *Enochrus nigritus*, *Enochrus affinis*, *Laccobius revelierei* (all newly recorded for Tunisia), and *Laccobius orientalis* and *Berosus spinosus* (newly recorded for North Africa) [42].

An updated checklist of the aquatic adephagan Coleoptera includes a total of 90 species, of them 57 were sampled in the study area (three Gyrinidae, six Haliplidae, one Paelobiidae, one Noteridae, and 46 Dytiscidae). *Hydroglyphus major* is recorded for the first time from Tunisia [43]. Dytiscidae have approximately 520 undescribed species. Gyrinidae represent a family of medium diversity, with an estimated 1000 species. Water beetles display their greatest diversity in the tropics except for Haliplidae and Helophoridae. Haliplidae are distinctly more diverse in the Holarctic Realm than in any of the tropical regions, although most tropical countries are still rather poorly examined [1].

3. Ecological traits of water beetles

3.1. Diversity and geographical pattern of water beetles

Species were categorized into three groups according to their adult phenology, following the approach of Valladares and Garrido [44]; permanent species (found over the course of the year), frequent species (encountered in three seasons), and seasonal species (occurring only during one or two seasons). The phenology of species is based on the presence of the adults since the capture of larvae is sporadic and requires an appropriate methodology [45]. The distribution of species in the studied areas in a transect from west to east took into account the differences in geological (landform localization) and hydrological (basin connectivity) characteristics. Four distributional categories of the water beetle species are distinguished according to the areas in which each species occurs (**Figure 1** and **Table 1**); 19 species occurring only in the Kroumir and Mogods Mountains (G1), mainly Hydraenidae and Hydroporinae (Dytiscidae) that are pollution-sensitive and rheophilous collected primarily from montane streams of the Aïn Draham region that are safe from any anthropogenic activities, 15 species also occurring east of the *Tell* mountain chain (G2), 19 species occurring only in the Bizerte and Cap Bon Peninsula (G3), and 70 species widespread throughout the study area (G4).

Beetles living in freshwater are strongly influenced by physicochemical and biological factors [46]. The chronology of appearance of the sampled species may be attributed to the flow of rivers, which may be temporary. The temporal appearance of species also is affected by a spatial variation that can hide the chronology of their emergence in each stream. Indeed, the study sites belong to different bioclimatic regions that largely influence water permanence, trophic factors, and hence, population dynamics (intensity of drift and migration and distribution of the fauna according to the availability of prey and competition).

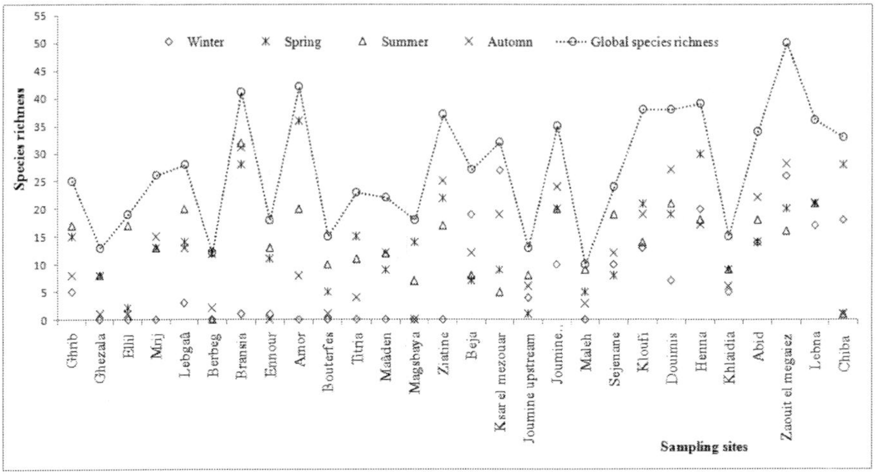

Figure 2. Seasonal and spatial variation in species richness of water beetles.

The species richness of water beetles follows seasonal fluctuation (**Figure 2**). The frequency and abundance of their adults attenuate considerably and even are absent when the environmental conditions become unfavorable in winter because of decreased water temperature, high turbidity, and the reduction of the aquatic vegetation as food, and refuge for the benthic community. The seasonal succession of species in temporary waters affects trophic structure, adaptations to drought, and traits common to most successful taxa, including highly flexible life cycle, temperature-dependent development, diapause or otherwise protected eggs, and high dispersal ability [46]. Among the main factors affecting community structure are runoff from agricultural areas, vegetation cover, and water chemistry [47]. Except for Noteridae, water beetles have terrestrial pupae. The life cycles of these species may include larval or adult terrestrial stages that minimizes the likelihood of their occurrence in the aquatic environments. In groups such as Dytiscidae and Elmidae, larvae and adults are aquatic. In other groups, such as Scirtidae, only larvae are aquatic. Hydraenidae and the hydrophilid (Hydrochinae and Helophorinae) have only adults as the aquatic stage [48]; they live as larvae in a dry cocoon in an excavated cavity above the water level, and can leave and return to it as adults [49], which minimizes the chance of capture. The overall phenology of the water beetle species reveals highest abundance and frequency in summer and in autumn. The phenological categories revealed a predominance of permanent species compared with frequent and seasonal species, with slight differences in abundance levels. Permanent and frequent species follow the overall phenological pattern, whereas the seasonal species show a spring maximum of abundance, which decreases toward a winter minimum [50].

Permanent species (52)—They may be considered the most eurytopic species in terms of environmental variables. Their majority are aquatic as larvae and imagoes (Gyrinidae, Haliplidae, Dytiscidae, and Elmidae). However, this more complete aquatic existence hides a variability of their existence in each river taken separately. The presence of each species can be used for the evaluation of its habitat preference [51].

The species exhibiting an autumnal peak of abundance are *Noterus laevis, Yola bicarinata,* and *Ochthebius viridescens. Haliplus lineaticollis, Peltodytes rotundatus, Laccophilus minutus, Hydrochus flavipennis, Berosus affinis, Helochares lividus, Hydraena leprieuri, L. orientalis,* and *Laccobius atratus* were most abundant in summer and autumn. Three species, *Hydroporus pubescens, H. rivularis,* and *Ochthebius dilatatus,* were most abundant in spring and in autumn, whereas only *Graptodytes flavipes* was maximally abundant in spring and in summer. *B. affinis* had a considerable population increase during spring, summer, and autumn, coinciding with a rise in water temperature and vegetation growth. The species displayed a phenology close to that recorded in the province of Palencia in Northern Spain with a peak of abundance in summer, but a light winter proliferation [44]. This phenology was identical to that observed by Aouad [52] in ecosystems of Morocco, with high abundance of larvae in spring. *H. lividus* and *Aulacochthebius exaratus* maintained similar abundance patterns during the annual cycle, similar to their phenology observed by Valladares and Garrido [44] in Palencia. *Agabus bipustulatus* is regarded by Ribera et al. [53] as the most abundant species of the genus year-round in the Pyrenees, whereas *Hygrobia hermanni* is maximally abundant during winter and spring. We found *Agabus didymus* the most abundant species year-round, and *H. hermanni* was uncommon during all seasons. According to Bertrand [54], *Hyphydrus aubei* is present with quasi similar seasonal abundances, with a collection of larvae during April in central and southern Europe and North Africa. We found this species slightly more abundant in spring and in summer. According to Millán [55], *N. laevis* is present year-round, but its abundance decreases during autumn in the south-eastern Iberian Peninsula. This is different from our observations that show an autumnal increase. Valladares and Garrido [44] mentioned that populations of *Hydroglyphus geminus* increase in summer and then decrease in autumn, whereas *Anacaena lutescens* is encountered only in winter. However, in the present study, we observed *H. geminus* more or less equally abundant year-round, and found *A. lutescens* all year but most abundant in spring and in summer. *H. flavipennis* is a permanent species with slightly higher abundance in summer and in autumn. According to Valladares and Garrido [44], it belongs to the group of frequent species, but it is absent in spring.

Frequent species (31 species)—they are mainly more localized montane species. These species exhibit low abundance associated with low frequencies and restricted distribution. The majority of the species exhibit similar seasonal abundance patterns. However, *Graptodytes* *fractus* and *Hydroporus feryi* present two peaks of abundance, in summer and in autumn. *Helophorus algiricus* is maximally abundant in spring, whereas *L. revelierei* is most abundant in summer. Sixteen species in this group were not captured in winter, which is characterized by a disturbance of microhabitats due to the flood effect affecting the majority of sites. Six species were not captured in summer when temporary streams become dry. Two species, *Helophorus asturiensis* and *Helophorus alternans*, were rare in autumn and in winter, became abundant in spring, and then were absent in summer. They are regarded as seasonal species by Valladares and Garrido [44], appearing only in winter and in spring with low abundance. Valladares et al. [45] noted that *H. lividus*, a permanent species in our study, and *Enochrus fuscipennis*, which have similar sizes, habitats, and prey type, exhibited peak abundance in spring and in autumn. This may be interpreted as an adjustment of their biological cycles to avoid the interspecific competition for shared trophic resources.

Seasonal species (40)—their absence from some sites may be accounted for sampling effort and/or scarcity. Their richness decreased from spring to winter, probably due to their ecological requirements. Abellán et al. [56] consider measurements based on richness, abundance, or evenness, which are usually used in the evaluation of the effects of environmental degradation on biodiversity, to be highly influenced by sample size, sampling effort, and the type or complexity of the habitat. Valladares and Garrido [44] observed that *Paracymus scutellaris* and *Dryops algiricus* exhibited the same phonological pattern in Spain, whereas *Berosus signaticollis* had clear peak abundance during these seasons. In the present work, the first species was found in spring and in summer only, the second species in summer and winter only in low numbers, and only a single specimen of the third species was taken (during spring). *H. atrata* was collected only in one site during September. It is present during four seasons, particularly in spring, in the Channel of Castile in Northern Spain [45]. In the same way, *Rhantus suturalis* was captured in a well during the spring, but it is permanent in the Pyrenees and has its population peak in winter [57]. *Hydrobius fuscipes* was collected only in April in our study, but in Spain, it is abundant in autumn and in spring and absent in winter [44].

The geographical distribution was analyzed by chorotype based on distributional patterns that are deduced from a comparative analysis of geographical ranges of species [58]. The fauna of Tunisia, as well as all of North Africa, is a heritage of Eurasian and Afrotropical elements. However, during the Pliocene, the isolation from Europe blocked the arrival of several European lineages. The desertification of the Sahara during the Holocene impeded the northward movement of Afrotropical species. These two facts explain the relative poverty of the water beetle fauna compared to less isolated zoogeographic regions in the world [32]. The chorological category corresponding to each species is given in **Table 1**. The most important chorotypes are Europeo-Mediterranean (19.2%), North African (15.8%), West-Mediterranean (12.5%), Turano-Europeo-Mediterranean (12.5%), and Mediterranean (8.3%). The number of endemic species is low, about 2.5% of the total fauna. For the species of the genus *Hydraena*, as for many other organisms of the chorotype North African (NA), the Sahara desert rather than the Mediterranean Sea forms a biogeographical limit. The majority of *Hydraena* species are North African. The main exchanges with Europe took place in the west where the Strait of Gibraltar is situated [32]. Abellan et al. [59] considered the biodiversity of freshwater systems to be endangered, especially in Mediterranean semiarid regions like our study area. Northern

Tunisia has a rich and endemic biota that is threatened by the development of surrounding land-crop irrigation. These freshwater habitats and species need more protection in order to preserve the biodiversity of the freshwater ecosystems of North Africa. Endemic species from Tunisia (En) comprises three species *H. kroumiriana, L. irmelae* and Dytiscidae (*Agabus africanus*). The distribution of *H. kroumiriana* is restricted to a small montane stream located in Northwestern Tunisia. It is threatened with extinction according to the categories and criteria established by the IUCN [18]. *L. irmelae* apparently has a small population since less than four adults were captured per site over the course of a year. *A. africanus* is more dispersed toward the east (Cap Bon Peninsula) but is still scarce.

The fauna of North Africa probably originated from the passage of Euroasian species to the African continent as a result of plate tectonics (in the Tortonian), leading to the connection of the two continents. The Mediterranean Sea and the Atlantic Ocean were later connected (in the Pliocene), thereby isolating the faunas of the two continents [20]. The origin of the water beetle fauna in Northern Tunisia reflects the history of the Mediterranean basin. During the secondary era, the coastal massifs of the Rif, as far as the Kroumir, were an emerged part of a continent or more probably an archipelago, the Tyrrhenid, which spread over what is now the western Mediterranean between Spain and Italy [32]. The Eocene and Oligocene transgressions reduced the Tyrrhenid to the Betico-Rifan massif separated from Europe by the North-Betican trough and from Africa by the South-Rifan trough and by some islets near the Kabylie. These lands remained emerged up to the present time. North Africa was joined to Eurasia at the end of the Miocene, and the Mediterranean, thus enclosed, dried up. During the Pliocene, a transgression covered the Tyrrhenid and the Strait of Gibraltar divided the Betico-Rifan into two parts [60]. These events provide a hypothesis explaining the chorological aspects of the current water beetle fauna of Tunisia and North Africa.

3.2. Altitudinal distribution of water beetles

The northern part of Tunisia comprises several orographic areas: the *Tell* (altitude ranges between 400 and 800 m), the *haut-Tell* and the *Dorsale* culminating at *Djebel Châambi* (1544 m), with a width of 40 km in the west and becoming narrow in the east toward the *Cap Bon Peninsula* [21]. As larger number of species are sensitive to this ecological parameter, the more significant are their role in the biotope [61]; an ecological study of the geographical distribution of water beetles in the mountains of Northern Tunisia was carried out with an analysis of the effect of altitude on the distribution of 123 species collected from 64 sampling sites. Species richness was analyzed at different altitudinal levels and the indicator species were determined by establishing their altitude profile in terms of reciprocal species-factor information (see Touaylia et al. [62]). The information related to the altitudinal gradient gave a score of $I(L; E) = 0.702$, while maximum entropy, which depends only on the number of altitudinal levels considered, was $H(L)\ max = \log_2 [NK] = \log_2 3 = 1.09$. The ratio between those two scores, which determines the sampling quality, was 0.644. This ratio shows that the altitudinal factor has been sufficiently sampled; however, some information is lost because of the high frequency (76.6%) of sampling sites in the first altitudinal level. The variation in aquatic beetle species richness at different altitude levels is shown in **Table 1**; overall, species richness decreases with increasing altitude. Species richness decreases within the different families with increasing altitude [62]. In an overall regional survey, we did not analyze the ecological features of all species because it would not be necessary to analyze features of species not having "useful information", and it would be difficult to work with such a large number of species. An indicator value is attributed to every altitude profile as high as the information score that it gives. The species with the highest I(L; E) scores are the most sensitive to the altitudinal factor and could be considered as altitude indicator species. Their high I(L; E) value highlights the important role of altitude in the distribution of species in the study area. Among the species whose reciprocal "species/altitude" information exceeds 0.04, those having a more significant altitude profile from the corrected frequencies and comprising a less variable spectrum of the altitudinal factor were selected as being representative (indicated with an asterisk). The ecological influence of altitude on the distribution of water beetles was demonstrated through a determination of indicator species linked to the altitudinal gradient. The assessment of the barycenter abscissa value is obtained from the set of information provided by the species' altitude profiles, highlighting those that give more information in relation to the considered factor (I (L; E) > 0.04). In this way, an ordering was established in the set of species, based on increasing "G" values, that is equivalent to classifying species along an altitudinal gradient. Species having similar barycenters are expected to possess the same ecological behavior, and the barycenter abscissa of the profile can be taken as an index of its ecological optimum in terms of the considered factor [63]. However, the frequency profile of the different species needs also to be taken into account. The analysis of the barycenter abscissa reveals that species having a low value are highly localized. Species whose profile shows an average barycenter have greater amplitude. Such phenomena are often very significant, as has been shown in similar ecological studies [64]. Taking into account their altitudinal preferences, the 30 representative species are categorized into five groups (**Table 1**):

·Species present along the entire altitudinal gradient—several altitudinally ubiquitous species of Hydraenidae, Hydrophilidae, Helophoridae, and Dytiscidae. *H. leprieuri* (7–588 m) is endemic in North Africa [32]; it lives in clear, preferably flowing, waters under detritus and stony substrata with rocks and gravel providing with it a homochromy for avoiding predators. *Limnebius pilicauda* (2–588 m)—a member of the Maghrebinian element— was captured at an altitude of 1300 m in Morocco [34]. It was collected in shady watercourses in the region of Aïn Draham as well as in lowland, lentic, vegetated streams with muddy and gravelly bottoms. *L. theryi* (16–588 m)—endemic in Algeria and in Tunisia [34]—is more abundant at high altitudes. Its broad ecological profile allows it to live in a variety of habitats. *B. affinis* is an ubiquitous specie, showing great ecological plasticity, occurring in a variety of aquatic ecosystems (lagoons, ponds, springs, rivers, and ditches) associated with muddy substrates. Its altitudinal distribution ranges from 5 to 1400 m [65]. It has been found associated with vegetation, sometimes in muddy and eutrophic as well as in clear waters [66]. It has a wide altitudinal distribution (2–588 m) and also high abundance and frequency. *H. algiricus* is a representative of the *obscurus* complex in North Africa. It occurs in a variety of shady places (5–2000 m) in Morocco [65]. Its altitudinal range is wide (6–631 m) with a greater abundance at high altitude sites. It was found along the banks of streams rich in organic matter. *H. feryi* shows a preference for highland sites, but rarely occurs in lowland areas; its wide altitudinal level (2–614 m) means it is anubiquitous species. It occupies the same habitats as *H. pubescens.*

Its highest abundance is in sites with bryophytes and mosses. *H. pubescens* was encountered on plains and low mountains in stagnant fresh or brackish waters [67]. It is more abundant in clear, well oxygenated, fast-flowing, but poorly vegetated waters (2–588 m). *Hydroporus tessellatus* is a typical species of subalpine levels. It has a wide ecological valence, occurring in lotic or lentic habitats [26]. It was captured at sites whose altitude ranges between 16 and 588 m. Its abundance increases considerably at higher altitudes, and it becomes scarcer at lower altitudes. *Stictonectes samai* is well represented in highland areas; in contrast, it was taken at a single lowland site. *Agabus brunneus* has been collected from flowing waters of small streams [54]. It was found at both lowland and highland sites (2–588 m), but it is not represented at mid-altitude sites. *A. didymus* is considered a montane species (645–1295 m), with greatest affinity for altitudes between 900 and 1200 m, where it is abundant [26]. It occurs in clear running waters of the Alps [54]. In our study, it was captured in the same habitat type up to 588 m. *Agabus nebulosus* lives in stagnant waters in the lower Pyrenees (>2000 m) [67]. It was captured in lentic (pond and irrigation canal) and lotic (river and high mountain) habitats [26]. In our collections, it was also found in lowland areas (16 m) with abundant vegetation and reached its highest altitude at 631 m. *Ilybius bedeli* (3–588 m) is scarce at lowland altitudes, but well represented at highland sites. It was recorded by Normand [12] and Vigna Taglianti et al. [58] from clear, flowing waters of mountainous part of Aïn Draham, Camp de la Santé, and El Feidja. *G. fractus* (2–714 m) is abundant mainly in the highest mountainous regions of Northwestern Tunisia, as also recorded by Normand [13]. *Graptodytes ignotus* is uncommonly distributed along an altitudinal gradient from 16 to 714 m. It has a tendency to sandy and gravelly sediments in clear, flowing waters of shady mountainous regions. *Graptodytes varius* (2–588 m) has highest abundance from streams of the Bni Mtir Dam basin. *Deronectes perrinae* is a highland species found in mountains (up to 631 m). However, one adult was collected at lowland (7 m), possibly carried there by river debris.

****Species present only in lowland areas***—*A. exaratus* was encountered in a small vegetated inlet stream with a substratum of gravel and sand at 265 m [26]. Its ecological requirements vary according to geographic area. It can tolerate a wide range of environmental conditions [65]. Its altitudinal range in the study area varied between 3 and 176 m. *Dryops sulcipennis* is found in two different kinds of habitats, bogs, and flowing waters [69]. It was collected at lowland sites (2–237m). *P. rotundatus* lives among filamentous algae in lakes, pools, ponds, rivers, marshes, brooks, and streamlets, and shows a preference for running water among vegetation of *Utricularia* and *Nuphar* [70]. It was encountered at sites located at altitude from 2 to 236 m.

****Species present only in lowland-midland areas***—*O. viridescens* was found in brackish as well as inland waters [71]. It lives mainly in lowland sites (2–255 m). It is associated with the roots of vegetation such as *Juncus multiflorus* and *Typha angustifolia*. *L. orientalis* can be considered lowland; it is very frequent in sites at low altitude (6–255 m). *H. lividus* is localized among mud and detritus on the edge of temporary ponds, lakes, lagoons, and rivers [66]. It was captured at altitudes ranging from 6 to 255 m. *N. laevis* (2 and 255 m) was found in stagnant waters with rich vegetation, but sometimes also in slowly flowing waters as mentioned by Vondel and Dottner [70]. *L. minutus* shares the same habitat type as *Laccophilus hyalinus*, but it shows a preference for stagnant waters and has been captured at altitudes up to 1500 m [54]. The altitude range of present collections lies between 2 and 484 m.

***Species present only in midland-highland areas**—*Hydraena testacea* lives in ponds, slow-flowing, oxygenated small streams, or other stagnant waters with a slimy, eutrophic substratum invaded by algae [26, 72]. Its distribution in Tunisia is limited to two closely approximate sites from the Aïn Draham forest (484–588 m), categorized as a threatened species at local scale. *Hydroporus obsoletus* lives exclusively in hollows and quiet stream edges of mountains [67]. It was collected at altitudes between 329 and 588 m. *Hydroporus analis* is typical of coastal brackish waters [67]. The habitat of our collection (484–588 m) is characterized by fast flowing, limpid, oligotrophic, unmineralized water, where it was recorded by Pederzani and Schizzeroto [16]. *Rhithrodytes numidicus* was recorded from small streams of Fernana and Aïn Draham Mountains [12]. Our captures are restricted to the Bni Mtir Dam basin (314–588 m). It was also collected in underground waters (wells) as a stygoxene epigean species.

***Species present only in highland areas**—*Agabus biguttatus* may be considered montane. In North Africa, it lives in forest streams (3200 m) [68]. Its occurrence in the study area is limited to several streams of the Bni Mtir Dam basin (Northwestern Tunisia) at altitudes between 563 and 631 m. It occupies a habitat type characterized by steep slopes, a substratum predominantly of stone and gravel, poor vegetation, and well oxygenated, fast-flowing waters. The mean value of the mutual information is higher when species have more affinity for several classes of the altitude. The importance of a descriptor corresponds to its effectiveness to select species in order to know the zoo-ecological groups among them [61]. Jacobsen [73] indicated that mean local and zonal family richness decreased by about 50% from sea level to 4000 m; local richness declined linearly from sea level to 1800 m.

The species richness of water beetles decreases with increasing altitude. This may be explained by the fact that some species present in lowland streams were not found at higher altitudes (**Table 1**). Furthermore, it can also be attributed to the fact that the majority of sampled sites were in the first altitudinal level. Also, there was a decrease in new species with the accumulation of new records. Five species were newly recorded in the mid-altitude level, whereas 25 species disappear in it. Sixteen species were added in the high altitude level to those of the mid-altitude level, with the disappearance of 18 species; seven species were new, and 69 were absent in the high altitude level in comparison with the low altitude level.

4. Conclusion

The water beetles in Tunisia are poorly studied not only in comparison with the European fauna but also with other zoogeographical areas. The present conducted survey aims at bettering the knowledge on this freshwater fauna. The species of richness of water beetles is updated by new records; it can be bettered through samples from central and southern Tunisia. This checklist seems to better the knowledge of the diversity of the water beetles' habitats and provides a solid basis for further research, focusing on macroinvertebrates in order to better direct monitoring conservation projects, and to asses the effects of anthropogenic activities on these fragile ecosystems. The fauna of Tunisia, as well as all of North Africa, is a heritage of Eurasian and Afrotropical elements. However, during the Pliocene, the isolation from Europe blocked the arrival of several European lineages. The desertification of the Sahara during the Holocene impeded the northward movement of Afrotropical species. These two facts explain the relative poverty of the water beetle fauna compared to less isolated zoogeographic regions in the world [32]. These aquatic insects are heterogeneous in their local and world distribution. Their assemblage is structured by physicochemical parameters. They include indicator species (water quality, altitude. etc.). Altitude could be considered among the physical factors that affect distribution of stream macroinvertebrate communities, but its effect is also combined with other environmental variables such as temperature, substratum, water flow, and stream geomorphology, particularly in streams extending along altitudinal gradients [74]. Water permanence and depth were considered by Williams et al. [75] among the main environmental variables explaining invertebrate assemblage structure. The present study was restricted to the northern part of Tunisia. The sampled sites ranged between 1 and 714 m, which is a rather limited altitudinal gradient. In many geographical areas, 714 m would hardly be considered "high altitude." This can make difficult a comparison with other areas in which the habitats typical of the Tunisian "lowland" are found above such an altitude. Therefore, the distribution of the species has to be related to the distribution of habitats, as the same species can be found at different altitudes in different areas, depending on where suitable habitats are found. Further sampling is required to confirm these results, especially in higher mountains in central and southern Tunisia, since those in central Tunisia rise to 1544 m.

Author details

Touaylia Samir

Address all correspondence to: toysam2010@yahoo.fr

Faculty of Sciences of Bizerta, Laboratory of Environment Biomonitoring, University of Carthage, Bizerta, Tunisia

References

[1] Jäch, M.A. & Balke, M. (2008). Global diversity of water beetles (Coleoptera) in freshwater. *Hydrobiologia*, vol. 595, 419–442.

[2] Abellán, P., Sanchez-Fernandez, D., Velasco, J. & Millan A. (2005). Conservation of freshwater biodiversity: a comparison of different area selection methods. *Biodiversity and Conservation*, vol. 14, 3457–3474.

[3] Ribera, I., Foster, G.N. & Vogler, A.P. (2003). Does habitat use explain large scale species richness patterns of aquatic beetles in Europe? *Ecography*, vol. 26, 145–152.

[4] Balke, M., Watts, C.H.S., Cooper, S.J.B., Humphreys, W.F. & Vogler, A.P. (2004). A highly modified stygobiont diving beetle of the genus *Copelatus* (Coleoptera, Dytiscidae):

taxonomy and cladistic analysis based on mitochondrial DNA sequences. *Systematic Entomology*, vol. 29, 59–67.

[5] Bedel, L. (1900). Reasoned catalog of the beetles of Tunisia. *Paris: national printing house.* xiv + 130pp.

[6] Seurat, L.G. (1921). Fauna of the continental waters of Berberia. Algeria: *Publication of the University, Faculty of Sciences Algiers.* 66p.

[7] Seurat, L. G. (1934). Aquatic fauna in the south and extreme south of Tunisia. *Annals of Natural Science of Zoology*, vol. 10 (17), 441–450.

[8] Seurat, L.G. (1938). Aquatic fauna of Southern Tunisia (South and extreme South). *Memory of the Society of Biogeography*, Vol. 6, 121–143.

[9] Peyerimhoff, P. (1924). New North African Coleoptera. Forty-eighth Note: insects of land and salt water, harvested by Messrs. Seurat and Gauthier Southern Tunisia. *Bulletin of the Entomological Society of France*, vol. 29, 158–161.

[10] Gauthier, H. (1928). Research on the fauna of continental waters of Algeria and Tunisia. *Ed. Minerva, Algiers:* 1 pl., 1 map, 419 pp.

[11] Omer-Cooper, J. (1930). Notes on the freshwater fauna of Southern Tunisia. *The Entomologist* 63, 251–255.

[12] Normand, H. (1933). Contribution to the catalog of Coleoptera of Tunisia. *Bulletin of the Natural History Society of North Africa*, vol. 24, 295–307.

[13] Normand, H. (1935). Contribution to the catalog of Coleoptera of Tunisia. *Bulletin of the Natural History Society of North Africa*, Vol. 26, 86–304.

[14] Normand, H. (1949). Contribution to the catalog of beetles of Tunisia. *Bulletin of the Society of Natural Sciences of Tunisia*, vol. 2, 79–104.

[15] Boumaiza, M. (1994). Research on the running waters of Tunisia. Faunistic, ecology and biogeography. *Ph.D. thesis, Faculty of Sciences of Tunis*, 429p.

[16] Pederzani, F. & Schizzeroto, A. (1998). Description of *Agabus* (*Dichonectes*) *africanus* n.sp. from north-west Tunisia and notes on the cohabiting species of Hydradephaga (Coleoptera Haliplidae, Gyrinidae & Dytiscidae). *Attidell' Accad. Roveret. degliAgiati*, vol. 7 (8B), 87–95.

[17] Löbl, I. & Smetana, A. (2003). Catalogue of Palaearctic Coleoptera, Vol. 1. *Strenstrup: Apollo Books*, 819 p.

[18] IUCN. (2006). Guidelines for Using the IUCN Red List Categories and Criteria. Version 6.2: Standards and Petitions Working Group of the IUCN SSC Biodiversity Assessments Sub-Committee.

[19] Touaylia, S., Bejaoui, M., Boumaiza, M. & Garrido, J. (2009a). Nouvelles données sur la famille des Hydraenidae Mulsant, 1844, de Tunisie (Coleoptera). *Bulletin de la Société entomologique de France*, vol. 114(3), 317–326.

[20] Bennas, N., Sáinz-Cantero, C.E. & Alba-Tercador, J. (1992). Preliminary data for a biggeographic study of the Betic-Rifle Massif based on aquatic beetles. *Zoologica Baetica,* vol. 3, 167–180.

[21] Ben Ayed, N. (1993). Tectonic evolution of the foreland of the alpine chain of Tunisia from the beginning of the Mesozoic to the Current. *State Thesis, National Office of Mines*: 282 p.

[22] Zielhofer, C. & Faust, D. (2008). Mid- and Late Holocene fluvial chronology of Tunisia. *Quaternary Science Reviews*, vol. 27, 580–588.

[23] Ben Jemaa, F., Houcine, I. & Chahbani, M.H. (1998). Desalination in Tunisia: past experience and future prospects. *Desalination*, vol. 116, 123–134.

[24] Houcine, I., Benjemaa, F., Chahbani, M.H. & Maalej, M. (1999). Renewable energy sources for water desalting in Tunisia. *Desalination*, vol. 125, 123–132.

[25] Bonada, N., Zamora-Muñoz, C., Rieradevall, M., Prat, N. (2005). Ecological and historical filters constraining spatial caddisfly distribution in Mediterranean rivers. *Freshwater Biology*, vol. 50, 781–797.

[26] Garrido Gonzalez, J., Diaz Pazos, A. & Regil Cueto, A. (1994). Aquatic fauna of the Foral Community of Navarre (Spain) (Col., Adephaga and Polyphaga). *Bulletin of the entomological Society of France*, vol. 99 (2), 131–148.

[27] Ferro, G. (1983). New interesting Hydraenidae of museum of natural history of Praga. II contribution (Coleoptera Hydraenidae). Bulletin and Annals of the Royal Belgian Society of Entomology, vol. 120, 73–80.

[28] Ferro, G. (1984). New interesting Hydraenidae of museum of natural history of Praga. III contribution *Bulletin and Annals of the Royal Belgian Society of Entomology*, vol. 120, 61–71.

[29] Ferro, G. (1985). Hydraenidae (Coleoptera Hydrophiloidae) of the Norte de Africa XV Contribution to the knowledge of Hydraenidae. *Bulletin and Annals of the Royal Belgian Society of Entomology*, vol. 121, 233–241.

[30] Ferro, G. (1986). Description of two new species of Hydraenidae (Col. Palpicornia) (XIX Contribution to the knowledge of Hydraenidae). *Bulletin of the Italian Entomological Society*, vol. 118 (8-10), 135–138.

[31] Berthélemy, C. (1964). Elminthidae from Western and Southern Europe and Africa Of the North (Coleoptera). *Bulletin of the Natural History Society of Toulouse,* vol. 99, 244–285.

[32] Berthélemy, C., Kaddouri, A. & Richoux, P. (1991). Revision of the genus *Hydraena* Kugelan, 1794 from North Africa (Coleoptera: hydraenidae). *Elytron*, vol. 5, 181–213.

[33] Jäch, M.A. (1991). Revision of the Palearctic species of the genus *Ochthebius* VII. The *foveolatus* group (Coleoptera: hydraenidae). *Koleopterologische Rundschau*, vol.61, 61–94.

[34] Jäch, M.A. (1993). Taxonomic Revision of the Palearctic species of the genus *Limnebius* Leach, 1815 (Coleoptera: hydraenidae). *Koleopterologische Rundschau*, vol. 63, 99–187.

[35] Kaddouri, A. (1986). Revision of the *Hydraena* of Morocco from Algeria and Tunisia (coleoptera, Hydraenidae). *Ph.D thesis. University Paul Sabatier*, Toulouse: 156p.

[36] Touaylia, S., Bejaoui, M., Boumaiza, M. & Garrido, J. (2010b). Contribution to the study of the Aquatic Coleoptera of Tunisia: Elmidae Curtis, 1830 and Dryopidae Billberg, 1820 (Coleoptera). *New Journal of Entomology*, Vol. 26 (2), 167–176.

[37] Bennas, N. & Saìnz-Cantero, C.E. (2007). New data on Coleoptera Aquatic plants of Morocco: the Elmidae Curtis, 1830 and the Dryopidae Billberg, 1820 of the Rif (Coleoptera). *New Journal of Entomology*, Vol. 24 (2), 61–79.

[38] Hansen, M. (2003). Helophoridae, pp. 36-41. In: Löbl I. & Smetana A. (ed.): Catalogue of Palaearctic Coleoptera, vol. 2. *Strenstrup: Apollo Books*, 942 p.

[39] Touaylia, S., Bejaoui, M. Boumaiza, M. & Garrido, J. (2009b). A study on *Hydrochus* Leach, 1817, species from Tunisia (Coleoptera, Hydrochidae). *Bulletin de la Société Entomologique de France*, 114(1), 11–16.

[40] Mart, A. & Erman, O. (2001). A Study on Helophorus Fabricius, 1775 (Coleoptera, Hydrophilidae) Species. *Turkish Journal of Zoology*, vol. 25, 35–40.

[41] Touaylia, S., Bejaoui, M., Boumaiza, M. & Garrido, J. (2009c). New data on the Helophoridae Latreille, 1802 species from Tunisia (Insecta, Coleoptera). *Nouvelle Revue d'Entomologie*, vol. 25(4), 317–324.

[42] Touaylia, S., Garrido, J. & Boumaiza, M. (2011a). A study on the family Hydrophilidae latreille, 1802 (Insecta, Coleoptera) from Tunisia. *Pan-Pacific Entomology*, vol. 87, 27–42.

[43] Touaylia, S., Garrido, J., Bejaoui, M. & Boumaiza, M. (2010). A contribution to the study of the aquatic Adephaga (Coleoptera: Dytiscidae, Gyrinidae, Haliplidae, Noteridae, Paelobiidae) from northern Tunisia. *The Coleopterists Bulletin*, vol. 64(1), 53–72.

[44] Valladares, L.F. & Garrido, J. (2001). Aquatic coleoptera of associated altitudinal wetlands of Castilla (Palencia, Spain): faunistic and phenological aspects (Coleoptera, Adephaga and Plyphaga). *New Journal of Entomology*, vol. 18 (1), 61–76.

[45] Valladares, L.F., Garrido, J. & Herrero, B. (1994). The annual cycle of the community of aquatic Coleopteran (Adephaga and Polyphaga) in a rehabilitated wetland pond: Le laguna de la Nava (Palencia, Spain). *Annales de Limnologie*, vol. 30(1), 209–220.

[46] Williams, D.D. (1996). Environmental constraints in temporary fresh waters and their consequences for the insect fauna. *Journal of the North American Benthological Society*, vol. 15(4), 634–650.

[47] Armin, B., Stefan, K., Barbara, P. & Albert, M. (2009). Abundance, diversity and succession of aquatic Coleoptera and Heteroptera in a cluster of artificial ponds in the North German Lowlands. *Limnologica*, vol. 40, 215–225.

[48] Tachet, H., Richoux, P., Bournaud, M. & Usseglio-Polatera, P. (2000). Freshwater invertebrates, systematics, biology, ecology. *CNDS Editions*: 581 p.

[49] Hutchinson, G.E. (1981). Thoughts on aquatic insects. *Bioscience*, vol. 31(7), 495–500.

[50] Touaylia, S., Garrido, J. & Boumaiza, M. (2011b). Chorological and phenologic analysis of the water beetle (Coleoptera, Adephaga and Ployphaga) fauna from Northern Tunisia. *The Coleopterists Bulletin*, vol. 65(3), 315–324.

[51] Valladares, L.F., Garrido, J. & Garcia-Criado, F. (2002). The assemblages of aquatic Coleoptera from shallow lakes in the northern Iberian Meseta: Influence of environmental Variables. *European Journal of Entomology*, vol. 99, 289–298.

[52] Aouad, N. (1988). The biological cycle and polymorphism of *Berosus affinis* (Coleoptera: Hydrophilidae) in Marocco. *Entomological News*, vol. 99(2), 105–110.

[53] Ribera, I., Isart, J. & Régil, J. (1995a). Autoecology of some species of Hydradephaga (Coleoptera) of the Pyrenees. I. Gyrinidae, Haliplidae, Noteridae and Hygrobiidae. *Zoologica Baetica*, vol. 6, 33–58.

[54] Bertrand, H. (1928). Larvae and Nymphs of Dytiscidae, Hygrobiidae and Haliplidae. 33 boards, 207 figures. *Ed. Paul Lechevalier*, Paris: 366p.

[55] Millán, A. (1991). The Coleoptera Hydradephaga (Haliplidae, Gyrinidae, Noteridae And Dytiscidae) of the Segura river basin, SE of the Iberian Peninsula. *PhD thesis. University of Murcia*. 567p.

[56] Abellán, P., Bilton, D.T., Millán, A., Sánchez-Fernández, D. & Ramsay, P.M. (2006). Can taxonomic distinctness assess anthropogenic impacts in inland waters? A case study from a Mediterranean river basin. *Freshwater Biology*, vol. 51, 1744–1756.

[57] Ribera, I., Isart, J. & Régil, J. (1995b). Autoecology of some species of Hydradephaga (Coleoptera) of the Pyrenees. I. Dytiscidae. *Zoologica Baetica*, vol. 6, 59–104.

[58] Vigna Taglianti, A., Audisio, P.A., Biondi, M., Bologna, M.A., Carpaneto, G.M., De Biase, A., Fattorini, S., Piattella, E., Sindaco, R., Venchi, A. & Zapparoli, M. (1999). A proposal for a chorotype classification of the Near Est fauna, in the framework of the Western Plearctic region. *Biogeographia*, vol. 20, 31–59.

[59] Abellan, P., Sanchez-fernandez, D., Velasco, J. & Millan, A. (2004). Conservation of freshwater biodiversity: a comparison of different area selection methods. *Biodiversity and Conservation*, vol. 14, 1–18.

[60] Dercourt, J., Zonenshain, LE, Ricou, LE, Kazmin, VG, Le Pichon, X., Knipper, AL, Grandjacquet, C., Sborschikov, IM, Boulin, J., Sorokhtin, O., Geyssant, J. , Lepierre, C., Biju-duval, B., Sibuet, JC, Savostin, LA, Westphal, M. & Lauer, JP (1985). Presentation of 9 1/ 20,000,000th palaeogeographic maps extending from the Atlantic to the Pamir for the period from Lias to the present. *Bulletin of the Geological Society of France,* Vol. 8.1 (5), 637–652.

[61] Daget, P., Gordon, M. & Guillerm, J.L. (1972). Profils écologiques et information mutuelle entre espèces et facteurs écologiques. In: *Grundfragenum Methoden in der Pflanzensoziologie. Junk Publ.* La Haya, pp. 121–149.

[62] Touaylia, S., Garrido, J., Bejaoui, M. & Boumaiza, M. (2011c). Altitudinal distribution of aquatic beetles (Coleoptera) in northern Tunisia: relationship between species richness and altitude. *The Coleopterists Bulletin,* vol. 65(1), 53–62.

[63] Dajoz, R. (1971). Precis of ecology. *Dunod,* Paris, p. 1–434.

[64] Daget, P. & Gordon, M. (1982). Analysis of the ecology of species in communities. *Coll. Of ecology. Masson,* Paris, p. 1–163.

[65] Bennas, N. (2002). Water Beetles Polyphaga du Rif (Northern Morocco): fauna, Ecology Biogeography. Ph.D *Thesis in Biological Sciences, University Abdelmalek Essaâdi, Faculty of Sciences of Tetouan:* 383p.

[66] Valladares, L.F. (1988). The Palpicornios aquaticos of the province of Léon (Coleoptera, Hydrophiloidea). *Ph.D Thesis, University of Léon.* 454p.

[67] Guignot, F. (1931-1933). The Hydrocanthares of France. Hygrobiidae, Haliplidae, Dytiscidae and Gyrinidae of continental France with notes on the species of Corsica and French North Africa. *Miscellanea Entomologica, Toulouse:* 1057p.

[68] Guignot, F. (1959). Revision of African Hydrocanthares, (Coleoptera Dytiscoïdea). First part. *Annales Royal Museum of Congo Tervuren,* vol. 70, (8), 1–313.

[69] Olmi, M. (1972). The Palearctic species of the genus *Dryops* Olivier (coleoptera: Dryopidae). *Bollettino dei Musei di Zoologia, Universita di Torino,* vol. 5, 69–132.

[70] Vondel, B.J. & Dottner, K. (1997). Insecta: Coleoptera: Haliplidae, Noteridae, Hygrobiidae. *SüBwaserfaunavon Mitteleuropa.* 147 p.

[71] Jäch, A.M. & Delgado, J.A. (2008). Revision of the Palearctic species of the genus *Ochthebius* Leach XXV. The super species *O.* (s.str.) *viridis* Peyron and its allies (Coleoptera: Hydraenidae). *Koleopterologische Rundschau,* vol. 78, 199–231.

[72] Mulsant, M. (1844). Natural history of the Coleoptera of France. Of house. Paris: 197p.

[73] Jacobsen, D. (2004). Contrasting patterns in local and zonal family richness of stream invertebrates along an Andean altitudinal gradient. *Freshwater Biology,* vol. 49, 1293–1305.

[74] Tate, C.M. & Heiny, J.S. (1995). The ordination of benthic invertebrate communities in the South Platte River Basin in relation to environmental factors. *Freshwater Biology*, vol. 33, 439–454.

[75] Williams, P. Whitfield, M. Biggs, J. Bray, S. Foxa, G. Nicolet, P. & Sear, D. (2003). Comparative biodiversity of rivers, streams, ditches and ponds in an agricultural landscape in Southern England. *Biological Conservation*, vol. 115, 329–341.

Chapter 2

Dung Beetles of Chile, with Emphasis in La Araucania Region

Ramón Rebolledo Ranz, Ricardo González Jiménez, Mario Elgueta Donoso, Rubén Palma Millanao, Vivian Medel Meza and Mauricio Reyes Schencke

Additional information is available at the end of the chapter

http://dx.doi.org/10.5772/67302

Abstract

Dung beetles are insects that provide a large-scale ecosystem service worldwide through their role in the decomposition of manure from livestock, thereby providing a series of environmental services, such as nutrients recycling, control of internal parasites of livestock whose eggs are in the feces, soil aeration, spreading of seeds and maintenance of ecological balance. Dung beetles are broadly classified according to their nesting behavior in three categories as telecoprids, paracoprids and endocoprids. Telecoprids are the rollers that make balls from feces and roll them into the ground; paracoprids are the tunnellers that bury the dung balls at different depths, forming galleries in the ground below or next to the food source and endocoprids, who are the dwellers that raise their larvae inside feces. There are 10 native species of dung beetles recorded in Chile, apart from 10 species of Aphodiinae, plus two introduced species, such as *Onitis vanderkelleni* and *Onthophagus gazella*. Dung beetles species were prospected in La Araucania Region and registered *Homocopris torulosus*, *Frickius variolosus*, *Podotenus fulviventris* and *Aphodius pseudolividus*. We found that species from genus *Homocopris*, *Podotenus* and *Aphodius* were distributed from 0 to 2000 m above sea level, while *F. variolosus* was distributed over an altitude of 350 m.

Keywords: dung beetles, Scarabaeidae, Geotrupidae, Chile

1. Introduction

Dung beetles (*Coleoptera*: *Scarabaeidae* and *Geotrupidae*) are well known because of their feeding habits, consisting in collecting the dung from various animals. In some countries, they are called scarabs [1–3]. Dung beetles have a worldwide distribution with a range of about 8500 species. They are present in every kind of environment such as deserts, forests, savannas, tropical or cold grasslands and urban areas [4–8]. Dung beetles appeared on Earth during the late Cretaceous, where they began taking advantage of the huge amount of accumulated manure and the absence of other insects or organisms that could play the role as decomposers (see Refs. [5, 8–10]).

The feeding habits of these insects bring significant environmental benefits and ecosystem services, such as recycling of soil nutrients, stimulation of plant development, spreading of seeds, biological control of parasite load in manure and indirect help in pollination [10–14].

In grasslands, the insects activity generates an increase in the rate of manure decomposition, allowing nutrient recycling, and also, the reuse of grassland, once feces have been cleared [15, 16]. The introduction of dung beetles has been used as a grassland management decision in other parts of the world. For example, in Eastern Island, the species *Onitis vanderkelleni* (Lansberge)and *Onthophagus gazella* (Fabricius) were introduced for grassland management purposes, and the last was introduced in Australia and other countries, too [17, 18]. In America, *O. gazella*, native to sub-Saharan Africa, is now introduced in the United States of America, Mexico, Guatemala, Nicaragua, some Caribbean islands, Colombia, Venezuela, Brazil, Peru, Bolivia, Paraguay Uruguay and Eastern Island and has not been recorded from continental Chile.

Dung beetles have received great attention in recent years due to their higher agro-ecological importance in crops and environmental management, especially due to their role in increasing soil fertility and parasites control in cattle. So much so that it is estimated that biological activity of dung beetles reaches an economic contribution to agriculture of US$380 million per year in the USA [19–21]. Dung beetles are also considered key important due to ecosystem services they provide worldwide by eliminating the remains of manure, derived from livestock production systems. These services are directly related to improving yields and crop sustainability; in addition, seeds dispersion become significant importance in reforestation, restoration and conservation of forests [19, 22–24].

Dung beetles have had great religious importance since ancient times, at such a point that have been considered as religious symbols in rituals or, as deities. Besides, beetles have been used as food, in medicine, and have played important roles in traditional culture and folklore (see Refs. [25–28]). Egyptians emphasized at that point, who observed dung beetles behavior and their ecology over 5000 years, and even compared a beetle with their deity, god Khepri. Beetles were considered sacred by Egyptians due to the role they played in the renewal, transformation and resurrection of life. There were four aspects that determined the relationship between facts and the biological-theological explanation [29]: (1) beetles looking for droppings, something to what the Egyptians attributed a sacred character; (2) a beetle rolling a ball

http://dx.doi.org/10.5772/67302

of dung and burying it in the ground; (3) the fact that most of their metamorphosis occurred underground; and (4) the fact that eggs hatch and restart the life cycle.

Dung beetles have been used as bio-indicators during the last three decades and also in biological models [23, 28, 29]. However, productive practices in croplands, prairies and even forests have lowered significantly the biodiversity of dung beetles, mainly due to the negative effects of chemical control against weeds and pests, such as herbicides and pesticides, as, for example, when controlling the horn fly (*Haematobia irritans*), or internal parasites of cattle (see Ref. [24]).

A decline in the population of dung beetles in prairies has great importance in intensive livestock production systems, where the greatest accumulation of manure cannot be eliminated by native insects. In those cases, as a management measure, entomologists propose the introduction of exotic dung beetle species. However, previous to such decisions, it becomes necessary to carry out studies for assessing possible losses of native species, as it has already happened in other parts of the world, such as in Colombia where the release of *O. gazella* would have produced smaller populations of native dung beetles [30–32]. The exotic *O. gazella* was released in Texas, USA in 1972, and since that year, the species have been widely spread across most of the countries of South America. This fact would indicate this species is affecting native populations of native dung beetles [33].

There are 10 native species identified in the literature as native species of dung beetles in Chile, four paracoprids and five telecoprids with no records of endocoprids [34]. They are *Frickius costulatus, Frickius variolosus, Taurocerastes patagonicus, Homocopris torulosus, Homocopris punctatissimus* and 10 species of genus *Aphodius*, plus two introduced species, the *O. vanderkelleni* and *O. gazella* [29, 33]. There are two species from subfamily Aphodiinae in La Araucanía, *Podotenus fulviventris* and *Aphodius pseudolividus* [34, 35]. The native paracoprids and telecoprids are detailed in **Table 1**.

Feeding class	Family	Species
Paracoprids	Geotrupidae	*Frickius costulatus* (Germain, 1897)
		Frickius variolosus (Germain, 1897)
	Scarabaeidae	*Homocopris torulosus* (Eschscholtz, 1822)
		Homocopris punctatissimus (Curtis, 1844)
Telecoprids	Scarabaeidae	*Megathopa villosa* (Eschscholtz, 1822)
		Scybalophagus rugosus (Blanchard, 1843)
		Tesserodoniella elguetai (Vaz de Mello and Halffter, 2006)
		Tesserodoniella meridionalis (Vaz de Mello and Halffter, 2006)
	Geotrupidae	*Taurocerastes patagonicus* (Philippi, 1866)
Endocoprids	Scarabaeidae	No records and there is not any species reported uncertain. *Podotenus (Podotenus) fulviventris* (Fairmaire and Germain, 1860)

Table 1. Native species of dung beetles in Chile according to nesting behavior.

There are few studies on both native and introduced dung insects in South Central Chile. One study in horse droppings was focused on the biology of *H. torulosus* (previously known as *Pinotus* or *Dichotomius torulosus*) [30, 35–38], *F. costulatus* [31], *A. pseudolividus* and *Megathopa villosa* [33]. In Los Rios region, there are some studies carried out during the autumn season to sample the presence and activity of dung beetles over grasslands [31, 34, 35]. Currently, there are no published records on the distribution and characterization of the environments in which it is possible to find these insects.

2. Methodology

The sample was conducted in five agro-ecological areas of La Araucanía Region covering approximately 3,184,200 ha (see Ref. [39]). La Araucania Region is located in Southern Chile and corresponds to a transitional zone between a dry Mediterranean and climate. As most of South Central Chile, the region is divided into five different landscapes: (1) coastal rain fed area (CRF) that includes the coast (west side of coastal range) and the coastal range; (2) the interior rain fed area (IRF) that includes the dry east-side of the coastal range; (3) the central plain (CP) that includes the flat lands of the central valley; (4) the pre-Andean area (PA) that includes lands of the piedmont of the Andes Mountains; and (5) the vo1canic Andean area (VA) that corresponds to Andes Mountains. The Pluviometry varies from west to east, with notorious lower precipitations in the east side of the coastal range. Temperature decreases from west to east in the measure altitude increases to up in the Andes. The differences in temperature and altitude determine some vegetation differences. Therefore, in the coastal range and its east side will be dryer with sclerophyll vegetation; the central valley will be a transition between sclerophyll and rainy temperate and evergreen forests, with these temperate forests covering until the piedmont of the Andes and, with changes in species distribution, leaving space to species better adapted to low temperature. At higher altitude, forests of Chilean monkey puzzle trees (*Araucaria araucana*), lenga birch (*Nothofagus pumilio*) or Ñirre birch (*Nothofagus antarctica*) take place replacing the species of the temperate forests.

In each agro-ecological area described above, we chose between 5 and 15 sampling areas based on the presence in pastures of manure from cattle and horses, and within native and exotic forests. The sampling areas were chosen and marked over the agroecologic map of the Araucanía Region [39].

At the same time, in each of these sampling areas, we chose as many as possible sampling points, where we sampled and measured the presence and abundance of insects in, over and around feces, and prepared a list of dung beetles. The sampling process was conducted during spring, summer and autumn, for 3 years (2008–2010). The observations were made during 30 minutes at each sampling place, and then samples with organic material were taken for analyses to laboratory. Once in the laboratory, the samples were compared against the material of reference (specimens) that are kept at Museum of Entomology, to assure a correct identification and classification of sample specimens. In a parallel procedure, the classification keys were examined in the literature. Insect samples were deposited in the Entomological collection at the Museum of Entomology.[1] Each sampling place was georeferenced using a GPS devise "Garmin".

In addition to the sampling areas, we had installed two light traps. Such light traps included a container to receive the insects. The traps were 1.1 m high and weighted approximately 5 kg each. The traps contained an ultraviolet illumination tube of 43 cm long and 20 W.

One in a farm located in the urban border of Temuco city, southern Chile and the second in the Experimental and Model Farm Maquehue, located at 38°50′27.20S 72°41′34.32″W at 12 km far from Temuco city. Both traps were set up for 10 years (year 2000–2010). The samples from each light trap were collected on daily basis and taken to the laboratory, for analysis and specimen classification.

In relation to the scientific names, we followed the classification for the Scarabaeoidea of South America and the most recent taxonomy used in the literature [34, 35].

3. Results and discussion

In our prospection performed in the sampling process (2008–2010) and during years 2000 and 2010 (light traps), we found four species of dung beetles in La Araucanía, which are detailed in **Table 2**.

Feeding class	Family	Species
Paracoprids	Geotrupidae	*Frickius variolosus* (Germain, 1897)
	Scarabaeidae	*Homocopris torulosus* (Eschscholtz, 1822)
Endocoprids	Scarabaeidae	*Podotenus fulviventris* (Fairmaire and Germain, 1860)
		Aphodius pseudolividus (Balthasar, 1941)

Table 2. Native species of dung beetles sampled in La Araucanía, Chile, period 2000–2010 according to nesting behavior.

We did not register the rest of dung beetles described in the literature nor the exotic ones. Apart of those detailed in **Table 2**, we would have expected to find in La Araucanía Region, the native species *H. punctatissimus* (Curtis, 1844), a species recently revalidated, *M. villosa* (Eschscholtz, 1822), and *F. costulatus* (Germain, 1897) [29, 33, 35]. In regards the introduced species *O. vanderkelleni* (Lansberge) and *O. gazella* (Fabricius), we did not succeed at registering them, in spite the fact their presence have been previously reported for Easter Island, an insular Chilean territory, and cited for different parts around the world [34, 37–40].

From the four species recorded in the study, *F. variolosus* and *H. torulosus* were found present in most of the samples along the year, while *P. fulviventris* always was recorded in early spring

[1]Laboratory of Applied Entomology at the Faculty of Agricultural and Forest Sciences, University of La Frontera, Chile.

and in huge amounts, with approximately 1800 individuals per feces, from both cattle and equine. However, *A. pseudolividus* was always recorded in the period from late February until mid-May to June, with counts of 4000 individuals per feces from cattle and equine. According to our literature review, there is no other record on this behavior; therefore, the current sample observations are the very first time this behavior is observed.

Regarding the seasonal flight of the two species *Aphodius* registered in this study (**Figure 1**), it is seen that, adult specimens of *A. pseudolividus* appeared between February and April, with a higher population peak in March. However, the adult specimens of *P. fulviventris* appeared in large quantity, in feces from cattle and equine, only between September and October.

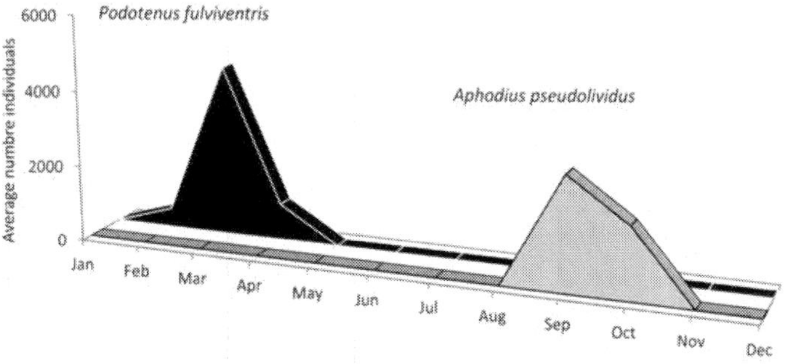

Figure 1. Flight and seasonal abundance of adults *of A. pseudolividus* and *P. fulviventris.*

Results indicate these species have completely unknown behavior in regards their univoltine life cycle. At least, one can know where they are when adults do not appear. It is interesting the fact that adults of both species coexist with adults of *H. torulosus* and *F. variolosus*. Apparently, there would be competition for the resource (feces) by having each species, a different feeding strategy (Paracropid and endocropid).

Regarding the samples collected with the light traps (**Figures 1** and **2**), specimens of *A. pseudolividus* and *H. torulosus* showed to be strongly attracted to light. However, it must be noted that specimens of *P. fulviventris* were not captured by these traps, but by manual collection. Adult specimens of *H. torulosus* showed to have a seasonal flight over the year, with a decrease in population in winter (July), when no adults flight, and an increasing population from August onwards, reaching a higher peak of individuals in late summer (February–March). It was difficult to determine the quantity of generations per year, given the fact that adults flight almost the whole year; however, we assume that there were present at least two generations a year (bivoltine). This situation should be validated and checked in the field in future research.

It was found that *H. torulosus* was sharing the food resource with *F. variolosus, P. fulviventris* and *A. pseudolividus* as adults. It was usual to find specimens of *H. torulosus* and *F. variolosus* together in the same dung, from equine and cattle. The number of individuals varied greatly from one sample to another, at sampling altitude near 500 m, was the two species are sympatric (**Figure 3**).

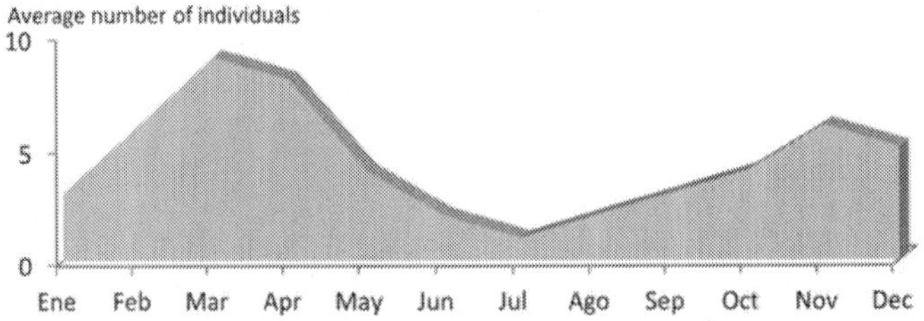

Figure 2. Flight and seasonal abundance of *Homocopris torulosus,* 10-year average 2000–2010 of monthly average registered by light traps at Maquehue, Temuco.

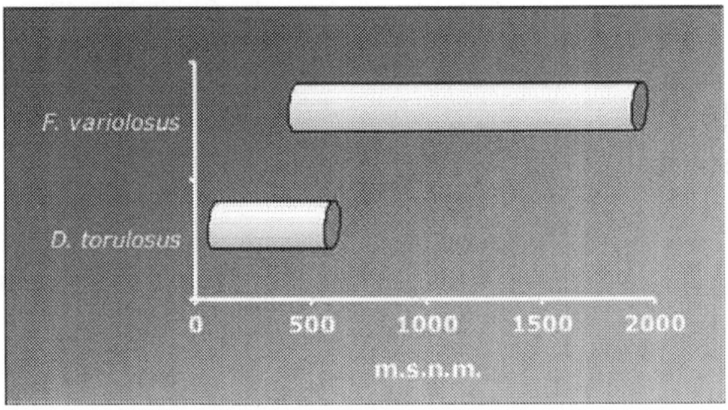

Figure 3. Altitudinal distribution pattern of *H. torulosus* and *F. variolosus.*

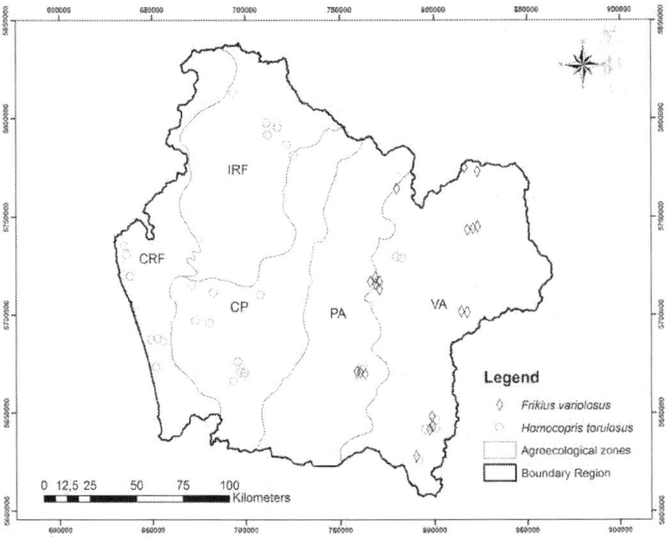

Figure 4. Distribution of *H. torulosus* and *F. variolosus,* across agro-ecological areas (CRF—coastal rain fed; IRF—interior rain fed; CP—central plain; PA—Pre Andean; VA—Vo1canic Andean).

Regarding the distribution of species in La Araucanía Region (**Figures 4–6**), it should be noted that according to the sampling methodology, where the observations were made at different times, we were able to describe the sample distribution of *H. torulosus* and *F. variolosus* over time.

Figure 5. Adults of *Homocopris torulosus*.

Figure 6. Adults of *Frickius variolosus*.

In **Figure 4**, we can see that both species are well distributed across the region, but *H. torulosus* is distributed between 0 and 850 m above level and *F. variolosus* was recorded as present between 400 and 2000 m above sea level (**Figure 3**). We registered between 1 and 50 specimens per dung (**Figure 2**) with a marked tendency of insects to prefer horse feces, in occurrence and abundance. On the other hand, *F. variolosus* was registered at altitude higher than 500 m. We could not identify a preference of *F. variolosus* for dung from a particular animal. Sample indicated that specimens of *F. variolosus* preferred sectors in the piedmont of the Andes. The registered abundance of *F. variolosus* was from a couple up to 250 of specimens per dung, which coincides with previous reports (see Ref. [29]).

In regards the preference for a type of vegetation, samples were taken in both native forest and forestry plantations and, according to registered data, both specimens of *H. torulosus* and *F. variolosus* preferred native forest to plantations.

4. Conclusions

According to the present study, we found evidence that dung beetles are well distributed in La Araucania region, where we were able to collect specimens from four species. They are *H. torulosus, F. variolosus, P. fulviventris* and *A. pseudolividus*.

All the species registered in this study are native. It is worth to mention that we did not find any specimen of those introduced species two decades ago. Besides, we could not find any record of previous studies that have found any specimen from such introduced species, which help in confirming our findings.

Also, we could not find any evidence, as mentioned in the literature, that dung beetles are present in higher frequency in manure located in native forests. We only could confirm that dung beetles are abundantly present in manure, independently of the type of cattle or type of forest (native or exotic) or vegetation cover, that is, they are present in forests, grasslands or any vegetation cover where there is manure that have been directly deposited over the ground by cattle. It is worth to note that dungs from wild animals like puma or foxes were not checked as part of this study.

The distribution of *H. torulosus* across La Araucania regions goes from 0 to 2000 m above sea level, while *F. variolosus* is distributed from 350 m and higher. In this study, we have registered for very first time the behavior of adults' specimens of *A. pseudolividus* that showed a higher abundance of higher than 1500 individuals per feces in the period from end of summer until late mid-Autumn (February until mid-May).

Finally, in the particular case of *P. fulviventris*, on who there is no previous published records on its habits, this study constitutes the first publication describing aspects on its biology, for example, that adults appear in very high densities in September and that they share the good substrate with *F. variolosus* and *H. torulosus*, and apparently, without any competition problems for the food resource.

Author details

Ramón Rebolledo Ranz[1]*, Ricardo González Jiménez[1], Mario Elgueta Donoso[2], Rubén Palma Millanao[3], Vivian Medel Meza[1] and Mauricio Reyes Schencke[1]

*Address all correspondence to: ramon.rebolledo@ufrontera.cl

1 Faculty of Agricultural and Forest Sciences, University of La Frontera, Temuco, Chile

2 Natural History Museum, Santiago, Chile

3 Faculty of Agrarian Sciences, University of Talca, Talca, Chile

References

[1] Halffter, G. and Edmonds, W.D. The nesting behavior of dung beetles (Scarabaeinae). An ecological and evolutive approach. Instituto de Ecología México. 1982. México D.F., Man and the Biosphere Program UNESCO, 177 p.

[2] Hanski, I. The dung insect community. From saprophagy to coprophagy. In Hanski, I. and Camberfort, Y. (eds). Dung beetle ecology. Princeton University Press, Nueva Jersey. 1991. pp. 5–21.

[3] Camberfort, Y. From saprophagy to coprophagy. In Hanski, I. and Camberfort, Y. (eds). Dung beetle ecology. Princeton University Press, Nueva Jersey. 1991. pp. 22–35.

[4] Camberfort, Y. and Hanski, I. Dung beetle population biology. In Hanski, I. and Camberfort, Y. (eds). Dung beetle ecology. Princeton University Press, Nueva Jersey. 1991. pp. 36–50.

[5] Davis, A.L. Community organization of dung beetles (Coleoptera: Scarabaeidae): Differences in body size and functional group structures between habitats. African Journal of Ecology. 1996, 34:258–275.

[6] Richards, O. and Davies, G. Imms's General Book of Entomology. Volume 2: Classification and Biology (Spanish Version). Omega, Barcelona, Spain. 1983. 998 p.

[7] Galante, E. and Marcos, M. Detritivores, coprophagous and necrophagous (In Spanish). Bol. S.E.A. 1997, 20:57–64.

[8] Martin, F. Sacred Beetles (In Spanish). Sociedad Entomológica Aragonesa. 1997, 20:327–330.

[9] Edwards, P. and Aschenborn, H. Patters of nesting and dung burial in Onitis dung beetles: Implications for pasture productivity and fly control. Journal of Applied Ecology. 1987, 24:837–851.

[10] Lastro, E. Dung beetles (Coleoptera:Scarabaeodae and Geotrupidae) in North Carolina pasture ecosystem. Thesis. MSc in Entomology, North Carolina State University. 2006. 121 p.

[11] Losey, J. and Vaughan, M. The economic value of ecological services provided by insects. Bioscience. 2006, 56:311–323.

[12] Nichols, E., Larsen, T., Spector, S., Larsen, T., Amezquita, S. and Favila, M. Ecological functions and ecosystem services provided by Scarabaeinae dung beetles. Biological Conservation. 2008, 141:1461–1474.

[13] Behling, C. Contribution of African dung-beetles to improving soil fertility (In Spanish). 2016. Available at www.AVPa.ula.uc/Congresos/Seminario_pastoX/Conferencia/a17-Cesar/2. Behling.pdf/. Retrieved on 7 of June 2016.

[14] Rojas, S. Biological control of pests in Chile (In Spanish) INIA Books Collection Number 12. Institute of Agricultural Research (In Spanish, Instituto de Investigaciones Agropecuarias, INIA), La Cruz, Chile. 2005. 123 p.

[15] Arriaga-Jiménez, A., Lumaret, J. and Halffter, G. Coprophagous beetles as bio-indicators of the conservation state of protected wild areas at the east of the transverse volcanic system (In Spanish). 2015. Available at http://www.mufm.fr/sites/mufm.univ-toulouse.fr/files/evenement/symposium/ponencias/alfonsina_arriaga_jimenez.pdf. Retrieved on 5 of June 2016.

[16] Howden, H. Geotrupinae (95-106). In Morón (Ed.). Atlas of beetles of Mexico (Coleoptera: Lamellicornia: Scarabaeidae, Trogidae, Passalidae y Lucanidae) (In Spanish). Vol. II. 2003. 227 p.

[17] Howden, H. Notes on Bolboceratini from Northeastern México with description of a new Bolbocerosoma Schaeffer (Coleoptera: Scarabaeidae: Geotrupinae). Folia Entomológica Mexicana. 2005, 44(2):213–218.

[18] Crespo, G. Factors determining nutrient recycling in permanent grazing lands, advances in their modelling (In Spanish). Revista Colombiana de Entomología. 2015, 49(1):1–10.

[19] Martínez, I., Cruz, M., Montes, E. and Suarez, T. The role of dung beetles in grazing lands (In Spanish). Veracruz, México. 2011. 71 p.

[20] Basto, G., Rodriguez, R., Delfín, H. and Reyes, E. Dung-beetles (Coleoptera:Scarabaeidae: Scarabaeinae) in cattle ranches of Yucatán, Mexico (In Spanish). Revista Mexicana de Biodiversidad. 2012, 83:380–386.

[21] Barriga, J. Coleoptera Neotropical. 2010. Available at http://www.coleoptera.neotropical.org/paginas/3nz_familias/SACARABAEOIDAE/1sp/Geotrupidae/Frickius costalatus.html. Retrieved on 15 of June 2016.

[22] Bustos, F. and Lopera A. Bait preferences of coprophagous beetles (Coleoptera: Scarabeidae) in a remnant dry tropical forest at the north of Tolima (Colombia) (In Spanish). Escarabajos de Latinoamérica: Estado del conocimiento. Vol. 3. Monografías Tercer Milenio, SEA Zaragoza, España. 2003. pp. 59–65.

[23] Hernández, B., Maes, M., Harvey, C., Vílches, S., Medina, A. and Sánchez, D. Abundance and biodiversity of coprophagous beetles and diurnal butterflies in a rancher landscape of Rivas, Nicaragua (In Spanish). Agroforestería en las Américas. 2003, 10(39–40):93–102.

[24] Morales, C., Guevara, F., Espinosa, J., Gómez, C., Llaven, J. and Hernández, R. Dung-beetles (Coleoptera: Scarabaeidae) of CUTT San Ramón, Villaflores, Chiapas, Mexico (In Spanish). Entomología Mexicana 2014, 1:1129–1132.

[25] Scholtz, Ch., Davis, A. and Kryger, U. Evolutionary biology and conservation of dung beetles. Pensoft Publishers, Sofia, Bulgaria. 2009.

[26] Jones, A.G., Forgie, S., Scott, D. and Beggs, J. Generalist dung attraction response in a New Zealand dung beetles that evolved with an absence of mammalian herbivores. Ecological Entomology. 2012, 37:124–133.

[27] Guiller, P. and Doube, B. Experimental analysis of inter and intraspecific competition in dung beetle communities. Journal of Animal Ecology. 1989, 28:129–142.

[28] Horgan, F. and Fuentes, R. Asymmetrical competition between neotropical dung beetles and its consequences for assemblage structure. Ecological Entomology. 2005, 30:182–193.

[29] Noriega, J., Moreno, J. and Otavo, S. Fifteen years after the arrival of the coprophagous beetle Digitonthophagus gazella (Fabricius, 1787) (Coleoptera: Scarabaeidae) to Colombia: Invasion process and possible effects of its settlement. Biota Colombiana. 2011, 12:35–43.

[30] De la Vega, C., Elizalde, H., González, M. and Reyes, C. Dung-beetles for cattle ranching in the Aysen Region, Chile (In Spanish). Instituto de Investigaciones Agropecuarias, Coyhaique. Boletín INIA N° 295. Chile. 2014. 84 p.

[31] Joseph, C. The Pinotus torulosus Eschsch. Revista Chilena de Historia Natural. 1929, 33:31–46.

[32] Peña, L. The Frickius costulatus Germain 1897, (Coleoptera: Scarabaeidae). Revista Universitaria Santiago. 1968, 53(31): 95–96.

[33] Zunino, M. and Barbero, E. Beetles, cattle, grasslands: Some deontological considerations (In Spanish). Folia Entomológica Mexicana 1993, 87: 95–101.

[34] Ovalle, M. and Solervicens, J. Some observations on the biology of Megathopa villosa (In Spanish). Boletín Museo Nacional de Historia Natural de Chile. 1980, 37:235–246.

[35] Ripa, R., Rojas, P. and Velasco, G. Releases of biological control agents of insect pest on Easter Island (Pacific Ocean). Entomophaga. 1995, 40:427–440.

[36] Smith, A.T.B. and Skelley, P.E. A review of the Aphodiinae (Coleoptera: Scarabaeidae) of South America. Zootaxa. 2006, 1458.

[37] González, M. Autumn presence of paracoprid native dung-beetles (Sacarabaeidae: Scarabaeinae) in second-growth native forests and associated wild-prairies (In Spanish). Tesis Ingeniero Agrónomo, Universidad Austral de Chile. 2010. 81p.

[38] Klemperer, H. Brood ball construction by the non-brooding Coprini *Sulcophanaeus carnifex* and *Dichoromius torulosus* (Coleoptera: Scarabaeidae). Ecological Entomology. 8(1):61–68.

[39] Rouanet, J., Romero, O., Demanet, R. Agroecolocial áreas in the IX Region, Chile. Description (In Spanish). IPA Carillanca. Instituto de Investigaciones Agropecuarias, Temuco, Chile. 1988, 7(1):18–23.

[40] Simmons, L. and Ridsdill, S. Reproductive competition and its impact on the evolution and ecology of dung beetles. In Simmons, L.W. and Ridsdill-Smith, T.J. (eds.). Ecology and evolution of dung beetles. First Edition. Wiley-Blackwell, Oxford. 2011.

Chapter 3

Weevil Borers in Tropical Fruit Crops: Importance, Biology and Management

Luis Martín Hernández Fuentes,
Alvaro Castañeda Vildozola and
Mario Alfonso Urías-López

Additional information is available at the end of the chapter

http://dx.doi.org/10.5772/66635

Abstract

Weevils are an economically important group of Coleopteran insects of the family Curculionidae. This is the largest insect family in the superfamily Curculionoidea. They may be found almost everywhere and more than 3000 species in near of 500 genera occur in North America. Most of them are plant feeders and others are key pests. These weevils use the snout to feed on plant tissues and notch egg-laying sites on it. Adults drill holes and feed in seeds, fruits and other reproductive parts of the plants. Some of the most notable examples of weevils include *Conotrachelus* spp. on avocado and guava, *Optatus palmaris* on anonaceous fruits, *Heilipus lauri* on avocado, *Hypothenemus hampei* on coffee berry and others. The presence of some of these species requires establishing measures of restriction when the product is for exportation. Management practices and postharvest treatments are required to ensure that the fruits will be free of larvae. In this chapter the main species of weevils in the most important tropical fruit are included, such as avocado, coffee, guava and anonas fruits. Weevils of economic and quarantine importance are considered.

Keywords: weevils, tropical fruit, management, importance

1. Introduction

The members of the Curculionidae family are called weevils, since most of them have more or less prolonged head anteriorly as a peak or nose; this term is less appropriate (and rarely used) for Platypodinae and Scolytinae, because there is only little development of the peak in these two subfamilies.

Borers and weevils, including bark and ambrosia, are the members of this family most commonly found. This family shows considerable variation in size and shape. The peak is well developed in most species, with capitate antennae appearing in the middle of the peak. In Scolytinae, Platypodinae and some Cossoninae, there is no peak. The Curculionodae species present a complete metamorphosis (egg, larva, pupa and adult); the family has great economic importance by the number of pest species.

The male external genitalia comes in two types: Ortocera or Gonatocera—in the first one, plates of the aedeagus are known as tecto and the pedon; in the second one, only it remains the pedon; and in some families, there is a clear reduction of the tecto. The antennae are presented in two types—straight or elbowed—in addition some appear to be geniculate [1].

Almost all weevils are phytophagous and among these, there are many important agricultural and forest pests. Almost any part of the plant can be attacked, from the roots to the aerial parts; usually the larvae feed within plant tissues and adults make holes in the fruits, nuts and other parts [2], except for myrmecophilous, saprophagous and predatory species [3].

As mentioned most of them are herbivorous and economically important pest because they affect crops by reducing yields or affect stored products. They can actually cause large losses since the female makes holes somewhere in the plant, either branches or fruits to oviposit. In other cases, females lay eggs on the surface and when the larvae emerge, they pierce the fruit or branches, feeding on plants tissues; in the case of ambrosial weevils, they feed on fungi, which are cultivated within the plant, causing direct and indirect damages.

We considered the main species of weevils in the most important tropical fruits, for example, avocado, coffee, guava and anonas fruits. Weevils of economic and quarantine importance are considered. We talk about topics like biology, importance, damages and suggestions to their control.

2. Big avocado seed weevil, *Heilipus lauri* Boheman

2.1. Description

The egg of *H. lauri* is oval; its surface is finely reticulate with pentagonal forms. It is 1.40 ± 0.06 mm long and 0.87 ± 0.03 mm wide. At the moment of the oviposition, the chorion is bright white; during its embrionary development, its color changes from clear brown to dark brown. The forth instar larvae reach 24.21 ± 1.49 mm length. The body is robust, curved with color opaque white; the measurement of the head is 1.87 ± 0.06 mm wide and color clear brown and it is not retracted in the prothoracic segment. The pupae are the exarate type with color creamed white and 16.97 ± 1.25 mm in length. Adult (**Figure 1**) with opaque integument is black reddish in most of the surface of the body, except legs which are reddish. The length of the body, excluding the head (rostrum), is 14.77 ± 0.87 mm in the case of females and 13.78 ± 0.76 mm in males. The head (rostru m) reaches 7.29 ± 0.67 mm in the females and 5.32 ± 0.28 mm in the males. It is curved in the females and slightly more thick than in the males; in malesthe rostrum is short and straight. A distinctive characteristic of this species is the pair of irregular enlarged spots, formed by copactation of small opaque orange oval scales. The first pair is the biggest and it is located 2/5 at the base of the elytra and the second one, 1/5 of the apice, located almost over the periapical callus [4].

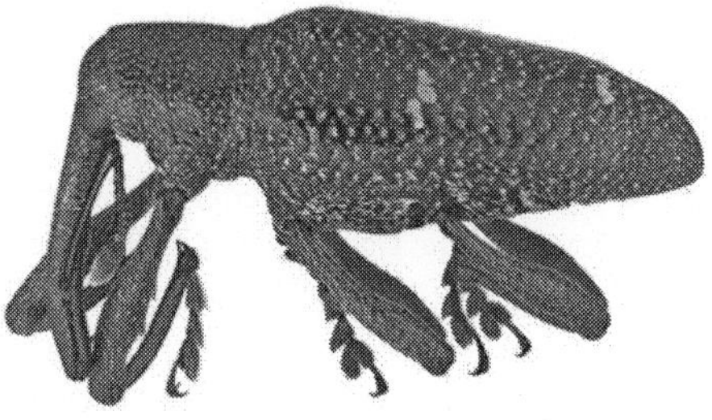

Figure 1. Male of *Heilipus lauri,* big seed borer of avocado.

2.2. Distribution

Heilipus lauri is a species with broad altitudinal adaptation, from 594 to 1900 over sea level. Its distribution includes Mexico [5, 6], Honduras (D. Cave 2013, *in litt.*) and Colombia [4]. Its presence in Mexico includes the states of Morelos, State of Mexico, Puebla, Hidalgo, Oaxaca, Puebla and Veracruz [2, 5, 7].

2.3. Host plants and biology

According to [8] in the State of Mexico, *H. Lauri* causes damage in avocado Hass, Fuerte and Colin V-33. Medina [9] documented that in the Morelos state. *H. lauri* has a preference for fruits of criollo avocado, *P. americana* var. drymifolia and Choquette cultivars Hass and Fuerte. Recently [7] added to *P. americana* var. *american* as a new host. Castañeda Vildozola [5] reported the presence of *H. lauri* in fruits of *Persea schiedeana* Nees in Huatusco and Zongolica, Veracruz. The information from these authors represents the first report of *H. lauri* damaging a different fruit of that of avocado species. The egg incubation period is 10.87 ± 0.45. In Ixtapan de la Sal Mexico State, females ovipositing have been recorded from March to September. This behavior is closely linked to the phenology of avocado in the region. Larva: It is found in a cotyledon where it feeds and develops its larval stage; its life cycle completes in 48.51 ± 2.30 days. A typical feature of *H. lauri* is that it develops a single larva per fruit and even two; these do not destroy the seeds, when they are used. In Ixtapan de la Sal, the presence of larvae of *H. lauri* occurred from April to August. Pupa: Close to pupation, the larvae form a pupation chamber inside one cotyledon, lodge and stand still to be transformed into pupa. The pupa occurs from late August to September. Adult: The longevity of adults in laboratory conditions was 309.55 ± 86.72 days. Adults have diurnal habits; their presence is very noticeable during fruiting avocado period from 09:00 to 19:00 h. Field emergence of new adults occurred from September to April 2005. *H. lauri* biology is closely linked with the phenology of avocado in

Ixtapan de la Sal. The availability of food throughout the year (fruits and foliage), avocado genetic diversity and environmental conditions play a role in favor of the presence of this insect [10].

2.4. Damage

Females drill and lay eggs directly in growing fruits (**Figure 2**) causing its injury inside; once the larvae emerge, they feed through the pulp to reach the seed where it completes its life cycle. When heavy infestations are recorded and no management is used, infestations of larvae can damage up to 67% of the fruits; in orchards with management of the pest, it has been quantified losses of 3% [9].

Figure 2. Damages in fruits by *Heilipus lauri*, big seed borer avocado.

3. Small avocado seed weevil, *Conotrachelus perseae*

3.1. Description

Egg: This is elliptical; the chorion is semitransparent when oviposited and acquires a graying color when the larva is next to emerging; it is less than 1 mm in length [1]. Larva is yellowish white and dark cephalic capsule. Prothorax has a suture open V-shaped and three ventral setae. The mesothorax and metathorax present two dorsal lobes; these segments in the posteroventral region have a dorsal setae. All abdominal segments, except the last one, have three dorsal lobes and three ventral setae. Well-developed larvae reach a length of 6 mm [11, 12]. In pupa, the prothorax has five pairs of lateral setae and three pairs forming two rows on the sides. The mesothorax has two pairs of setae located within a lobe and in the middle part is the other rounded lobe. Abdominal segments are characterized by two pairs of setae in the ventral part of each segment; the eighth and ninth segments present a pair of lateral setae [12]. Adults (**Figure 3**) are dark brown; prothorax, in dorsal view, is strongly constrained in the apical portion [3, 13–15]. Humeral region of the elytra of *C. perseae* is wider than the base of prothorax.

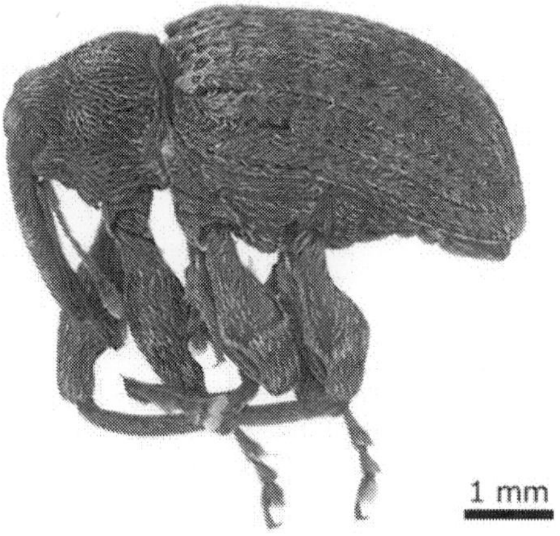

Figure 3. *Conotrachelus perseae*, small avocado seed weevil.

3.2. Distribution

Conotrachelus perseae was first described from specimens collected in Guatemala [14]. This pest has been reported in Honduras, Costa Rica and Mexico [15, 16]. In Mexico, its presence has been documented in Chiapas Guanajuato, Michoacan, Puebla Hidalgo, State of Mexico and Veracruz [17, 18]. *C. perseae* adults are nocturnal and stay hidden during the day. Damaged fruits show small superficial holes. The larvae feed in seeds and the pupae are formed out of the fruit, commonly on the floor.

3.3. Host plants and biology

Both weevils have been collected in "Criollo" avocados, Hass and Fuerte [3, 19]. Castañeda Vildozola [17] added *Persea floccosa* as a new host of *C. perseae* in Mexico. The egg incubation period occurs between 4 and 13 days after oviposition. The duration of the larval stage of *C. perseae* is 15 to 30 days. Once the larva has completed its development inside the fruit, they migrate to ground where pupation occurs, reaching a depth of 3 to 11 cm. Pupal stage lasts 35.5 days. The longevity of adults is 111 days for male and 140 days for females. Adults are nocturnal; during the day they remained hidden in the stems and leave bent leaves and inflorescences [2, 12, 19]. In Tacambaro, Michoacan [8, 20], he concluded that the presence of the different stages of development occurs as follows: egg from late January to late

March, larvae from February to April, pupa from March to early May and adult from July to mid-September.

3.4. Damage

Adults damage fruits when they reach 2 cm diameter; fruits are perforated during oviposition. The larvae feed the pulp of the seed. Once concluded the larval stage, the fruits fall to the ground because of the destruction of the seed [12] causing loss of 85% if control measures are not applied.

4. Avocado branch weevil borer, *Copturus aguacatae*

4.1. Description

The egg is oval in shape, measuring 0.5 mm long and 0.3 mm wide; early oviposited eggs are translucent and then change to white. Larva. A well-developed larva reaches 12 mm in length and is milky white. The head capsule is retracted in the first thoracic segment. The prothorax is ossified in its highest zone and covered with small scales and laterally presents an oval spiracle. In the predorse and postdorse of the mesothorax and metathorax are present two dorsal folds. On the first abdominal segment, there are three dorsal folds and the segments II to VII with four folds. VIII to IX segments have no dorsal folds. All abdominal segments have setae varying in number, size and arrangement. The pupa is exarata and measures 5–8 mm long and 2–2.5 mm wide [9]. They are creamy white. The rostrum reaches the metathoracic coxae. Adults are romboidal shaped, measuring 3.77–5.0 mm long and 2.0–2.5 mm wide [21]. They are black to reddish black with small white scales, red, orange, or black. The females are bigger than the males. They have an almost triangular prosternal ridge between coxae.

4.2. Distribution

The presence of the avocado branch borer has been registered in the states of Morelos, Michoacan, Puebla, State of Mexico, Queretaro, Guerrero, Nayarit and Oaxaca [9, 22, 23].

4.3. Host plants and biology

The avocado is reported as only host [9, 23]. Medina [9] reported that in the state of Morelos, *C. aguacatae* cycle lasts 200 days; egg has an incubation period of 10–12 days; the larval state lasts 120 days with five instars, each lasting 20–24 days; pupae last 15 days; and the adult has a life span of 45 days. Leos-Rodríguez [24] concluded that the biological cycle of *C. aguacatae* in Uruapan, Michoacan, had a duration of 169–192 days, the incubation period of the egg was 10–12 days, the larval stage was completed in 108–117 days and the pupa in 17–19 days and adult longevity was 5 days. The adults are diurnals; often they walk on branches with quick movements. In Ziracuaretiro, Michoacan, the presence of adults occurred from June to November; the egg state was recorded from September to June. The larvae were recorded in the months of September to June and pupae were recorded in May and July [23].

4.4. Damage

Damage is caused by larvae. The branches affected are those with greater exposure to the sun. Flowers and fruits are not damaged by the larva. The branches are affected from the epidermis to the core; damage is observed on large surface areas covered with lumps of crystallized wise. A consequence of damage, branches in production can be disrupted by the weight of the fruit [23].

5. Control strategies for *Heilipus lauri, Conotrachelus perseae, C. aguacatae* and *Copturus aguacatae* in Mexico

The importance of these species in Mexico is that they are classified as quarantine importance insects; fruit movement to avocado regions free of this pest is not allowed. This restriction is also valid for international market, specifically in the United States where the marketing of Mexican avocados were prevented since 1914 until November 1997 [25, 26].

Mexican Official Standard NOM-066-FITO-2002 has defined areas as free of seed borers and phytosanitary areas under control. The first step to prevent the spread of weevils to free areas of the pest is that the government of Mexico through the SAGARPA considers more rigorous sampling at points of entry and exit of agricultural products and makes strong awareness campaigns for people to be careful with fruit that moves from one place to another [27]. It is important to continue the implementation of studies to learn more about these insects of which is much talk, but little is known.

5.1. Control methods

5.1.1. Cultural control

It is advisable to remove the fruits with signs of damage by seed borers and prune the branches affected by this pest. All these materials should be burned for the control of eggs and immature stages; even though it is a late form of control, it is certain that it will be effective for the next cycle of the crop; less damage will be present, if control is supplemented by other measures. For *Conotrachelus perseae* it is recommended to track the ground because larvae next to pupation leave the fruit and stay on the floor to form cell pupation; this practice exposes larvae to desiccation or predators as birds.

5.1.2. Chemical control

In Mexico, some products are approved to be used in avocado orchards; an example is permethrin at doses of 200–300 cc in 100 L of water. The organophosphate insecticides, such as methyl parathion and malathion, are recommended for avocado pest control [9]. In the case of *H. lauri* and *Copturus aguacatae*, for their diurnal habits, the recommendation is to apply at early hours in the morning; for *Conotrachelus perseae* it should ideally apply at night as these insects are active from start of the night until dawn [19].

5.1.3. Legal control

It requires a local legal framework (in the stated affected by the presence of weevils) to attend a plant management program in backyard orchards which are reservoirs of this group of insects. Integrated management campaigns, which according to the NOM-066-FITO-2002 standard is to establish requirements and specifications for the phytosanitary handling and movement of plants and fruits of avocado. This standard indicates that avocado weevils are considered as an obstacle to export fruits; in consequence this is a limitation for exporting Mexican avocados to the US market. This situation remains latent, as producers of avocado in California and Florida are still dissatisfied with market opening arguing great risk of introducing weevils and other insects which may affect avocado orchards in these US states [26].

6. Guava weevil, *Conotrachelus dimidiatus*

6.1. Description

According to [28] adult presents elytra with discontinuous ridges and is less developed than *C. psidii*; the mesosterna is concave between mean coxae where few shallow spots occur; the metasternum has points more or less deep and evenly distributed and the crest is continuous in the range five; unlike *C. psidii* C, *C. dimidiatus* presents bifid nails.

6.2. Distribution

In Mexico, it was reported by misidentification of the species *C. psidii* [29]; years later [30] reported that the species belonged to *C. dimidiatus*, which is distributed in Mexico, Guatemala and Honduras [31]. This species causes major damages to the crop and is widely distributed in the producing areas of Mexico.

6.3. Host plants and biology

In Mexico, two species of *Conotrachelus* damaging guava are reported; however, *C. dimidiatus* is the species that causes the greatest damage [32]. The insect life span is one year; the egg incubation period is 6–9 days; the larva goes through a period of diapause for several months; however, on average it takes 51 days, the pupae 30 days and adult 75 days.

6.4. Damage

Once adults emerge from the soil, they fly to guava trees to feed; females lay eggs on the middle of developing fruits (**Figure 4**); after hatching, the larvae penetrate into the fruit to feed. Damaged fruits develop kidney shaped, mature and fall prematurely (**Figure 5**). In Calvillo, Aguascalientes, Mexico, the infestation of fruits averages 37.4%, which is higher in the lower portion of the tree and during the rainy season, the fruits are more susceptible when they are young, developing two to four cm of polar diameter [33]. The larvae feed the pulp, causing destruction and blackening of this and seeds (**Figure 6**). If control practices are not done or treatments are performed at inopportune seasons, damages could be extreme [22].

http://dx.doi.org/10.5772/66635

Figure 4. Damages in fruits by adults of *Conotrachelus dimidiatus*.

Figure 5. Damaged fruits by *Conotrachelus dimidiatus*.

Figure 6. Damaged fruits by larvae of *Conotrachelus dimidiatus*. Note the blackening of the pulp.

6.5. Control methods

In the orchards, control is based considering the behavior and biology of the insect, according [22, 33] to the presence of adults they have relationship with the rainy season, which are the ones to initiate the damage and infestation of fruit. Based on sampling, using a network under the tree before the start of rainy season, the detection of an adult/tree and observation of fruits with oviposition, applications of chemical insecticides, repellents and entomopathogenic are needed. Local tests to evaluate the effectiveness of these products are recommended; control is complemented with the destruction of damaged fruit that remains attached to the tree. When the fruits are harvested to export to the United States, they are treated by irradiation to eliminate, among other pests, guava weevil. According to [33, 34], the use of chemical attractant traps can be used to sample and detect weevil populations. Biological control is still not a practice used against weevil; we have found parasitoids, predators of larvae and prepupae; however, it is an activity that should be investigated and expand their knowledge in the future. Similarly, the identification of chemical attractants and pheromones as an alternative control should be integrated into weevil control.

7. Coffee berry borer, *Hypothenemus hampei*

7.1. Description

According to the description [35], the egg is bright white at the beginning to become opaque later. Chorion surface is smooth, measured 0.64 mm long and 0.32 mm in diameter. The larva is white, wormlike and fully developed measuring 1.39 mm in length and has well-developed sclerotic jaws. The pupa is white at the beginning and becomes coffee later; sexual dimorphism exists in relation to size, which is also manifested in adulthood; the female pupa measures 1.18 mm long and 0.88 mm male. The adult is 0.5–0.8 mm in length.

7.2. Distribution

The coffee berry borer (CBB) is native to equatorial Africa and is currently distributed virtually in every country where coffee is grown [36].

7.3. Biology

These insects remain within the fruit much of its life cycle, as egg, larva, pupa and adult stages. The female only leaves to infest new fruit; males remain inside until death making difficult its control [37].

7.4. Damage

In some countries the infestation levels can be up to 90%; in Nayarit, Mexico, it has been recorded up to 70% in orchards with minimum crop management. The female drills close to physiologically mature fruits; inside it lays eggs and, when the larvae emerge, feeds the endosperm of the seed; and it decreases quality and yield.

7.5. Control methods

Due to the wide distribution and importance of the coffee berry borer and the conditions for growing coffee, its control requires a large integrated approach, in which the use of traps with attractants and the application of entomopathogenic microorganisms and parasitoids predominate over the application of insecticide chemicals. The integrated management of the coffee berry borer starts with sampling and determining the economic threshold of action, in Colombia, for example, [38] recommends a maximum fruit infestation of 2% to initiate control activities. The use baited traps for catching and sampling of coffee berry borer is a widespread technique and contributes to both detection and reducing pest significantly. To this respect [39] recommend installation of 22 traps each 10 ha with a mixture of ethanol-methanol 1:3 (v/v) with 1% benzoic acid, as an optimal, effective and inexpensive amount for integrated management of the CBB. The widespread use of *Beauveria bassiana* against adults of the CBB and the presence of native entomopathogenic fungi must be taken into account because of the results obtained [20] to observe inhibitory effects of the development of this fungus by applying insecticides such as chlorpyrifos, endosulfan and disulfoton. The use and establishment of parasitoids with classical biological control method have been successful in all places where it has been implemented; the parasitoid *Cephalonomia stephanoderis* is widely distributed and it has remained despite applications of chemical insecticides. In some cases this can control 94.8% of the CBB [40].

8. Annonaceae weevil, *Optatus palmaris*

8.1. Description

According to [41], the adult presents strops separated at the apex by a visible keel, middle lobe of the pronotum with a short tubercle on each side and elytra with points of inter-stretch marks forming transverse grooves; the forelimbs are longer in males, with more dilated and bristly shanks (tarsus). The length of prothorax + elytra is 6.5–9.3 mm in males and 6.8–8.6 mm in females. It has a wide body, is rhomboid, has blackening layer, and is covered with fine decumbent scales. The head and face are 7.6–9.0 mm long in males and 7.5–8.2 mm in females and 1.4–1.8 mm wide in male and 1.5–1.7 mm in females, sometimes as long as the prothorax length, inserted at the top of the head in lateral view and slightly curved, reducing its size toward the apex, with lateral grooves in the middle basal; the face (rostrum) of the female is less sculpted than in the male, almost cylindrical, with scores and smaller scale and more limited compared to the male. Middle inserting antennae are 0.53–0.55 in males and 0.52–0.56 mm in females. Antennae with scapum are 0.34–0.38 and 0.33–0.35 times the length of the head (rostrum) or peak in males and females, respectively.

Thorax: Prothorax sometimes is wider than longer. Pronotum is moderately convex; medium lobe is pronounced, subtruncate, slotted dorsally and moderately emarginate and with a short tubercle on each side. The lateral region of prothorax is furrowed near the tip; the lower anterior margin is sub-straight. Mesepimeron is not prominent in the anterior region but visible in dorsal view. Scutellum (1.4–2.3 mm in male) sometimes is as wide as long with the anterior margin sub-straight and acute apex. Elytra (1.8–2.0 in males, females 2.0–2.2) is longer than the prothorax, (1.1 mm male) sometimes as long as broad; subtriangular; humerus is rounded and is present dorsolaterally; pre-apical slang developed; find superficial stretch marks. Description of larval and egg is missing, being a pest of recent detection and increasing their damage, are required to perform morphological studies and evaluate alternative control.

8.2. Distribution

Optatus palmaris spread in Mexico, Guatemala, Costa Rica, Peru, Brazil, Bolivia, Colombia, Paraguay, French Guyana, Ecuador, Honduras, Panama, Venezuela, Trinidad and Tobago and Argentina [42, 43]. In México it is present in the states of Michoacan, Guanajuato, Oaxaca and Nayarit [31, 34].

8.3. Host plants and biology

The genus *Optatus* is closely related to fruits of Annona [43]. *O. palmaris* in Mexico is associated to *A. diversifolia, A. cherimola* and *A. muricata* [28, 34]. According to [34], the egg incubation time is 5.36 ± 0.69 days with variation of 4 to 8 days at 24°C and relative humidity of 72 ± 2%. The larva lasts 73.5 ± 3 days, with an interval of 54 to 93 at 24.9°C and relative humidity of 81.3%; its pupal period lasts 25.1 ± 1.6 days in the range of 17–41 at 25°C and relative humidity 81.5 ± 0.12%; the adult lives 34–150 days, with an average of 112 ± 23.51 at 26.76 ± 0.03°C and relative humidity of 67 ± 0.2%. *O. palmaris* biological cycle lasts on average 216 ± 28.7 days, with an interval of 109–292 days.

Adults take refuge in the foliage of their host and feed on vegetative buds, flowers and fruits. They choose fruits close to physiological maturity where they form groups to feed, lay eggs and copulate. It is possible that the adult releases an aggregation pheromone; this may be influenced by volatile compounds of fruits during its developing stage, maybe because in

http://dx.doi.org/10.5772/66635

young vegetative buds and flowers, they appear very sporadic and solitary. The adults of *O. palmaris* make circular holes in the fruits that later become necrotics; they can create a sort of chamber where the female lays eggs. The female feeds and lays eggs; the male is dedicated exclusively to copulate and scarcely takes food during copulation time. After this period, adult migrates to the top of the tree for resting and feeds on young vegetative buds and flower buds; after 24 h it looks for females and continues copulating. These activities are done 15 h per day.

8.4. Damage

Optatus palmaris causes damage in chirimoya (*Annona cherimola*) [41, 42]; it has also been found in soursop, *Annona muricata* L. and ilama (*Annona diversifolia* L.). The larvae feed on the pulp and seeds of the fruit; in the final of the larval stage, it leaves the fruit and pupae in the soil [42]. Adults make holes in the fruits of soursop and chirimoya when they are feeding (**Figure 7**) or lay eggs (**Figure 8**); when there are no developed fruits, they feed on the petals and pedicel of small fruit, causing its fall. Damage of the species of the genus *Optatus* in *A. cherimola* corresponds to a pattern of drawings, similar to the letters "C" or "O," which is a group of holes caused by adults when they feed; in places where the larvae feed, first necroses and slight watery discharge are observed; feeding causes a small hole deepens to seed; when moves from the fruit, it leaves a hole of 2–3 mm in diameter [42]. In soursop, *O. palmaris* adults can damage 38% of the total area of the fruit and it is possible to find an average of six larvae per fruit. The fruits most affected are those close to the harvest [34].

Figure 7. Damaged fruits of soursop by *Optatus palmaris*.

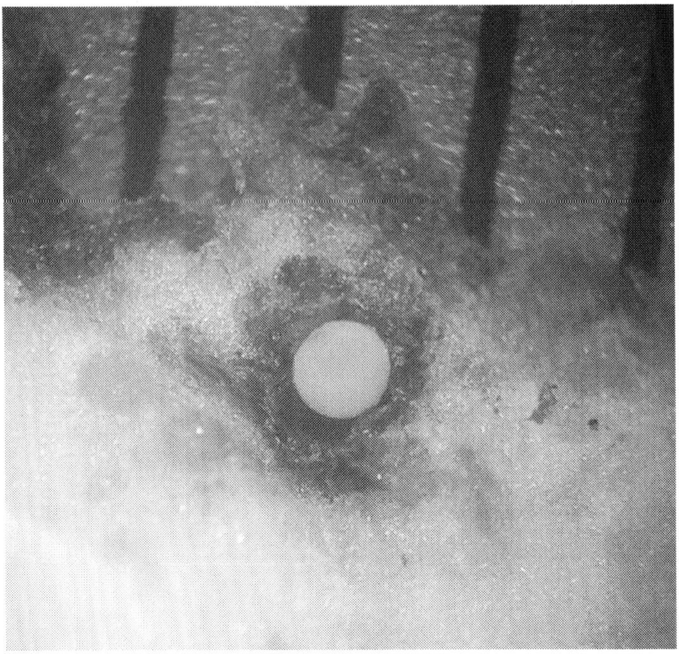

Figure 8. Egg of *Optatus palmaris*. Space between the black lines is a millimeter.

8.5. Control

Because of recent detection as Annonaceae pest, considering aspects of its biology and habits, control methods have not been evaluated. It has been observed that it is more frequently damaging in the producing annonaceae areas in Mexico; it is necessary to consider this pest as main insect pest of these crops and evaluate different methods of control and search of natural enemies: parasites, predators and entomopathogenic fungi. Based on their mating and feeding habits recorded by [34, 42], the use of volatile compounds appears to be a good alternative management; however, further studies are required in this regard.

Author details

Luis Martín Hernández Fuentes[1]*, Alvaro Castañeda Vildozola[2] and Mario Alfonso Urías-López[1]

*Address all correspondence to: lmhf01@yahoo.com.mx

1 National Institute for Research in Forestry, Agriculture and Livestock, México

2 Autonomous Mexico State University, Mexico State, Mexico

References

[1] Leitao B. M. L. Phylogenetic systematic of the tribu Neotropical Optatini Champion, 1907 (Coleoptera, Curculionidae, Baridinae). Instituto Nacional de Pesquisas da Amazonia. Universidade Federal do Amazonas. Programa Integrado de Pos-Graduacao em Biologia Tropical e Recursos Naturais, Manaus, Amazonas. 2007.

[2] Domínguez J. V. Morphology of small borer seed *Conotrachelus perseae* Barber. Campus Montecillo, Postgrado en Fitosanidad, Entomología y Acarología, Colegio de Posgraduados, Montecillo, Texcoco, Estado de México. 2006.

[3] Muñiz V. R. Study morphological of two species of *Conotrachelus,* that are pests of the avocado (Persea gratísima Gaertn) in Mexico. Rev. Soc. Mex. Hist. Nat. 1970. 31:289–337.

[4] Castañeda Vildozola A. Bioecology of the big borer seed of the avocado *Hilipus lauri* Boheman (Coleoptera: Curculionidae) in the region Central de Mexico. Colegio de Posgraduados en Ciencias Agrícolas, Montecillo, Texcoco, Edo. De México. 2008.

[5] Castañeda Vildozola A., Equihua-Martínez A. and Peña J. E. Avocado weevils of the genus *Heilipus,* pp. 35–47. In: Peña J. E. (ed.), Potential Invasive Pests of Agricultural Crops. CAB International, London, U.K. 2013. 464 pp.

[6] García A. P. *Hilipus lauri* Boheman a borer of the avocado seed in Mexico. Escuela Nacional de Agricultura, Chapingo, México. 1962.

[7] Palacios-Torres R. E., Castañeda Vildozola A., Sánchez-Pale J. R., Franco Mora O. and Vargas-Rojas L. Interaction of Two Avocado Fruit Insect Borers in the Same Agroecological Area in Mexico. Pan-Pacific Entomol. 2016. 92(2):100–103.

[8] Salgado-Siclán M. L., Bautista-Martínez N. The big borer seed of the avocado in Ixtapan of the salt, Mexico State. In Memoria. Fundación Salvador Sánchez Colín CICTAMEX, S.C. Coatepec Harinas, Estado de México. 1993. pp. 225–231.

[9] Medina Q. F. Incidence of the big borer seed of the avocado *Heilipus lauri* Boheman (Coleoptera: Curculionidae) in Tepoztlan, Morelos. Facultad de Ciencias Agropecuarias. Universidad Autónoma de Morelos. 2005.

[10] Castañeda Vildozola A., *Persea schiedeana* (Lauraceae), new host of *Heilipus lauri* Boheman (Coleoptera: Curculionidae) in Veracruz, Mexico. Neotrop. Entomol. 2009. 39:914–915.

[11] Díaz C. F. Biology and control of the small borer seed (Conotrachelus perseae) Barber in the region of Toliman, Queretaro. Escuela Nacional de Agricultura. Chapingo, México. 1972.

[12] Llanderal De la Peña L. L., Ortega H. A. Biology, habits and identification of the small borer seed of the avocado (*Conotrachelus perseae Barber*) in Ziracuaretiro, Michoacán. Facultad de Agrobiología "Presidente Juárez". Universidad Michoacana de San Nicolás de Hidalgo, Uruapan, Michoacán. 1990.

[13] Barber H. S. Avocado Weevils. Proc. Entomol. Soc. Wash. 1919. 21:53–61.

[14] Barber, H. S. Two New *Conotrachelus* from Tropical Fruits. (Coleoptera: Curculionidae). Proc. Entomol. Soc. Wash. 1923. 14:181–183.

[15] Whitehead D. R. Recognition Characters and Distribution Records for Species of *Conotrachelus* (*Coleoptera: Curculionidae*) that Damage Avocado Fruits in Mexico and Central America. Proc. Entomol. Soc. Wash. 1979. 81:105–107.

[16] González-Herrera A. Arthropods associated with the cultivation of the avocado (*Persea americana* Mill.) in Costa Rica. Proceedings of the World Avocado Congress V. Malaga, Spain. 2003. pp. 449–454.

[17] Castañeda Vildozola Á., Franco-Mora O., Reyes Aleman J. C., Ruiz-Montiel C., Váldez-Carrasco J. and Equihua-Martínez A. New Distribution Records of the Small Avocado Seed Weevil, *Conotrachelus perseae* Barber (*Coleoptera: Curculionidae*), in Mexico and Notes on Its Biology. Coleop. Bull. 2015. 69:267–271. 2015.

[18] Vázquez A. M., Cruz-López L. and Chamé-Vázquez E. R. First Record of *Conotrachelus perseae* (*Coleoptera: Curculionidae*) in Comitán, Chiapas, Mexico. Fla. Entomol. 2015. 98:1252–1253.

[19] Salinas C. A. Biology of the small borer seed of the avocado *Conotrachelus perseae* Barber in Tacambaro, Michoacan. Especialidad en Entomología y Acarología. Instituto de Fitosanidad. Colegio de Posgraduados, Montecillo, Texcoco, Edo. de México. 1999.

[20] Abreu M. S. Penetration of the Coffee Borer *Hypothenemus hampei* (Ferrari) (Coleoptera: Scolytidae) in Coffee Fruits Different Growing Stages and Selectivity of Insecticides and Fungicides to the Entomopathogen Fungi *Beauveria bassiana.* Universidad Federal de Vicosa. 2002.

[21] Kissinger, D.G. Description of New *Copturus* Pest of Avocado from Mexico (Coleoptera: Curculionidae, Zygopinae). Acta Zool. Mex. 1957. 2:1–8.

[22] Francia R. M. Distribution of seed borers *Conotrachelus aguacate* Barber and *C. perseae* Barbe (Coleoptera: Curculionidae) in the municipalities of Tacambaro, Tocumbo, Cotija, Susupuato and Ziracuaretiro, Michoacán. Campus Montecillo, Postgrado en Fitosanidad, Entomología y Acarología, Colegio de Posgraduados, Montecillo, Texcoco, Estado de México. 2008.

[23] Gasca C. L. Trials with traps adhesive of color and monitoring of *Copturus aguacatae* Kissinger, in Ziracuaretiro, Michoacan. Instituto de Fitosanidad, Especialidad en Entomología y Acarología. Colegio de Posgraduados. 2000.

[24] Gudiño J. L. M., García A. M. Biology and habits of the branches and trunks borers of the avocado (avocado Kissinger Copturus) in the region of Uruapan, Michoacán. Facultad de Agrobiología "Presidente Juárez". Universidad Michoacana de San Nicolás de Hidalgo, Uruapan, Michoacán. 1990.

[25] Leos-Rodríguez J. A.; Kido-Cruz M. T. and Valdivia-Alcalá R. Impact of the barriers phytosanitary in the trade of avocado between Mexico and the United States of North America. Rev. Chapingo Serie Hort. 2005. 11:99–103.

[26] Peterson E. V. and Orden D. Avocado Pests and Avocado Trade. Am. J. Agric. Econ. 2008. 90:321–335.

[27] Espinosa G. N. Validation and simulation statistics of methods for the sampling of *Copturus aguacatae* Kissinger, *Conotrachelus perseae* Barber and *C. aguacatae* Barber; pests of avocado (Persea americana Mill.), in Tingüindin, Michoacan, Mexico. Departamento de Parasitología Agrícola. Universidad Autónoma Chapingo, Chapingo, México. 2007.

[28] Salas A. M. D., O'Brien C. W. and Romero-Napoles J.. Curculionoidea (Insecta: Coleoptera) from the State of Guanajuato, México. Insecta Mundi. 2001. 15:45–57.

[29] Sánchez, S. S. New records of phytophagous insects to the State of Tabasco, Mexico. Folia Entomológica Mexicana 2000. 109:113–116.

[30] Sánchez, S. S. *Conotrachelus dimidiatus* Champion (Coleoptera: Curculionidae) the weevil of the guava (*Psidium guajava* L.) in Tabasco, Mexico. Boletin del Museo de Entomología de la Universidad del Valle, Colombia. 2011. 12 (2):17–18.

[31] O'brien C. W. and Wibmer G. J. Annotated Checklist of the Weevils (Curculionidae sensu lato) of North America, Central America and the West Indies (Coleoptera: Curculionoidea). Memoirs of the American Entomological Institute. 1982. 34:1–382.

[32] Gonzalez G. E. Control of insect pests. In guava, its cultivation in Mexico. Gonzalez G. E., Padilla R. S., M. L. Reyes, C. M. Perales and V. F. Esquivel (ed.). Libro técnico 1. Instituto Nacional de Investigaciones Forestales, Agrícolas y Pecuarias. Campo Experimental Pabellon. Aguascalientes, México. 2002. 181 p.

[33] Tafoya F. C., Perales S. E., González G. H. G., Calyecac C. Fruit Damage.Patterns Caused by Ovipositing Females of *Conotrachelus dimidiatus* (Coleptera: Curculionidae) in Guava Trees. Psyche Res. Artic. 2010. 4 p.

[34] Maldonado J. E., Hernández L. M. F., Luna G. E., Gómez J. R. A., Flores R. J. C., Santos M. O. Bioecology of Optatus *palmaris Pascoe* (Coleoptera: Curculionidae) in Annona muricata L Southwest. Entomol. 2014. 39(4):773–781.

[35] Rubio G. J. D. External morphology of the immature stages of *Hypothenemus hampei* Ferrari (Coleoptera: Curculionidae). Rev. U. D. C. A. Actualidad y Divulgación Científica. 2009. 12 (2):157–161.

[36] Le Pelley R. H. Pests of Coffee. Longmans, Green and Co. Ltd., London. 1968. 590 p.

[37] Vega F. E. Coffee Berry Borer, *Hypothenemus hampei* (Ferrari) (Coleoptera: Curculionidae: Scolytinae). 2004. pp: 959–960.

[38] Pablo Benavides, Carmenza Góngora and Alex Bustillo (2012). IPM Program to Control Coffee Berry BorerHypothenemus hampei, with Emphasis on Highly Pathogenic Mixed Strains of Beauveria bassiana, to Overcome Insecticide Resistance in Colombia, Insecticides - Advances in Integrated Pest Management, Dr. Farzana Perveen (Ed.), ISBN: 978-953-307-780-2, InTech, Available from: http://www.intechopen.com/books/insecticides-advances-in-integrated-pest-management/ipm-program-tocontrol-coffee-berry-borer-hypothenemus-hampei-with-emphasis-on-highly-pathogenic-mix

[39] Fernandes F. L., Picanco M. C., Fernandes M. E. S., Dangelo R. A. C., Souza F. F. and Guedes R. N. C. A New and Highly Effective Sampling Plan Using Attractant-Baited Traps for the Coffe Berry Borer (*Hypothenemus hampei*). J. Pest. Sci. 2015. 88:289–299.

[40] Bustillo, P. A. E. The role of biological control in the integrated management of the seed borer of the coffee, *Hypothenemus hampei* (Ferrari) (Coleoptera: Curculionidae: Scolytinae). Rev. Acad. Colomb. Cienc. 2005. 24(110):55–68.

[41] Hernández F. L. M., Gómez J. R., Andrés A.J. Importance, pest insect and diseases fungus of the cultivation of the soursoup. Libro Técnico Núm. 1. Instituto Nacional de Investigaciones Forestales, Agrícolas y Pecuarias. Campo Experimental Santiago Ixcuintla, Nayarit. México. 2013. 87 p.

[42] Corrales M. G. J. Observations on the biology and behavioral food of *Optatus* sp. (Coleoptera: Curculionidae) in *Annona cherimola* Mill. In: Resúmenes de *2do*. Congreso Centroamericano y del Caribe y 3ero. Costarricense de Entomología. Editorial San José ASENCO. San José, Costa Rica, 17–21 Julio. 1995. p. 6.

[43] Hernández F. L. M., Valdez J.C., Illescas C. P. R., Maldonado E. J., Orozco M. S. Morphological description of the aedeago of *Optatus palmaris* Champion (Coleoptera: Curculionidae). Southwest. Entomol. 2012. 37:225–229.

Chapter 4

Tabanids in South America

Ronald Rodrigues Guimarães,
Harlan Ronald Storti Rodrigues and
Ronald Rodrigues Guimarães Júnior

Additional information is available at the end of the chapter

http://dx.doi.org/10.5772/67108

Abstract

The text provides information on taxonomy, morphological data, distribution, and bionomy on most recorded species of tabanids in South America. The distribution parameters of species according to classification by biogeographical regions are used. An appendix indicating the main studies about tabanids according to the countries of their origin is still offered.

Keywords: insect vector, horsefly, Neotropical region, taxonomy, bionomy

1. Introduction

The species of family Tabanidae Latreille, 1802, commonly known in South America as "mutucas," "botucas," "mbutú," "colihuacho," and "moscas de los caballos", comprises more than 4400 worldwide species, absent only in the regions of higher altitudes and eternal snows [1, 2], with more than 1800 species present in the Neotropics [3]. They are the largest bloodsucking Diptera, reaching up to 25 mm, with a robust body and some with well-developed proboscis, an aspect that causes respect and fear. Females often require blood meal for maturation of eggs, at least after the first posture, so they are considered autogenous, partially autogenous, or nonautogenous [4]. Males are phytophagous, but females, always in search of blood, repeatedly attack humans, domestic and wild animals, among primates, rodents, alligators, snakes, turtles, and birds, especially during the drier seasons [5–8]. Tabanids are known worldwide for its painful sting and are mechanical and biological vectors of several helminths, viruses, bacteria, and protozoa, etiologic agents of diseases that can affect humans and wild and domestic animals [9–11]. Tabanids have the characteristics necessary for a good mechanical vector: interruption of hematophagism, high mobility, and large mouthparts that can carry blood [10]. The painful tabanid sting is recognized as a determining factor to interrupt blood meal. The sting causes reactions in the host, such as muscle tremors, tail movement, hit with the head, and kick in order to make the tabanid fly away [12]. But the presence of tabanid predators as the solitary sand wasps *Stictia punctata* (Fabricius) and *Stictia signata* (Linnaeus) (Hymenoptera: Crabronidae) plays an important epidemiologic role: the wasps catch tabanids and take them away, before and during the blood meal, and in this last case, causing interrupted hematophagism [13]. Around the world, more and better evidence has been gathered to assess the importance of tabanids in an epidemiological context. Several studies show a correlation between the time of increased activity of horseflies and the appearance of diseases in animals and man. The season of year in which the vectors are more common means increased health risk to animal populations and exposed human [9, 10, 14, 15]. In Neotropics, tabanids occur mainly in tropical rainforest, deciduous forest, wet savannah, and grassland meadows; they appear to be rare or absent in open savannah, oak forest, tropical dry forest, and seaside mangroves [16]. According to Raymond, there is greater number of species and greater possibility of finding rarer species in the ecotone areas [17].

The classification adopted by most current authors about Neotropical tabanids is the proposal by Mackerras [18], which divided the Tabanidae family into three subfamilies, Chrysopsinae Blanchard, 1840, Pangoniinae Rondani, 1856, and Tabaninae Latreille, 1802, mainly based on genitalia morphology.

The subfamily Chrysopsinae, with species that are an intermediate between Pangoniinae and Tabaninae, is divided into three tribes: Bouvieromyiini Séguy 1930 the more primitive, but with relatively specialized species; Chrysopsini Blanchard, 1840 with fairly structurally uniform species, and few genera but numerous species; and the tribe Rhinomyzini Enderlein, 1922 with only a very specialized species in the Neotropics. The subfamily Pangoniinae, in the Neotropical region, is divided into the tribes Mycteromyiini Coscarón and Philip, 1979, Pangoniini Rondani, 1856, and Scepsini Bequaert, 1930, all with more primitive as specialized species, and Scionini Enderlein, 1922, with only one primitive and anomalous species. The Tabaninae subfamily consists of three tribes, Diachlorini Lutz, 1909, Tabaninae Latreille, 1802, and Haematopotini Bequaert, 1930; the latter does not have representatives in the Neotropics. The Diachlorini tribe has primitive species and specialized species and the largest number of species in the Neotropics. Tabanini tribe also brings together both primitive and specialized species, but not as much as those belonging to tribe Diachlorini; most species is found only in one genus (*Tabanus* Linnaeus) [19, 20]. The terms "primitive" and "specialized" refer to the position of key characters: the most primitive are the ones that are closer to the conditions found in presumably ancestral forms, such as Nematocera, specializing more away from them, either by reducing and increasing the structural complexity.

2. Morphology of tabanids

Tabanids belong to the Suborder Brachycera, characterized by short antennae with three (up to five) segments and the adults emerge from the puparium by a T-shaped slot. The species of family Tabanidae have head wider than thorax and frons may have one or more callus; generally adjacent eyes (holoptics) in males and separated (dicoptics) in females; subcallus generally inconspicuous, but sometimes well-developed, smooth and shiny; antennal flagellum with first major flagellomere and 4–8 apical flagellomeres; maxillary palps with two segments; blood-sucking mouthparts with mandibles and stiliform maxiles of most females adapted to puncture the skin of the host; thorax with prominent notopleural lobes; legs with apical spurs in the median tibiae, and may be absent in hind. Wings with veins R_4 and R_5 limiting its apex; radial basal, basal medium, and discal cells large; posterior cubital cell usually closed near the edge of the wing; membranous wings with varying patterns. Male with gonocoxite fused with hiparium and single or partially divided gonostile; epandrium whole or divided; 10th tergite absent and flattened cerci; females usually with 10th tergite divided, 8th sternite is a shield-shaped enclosure and one-segmented [21, 22] (**Figure 1**). About the morphology of tabanids, Barretto's studies must be emphasized, mainly that external morphology of *Poeciloderas quadripunctatus* [23]. The studies of Iide, about the morphology of *Tabanus importunus* [24] and species of *Fidena* (*Fidena*) [25–28] are still used as a reference in morphology studies.

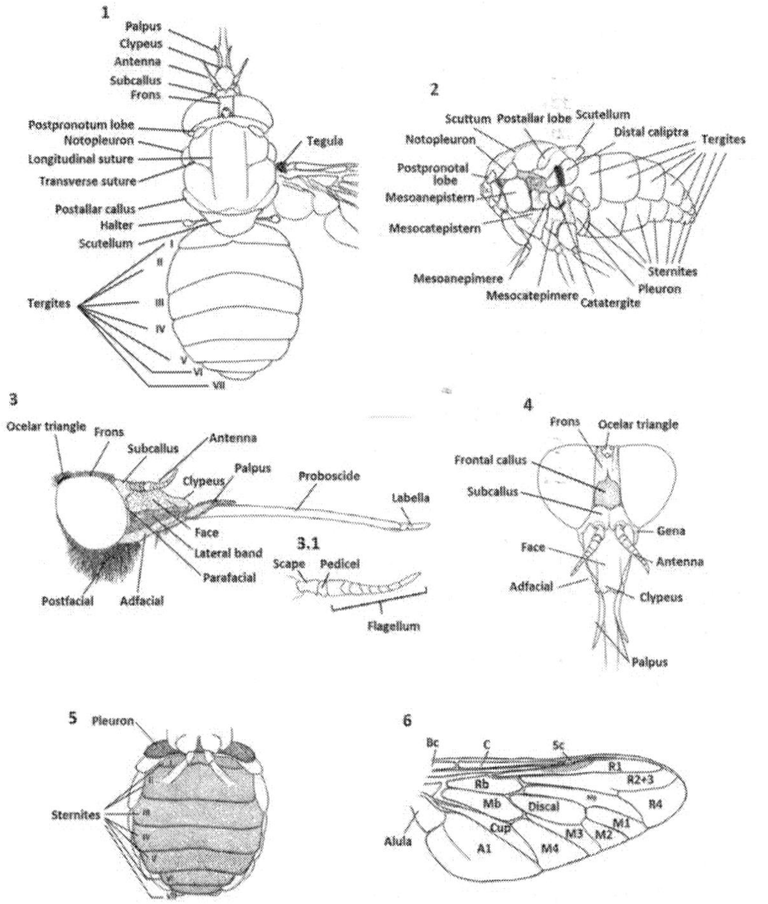

Figure 1. Main parts of a tabanid body (Genus *Fidena*). 1 – Dorsal view of body; 2 – Lateral view of thorax and abdomen; 3 – Lateral view of head; 3.1 Antenna; 4 – Frontal view of head; 5 – Ventral view of abdomen; 6 – Wing: Bc – Basal costal cell; C- Costal cell; Cup – Cubital posterior cell; Mp – Medial posterior cell; M1, M2, M3, M4 – Medial cells; R1, R2+3, R4 – Radial cells. According MacAlpine. Figures used with permission of Gorayeb.

3. South-American tabanids

Studies of tabanid species in South America began in the second half of the nineteenth century, with foreign researchers, some of which have never been on the continent. These studies were in full descriptions of native species, and were based on specimens deposited in private collections, museums, or European universities, sent by professional collectors [1]. These early records were made by Linnaeus, Scopolli, Strom, DeGeer, Fabricius, Thunberg, Meigen, Latreille, and Palisot Beauvois. More extensive studies on tabanofauna South America were performed by several authors. Wiedemann described a large number of species from South-American continent. Walker studied and described several species of South-American tabanid specimens deposited in British Museum and Saunders collections. Kröber was in Argentina and Bolivian Chaco regions, collecting dipterous between 1925 and 1926; he also studied the taxonomy and described several species of South-American tabanids from specimens deposited in museums and research institutes in Europe. It should be mentioned that an important study by Martins [29] on tabanids from Minas Gerais state, Brazil, located in provinces of Cerrado (Chacoan sub-region) and Parana Forest (Parana subregion) revealed the occurrence of 9 genera and 42 species of Pangoniinae, and 12 genera and 52 species of Tabaninae (the most common genera *Chrysops*, with 15 species and *Tabanus*, with 18 species). Descriptions of species carried out at this time are mostly considered insufficient to identify the species currently collected, if holotypes or paratypes are missing for comparison. Thus, many of these species were re-described, using more specific characters, which previously were not valued.

The first tabanid catalog of South America was published by Hunter in 1901 [30], naming 319 species; in separate listing, Hunter lists 64 South-American species described by Walker and another list with 62 species described by Wiedemann, Macquart, and Walker, but without information of locations from where they were collected, presumably from South America. Kröber, in 1934, published another catalog that included species of tabanofauna from South and Central America, Mexico, and the West Indies, which listed 861 South-American species [31]. In 1969, Fairchild [19] published an excellent study of the Neotropical tabanids, with key to genera and subgenera, containing information on the geographical distribution and morphology. Two years later, the same author published his catalog about tabanids from South of the United States, listing 707 species recorded in South America [32]. These Fairchild publications served as the basis for the manual to identify genera and subgenera published by Coscarón and Papavero in 1993 [33], as well as for a new catalog on tabanids from South of the United States, by Fairchild and Burger in 1994 [34]. More recently, in 2009, Coscarón and Papavero [20] published a new catalog of the Neotropics, including the species of Central America, southern part of Mexico and Baja California peninsula, southern Florida, all Caribbean islands, and South America. In the same year the authors also published a new illustrated manual for identification of the subfamilies, tribes, genera, and subgenera of Neotropical tabanids [35]. After the publications of Coscarón and Papavero [20] several species have been described in Neotropics and South America, giving rise to the addendum of 11 new taxa to the catalog [36]. And even after this publication, other species have been described from South America. *Pityocera (Pseudelaphella) ecuadorensis* Krolow and colleagues [37] was described from coastal zone of Arid Ecuador province; *Protosilvius gurupi* Rafael, Marques, and Limeira-de-Oliveira [38], *Muscotabanus rafaeli* Henriques and Krolow, *Pityocera (Pseudelaphella) pernaquila* Gorayeb and Krolow [37], *Elephantotus tracuateuensis* Gorayeb, *Dasybasis antillanca* González [39], all from Brazilian Amazonian Subregion. *Stenotabanus clavijoi* Gorayeb, Gómez and Velásquez-de-Rios was described from Venezuelan Amazonian Forest [40] and *Dasybasis collagua* González from Chilean Andean region [39]. *Dichelacera matogrossensis* Henriques and Krolow, *Pityocera (Pseudelaphella) barrosi* Gorayeb and Krolow [37], and *Pityocera (Pseudelaphella) gorayebi* Limeira-de-Oliveira and Krolow [37] were described from midwest Brazilian Cerrado and *Pityocera (Pseudelaphella) rhinolissa* Krolow and Henriques [37] from midwest Brazilian Cerrado and Bolivia (central Bolivian Plateau). *Dichelacera walteri* Guimarães, Gorayeb and Carvalho was described from southeast coast, Brazilian Atlantic Forest province [41]. Most studies on tabanids in South America are morphological, with few others about biology, behavior or seasonality, as will be seen by studying the main species of the genera represented in the South America.

South America has 11 major biomes: rainforest spanning the Amazon Forest and Atlantic Forest; the fields and southern savannas; the flooded fields (Pantanal); the montane camps; deserts and scrublands; tropical and subtropical conifers forests (Araucaria Forest); temperate forest; dry tropical forest; mangroves; Mediterranean shrub; and coastal areas of salt marshes. The tabanofauna is found in all biomes, except at higher altitudes, because of the restrictions imposed by low temperatures. In South America, tabanids are found in virtually all habitats and environments from the beaches of coastal areas, salt marshes, mangroves, salt lakes, Chilean and Peruvian deserts, southern grasslands, savannas and scrublands, rain forests, plains, up the slopes the mountains in the line of snow in the Andes [4]. Certainly the habitat of tabanids is also influenced by the food source, e.g., the arboreal fauna of the Neotropical mammals determines that a large number of tabanid species live in that habitat. Fairchild observed the preference of *Philipotabanus inauratus* Fairchild, *Stibasoma apicimacula* Fairchild, *Stenotabanus jaculatrix*, *Stenotabanus maruccii* (Fairchild), *Stibasoma fulvohirtum* (Wiedemann), *Tabanus defilippii* Bellardi, *Dichelacera crocata* Fairchild, *Catachlorops umbratus* Hine and *Stibasoma panamensis* Curran, by treetops of rainforest [42]. Generally, tabanids prefer defined habitats, although a few species widely distributed can be found in many environments, especially those altered by human activity, such as agriculture and livestock [19].

From many studies conducted over the years, it has been possible to map, at least to some degrees, the biogeographical distribution of main tabanids genera of South America; however, there is no study on biogeographic distribution that contemplate tabanids, except that of Fairchild [19]. To characterize the tabanid distribution in South America, a proposal has been elaborated bringing together the studies of Fairchild [19] and Morrone [43, 44], and their divisions of the Neotropical region in biogeographic subregions. The proposal of Morrone is based on previous studies of panbiogeography and cladistic analysis of insect fauna of Latin America [44]. Thus, an attempt to join the proposed biogeographical models and current knowledge about tabanids in South America is presented here.

According to Morrone [44], in a biogeographical context, South America is characterized as consisting of three regions: Neotropical, South American transition zone, and Andean regions (**Table 1**). **Neotropical region** extends from north-central Mexico to Argentina, comprises the

Region/transition zones	Subregion	Dominion	Provinces
Neotropical	Caribbean	North-western South America	Choco (1)
			Maracaibo (2)
			Venezuelan Coast (3)
			Trinidad and Tobago (4)
			Magdalena (5)
			Venezuelan Llanos (6)
			Cauca (7)
			Galapagos Islands (8)
			Western Ecuador (9)
			Arid Ecuador (10)
			Tumbes-Piura (11)
	Amazonian		Napo (12)
			Imeri (13)
			Guyana (14)
			Humid Guyana (15)
			Roraima (16)
			Amapa (17)
			Varzea (18)
			Ucayali (19)
			Madeira (20)
			Tapajos-Xingu (21)
			Para (22)
			Pantanal (23)
			Yungas (24)
	Chacoan		Caatinga (25)
			Cerrado (26)
			Chaco (27)
			Pampa (28)
	Parana		Brazilian Atlantic Forest (29)
			Parana Forest (30)
			Araucaria angustifolia Forest (31)
South-American transition zone			North Andean Paramo (32)
			Coastal Peruvian Desert (33)
			Puna (34)

Region/transition zones	Subregion	Dominion	Provinces
			Atacama (35)
			Prepuna (36)
			Monte (37)
Andean region	Central Chilean		Coquimbo (38)
			Santiago (39)
	Sub-Antartic		Juan Fernandez Islands (40)
			Maule (41)
			Valdivian Forest (42)
			Magellanic Forest (43)
			Magellanic Paramo (44)
			Malvinas Islands (45)
	Patagonian		Central Patagonia (46)
			Subandean Patagonia (47)

Table 1. Biogeographical classification of South America, adapted from Morrone [44].

Caribbean, Amazonian, Chacoan, and Parana subregions. **Caribbean subregion** from South America, with Northwestern South American dominion, comprises the provinces of Choco, Maracaibo, Venezuelan Coast, Trindad and Tobago, Magdalena, Venezuela Lannos, Cauca, Galapagos Islands, Western Ecuador, Arid Ecuador, and Tumbes-Piura, where some species of genera *Lepiselaga, Chrysops, Esenbeckia, Selasoma, Stibasoma, Dichelacera, Acanthocera,* and *Tabanus* occur. **Amazonian subregion** is the largest subregion of Neotropics, extending from Brazil, the Guyanas, Venezuela, Colombia, Ecuador, Peru, Bolivia, Paraguay, to Argentina. This subregion comprises 13 provinces: Napo, Imeri, Guyana, Humid Guyana, Roraima, Amapa, Varzea, Ucayali, Madeira, Tapajos-Xingu, Para, Pantanal, and Yungas. The tabanid species occurring in these provinces belong mainly to genera *Chrysops, Esenbeckia, Fidena, Boliviamyia, Catachlorops, Chlorotabanus, Cryptotylus, Dasychela, Diachlorus, Dichelacera, Lepiselaga, Leucotabanus, Phaeotabanus, Pityocera, Poeciloderas, Pseudacanthocera, Selasoma, Stenotabanus, Stibasoma, Stypommisa,* and *Tabanus.* There are four provinces assigned to **Chacoan subregion**: Caatinga, Cerrado, Chaco, and Pampa. There are no published records for tabanids in Caatinga province, except the fossil *Cretotabanus stonemyomorphus* Martins Neto and Santos [45]. Limeira-de-Oliveira informed that he had captured specimens of *Chrysops, Pityiocera, Catachlorops, Diachlorus, Dichelacera, Leucotabanus, Phorcotabanus, Poeciloderas,* and *Tabanus* in states of Piauí and Ceará, Brazil (personal communication). In other three provinces (Cerrado, Chaco, and Pampa), species of genera *Chrysops, Scaptia, Esenbeckia, Fidena, Acanthocera, Catachlorops, Chlorotabanus, Diachlorus, Dasybasis, Dichelacera, Lepiselaga, Leucotabanus, Phaeotabanus, Phorcotabanus, Stenotabanus, Stypommisa, Poeciloderas,* and *Tabanus* occur. **Parana subregion** comprises three provinces: Brazilian Atlantic Forest, Parana Forest, and *Araucaria augustifolia* Forest. The species reported to this subregion belong to genera *Chrysops, Esenbeckia, Fidena, Scaptia, Scepsis, Acanthocera, Catachlorops, Chlorotabanus,*

Dichelacera, Diachlorus, Lepiselaga, Leucotabanus, Phaeotabanus, Pseudacanthocera, Rhabdotylus, Stigmatophtalmus, Poeciloderas, and *Tabanus.* The **South American transition zone** extends along the highlands of Andes between Venezuela, northern Chile, and western Argentina, and comprises six provinces: North Andean Paramo, Coastal Peruvian Desert, Puna, Atacama, Prepuna, and Monte, where tabanids species of genera *Dasybasis, Esenbeckia, Fidena, Scione,* and *Tabanus* occur. The **Andean region** extends from central Chile and Patagonia, along the high mountain ranges of Venezuela, Colombia and Ecuador, through the coastal desert and Puna of Peru, Bolivia, northern Chile and Argentina, to Argentine-Chilean Patagonia. Andean region consists in three subregions: **Central Chilean subregion** with the provinces of Coquimbo and Santiago with species of *Veprius, Protodasyapha, Esenbeckia, Scaptia, Mesomyia, Mycteromyia, Dasybasis,* and *Tabanus.* **Subantartic subregion**, with the provinces of Juan Fernandez Islands, Maule, Valdivian Forest, Magelanic Forest, Magelanic Paramo, and Malvinas Islands, with tabanids species belonging to genera *Parosca, Silvestrielus, Scaptia, Acellomyia, Agelanius, Dasybasis, Scaptioides, Pseudoscione, Haematopotina, Poeciloderas* and *Tabanus.* **Patagonia subregion** is divided in two provinces: Central Patagonia and Subandean Patagonia, where species of genera *Scione, Scaptia, Acellomyia, Agelanius, Caenopangonia, Chrysops, Dasybasis, Haematopotina, Nubiloides, Protodasyapha, Scaptioides, Silvestriellus, Veprius,* and *Tabanus* occur (**Figure 2**). There is no doubt that there are species within genera that are ubiquitous, and there are genera with species more restricted to specific habitats. This issue was discussed by Fairchild in 1969 [19], but lacks a more current study on biogeography of tabanids in Neotropics.

Despite the tendency to turn the habitat as the determining factor for tabanid distribution, most authors use the politics division, by countries and their states and provinces, as the distribution paradigm. So, by the end of the chapter, the authors provide (as an Appendix), the main studies on tabanofauna of South-American countries.

Following, the list of subfamilies, tribes, genera, and more registered species of South-American tabanids, offering information on morphological characteristics, distribution, and the most important references of each taxa.

3.1. The subfamily Chrysopsinae (Blanchard, 1840)

This is underrepresented in number of species, except for those belonging to the genus ***Chrysops*** Meigen, 1803, which has a worldwide distribution. The females of this subfamily have a simple pointed genitalia style, simple caudal ends of spermathecal ducts, without cup-like expansions, ocelli, and eyes nearly always patterned with contrasting colored bands or spots. The subfamily is represented by two tribes in the tropical region: **Bouvieromyiini** Séguy, 1930 and **Chrysopsini** Blanchard, 1840 [19].

3.1.1. Tribe Bouvieromyiini (Séguy, 1930)

This comprises primitive and specialized species, with the following characteristics: first antennal segment hardly longer than width, antennae shorter than antero-posterior thickness of head, callus variable, narrower than frons, and eyes with a transverse band [19]. In South

http://dx.doi.org/10.5772/67108

Figure 2. Distribution of genera of tabanids according to occurrence of species in regions and subregions of South America. Neotropical Region: Caribbean, Amazonian, Chacoan and Parana subregions; South American Transition Zone; Andean Region: Central Chilean, Subantartic and Patagonian subregions. Adapted with permission of Morrone [43, 44].

America especially in Chile and Argentina, only the primitive species can be found. Other primitive species occur in eastern Nearctic region, temperate South Africa, Northeast Asia and Japan, and eastern and southeastern Australia. The more specialized representatives occur in the Old World tropical and southern Africa, Madagascar, Seychelles, Indonesia, New Guinea, and Australia with few species in India and Southern China. The South-American species of Bouvieromyiini tribe are species that occur in the temperate regions of central Chile and Southeast Argentina (Chacoan subregion), and belong to the single subgenus of tribe.

3.1.1.1. Genus Pseudotabanus (Ricardo, 1915)

This genus comprises only three species in the subgenus Coracella (Philip, 1960).

3.1.1.1.1. Subgenus Coracella (Philip, 1960)

Pseudotabanus (Coracella) araucana Coscarón, *Pseudotabanus (Coracella) carbo* (Macquart), and *Pseudotabanus (Coracella) rubricornis* (Kröber). The first species occurs in Chile and Argentina [46] and the latter two only in Chile [34]; *Pseudotabanus (Coracella) araucana* is considered by Coscarón and Papavero [20] as *araucanus*. Coscarón [47] reviewed the subgenera *Coracella*, as belonging to the genus *Mesomyia* Macquart 1850, which continued to appear [46] until the research publication by Fairchild and Burger [34], in which *Coracella* is considered as subgenus *Pseudotabanus*; in the same study, Coscarón [47] described *Pseudotabanus (Coracella) araucana* and provided a key to separate the three species.

3.1.2. Tribe Chrysopsini Blanchard, 1840

This comprises less restrictive species in habitats, and are separated from Bouvieromyiini by first antennal segment longer than width, near always at least twice as long as width, the third with basal plate and four annuli, antennae longer than width of head, callus generally as wide as high or wider, eyes speckled or with a specific pattern of spots and bars [19]. Most species occur in tropics in South America and Africa, but they are well represented in the Nearctic and Palearctic regions, but few eastern, Australian, or Chilean species. Only two genera are present in the Neotropics, ***Chrysops* Meigen**, 1803 and ***Silvius* Meigen**, 1820, but only the first is present in South America.

3.1.2.1. The genus Chrysops Meigen, 1803

It is represented worldwide and brings together 75 species in Neotropics, from Mexico to Argentina, of which 52 are South American [20]. In epidemiological and diagnosis of tabanofauna studies, the *Chrysops* species are more reported in South America. *Chrysops variegatus* (DeGeer) in Paraguay is a possible vector of the equine disease "mal de caderas," caused by *Trypanosoma evansi* [48]. It was the most abundant species observed in a survey performed in Aregua, central Paraguay [49]. Rafael and collaborators [50] captured *C. variegatus* on Maraca Island, Amazonian subregion, Guyana province. Bermúdez and Bermúdez described the larvae and pupae of *Chrysops variegatus* collected between March and April, in tropical area of high humidity and temperatures, associated with aquatic plants *Pondeteria sagitata* Presl and *Sagittaria* sp., in the livestock region of Mexico [51]. The species (as *variegata*) was the second-most collected species after *Lepiselaga crassipes* (Fabricius) in areas of ecotone between secondary forest and pastures, in northern Colombia, Caribbean subregion [52]. The species was also collected in eastern Amazon, in areas of primary forest and pasture [53] and in Brazilian northern Amazon, also in primary forest but at 1200 meters of altitude, border Brazil-Venezuela [50]. In a survey conducted during 1995–1996 in Central Amazon four species of *Chrysops* were found, and the most abundant species was *Chrysops formosus* Kröber; the other species were *Chrysops incisus* Macquart, *Chrysops ecuadoriensis* (Enderlein), and *Chrysops variegatus* (DeGeer) [54]. It also occurs in central Amazon, Brazil, in primary forest and "campinaramas" ground level, and in forest canopy, 40 meters high, Amazonian Subregion, Varzea province [55–57]; the species was also observed in transition zone between the savannah and the Brazilian Amazon forest [58] and in Caatinga (Limeira-de-Oliveira, in personal communication). *Chrysops variegatus* also occurs in coastal highlands of southeast Brazilian Parana subregion [59], but was not recorded in Brazilian southern Pampa [60]. Buestán captured *Chrysops variegatus* in coastal zone of Ecuador, Western and Arid provinces, and Caribbean subregion [61]. *Chrysops varians* Wiedemann is another wide distributed species and has been captured from savannah of French Guiana [17], in pasture area in western Amazon [53], in the transition zone between the savannah and eastern Brazilian Amazon forest [58], in highlands of southeast Parana subregion [62, 63], and in southern Pampa, province of Chaco subregion, in Brazil [60]. On Marambaia Island, southeast Parana subregion, Brazilian Atlantic Forest province, *Chrysops variegatus* and *Chrysops varians* were observed flying around the head and curling themselves on the hair of people who walk through forests or sandbanks [64, 65].

3.2. Subfamily Pangoniinae (Rondani, 1856)

This comprises the more primitive species of tabanids. They are characterized by ninth tergite undivided in both sexes, caudal ends of spermathecal ducts without cup-like expansions, usually with 7–8 annuli in third antennal segment, ocelli and hind tibial spurs present [19]. Species are distributed in four tribes, all represented in Neotropics [20].

3.2.1. Tribe Mycteromyiini (Coscarón and Philip, 1979)

This was created from genus *Mycteromyia* Philippi, 1865, characterized by elongated body, grayish or yellowish to brown, elevated ocelli at vertex, frons about as wide as high, no callus, but some rugosities above subcallus, face conically produced, wing elongated, accentuated clouds on crossveins, first posterior cell closed and petiolate [19]. The tribe is currently divided in four genera.

3.2.1.1. Genus Caenopangonia (Kröber, 1930)

This was recently placed within this tribe, with small to medium yellowish-brown species, dichoptic eyes in both sexes, widened frons, strong scutal vittae, wings with spotted crossveins, palpi small subcilindrical with reduced apical pits [66]. The genus comprises five species occurring in Central Chile and Midwest Argentina. The former genus comprised three species: *Caenopangonia aspera* (Philip), *Caenopangonia brevirostris* (Philipp) and *Caenopangonia hirtipalpis* (Bigot). Two new species from Chile, *Caenopangonia cerdai* Krolow, Henriques and González and *Caenopangonia coscaroni* Krolow, Henriques and González, were recently described and a key for identification of current species was provided [67].

3.2.1.2. The genus Mycteromyia (Philippi, 1865)

This appears in Coscarón and Papavero's catalog [20] and comprises three Chilean species: *Mycteromyia conica* (Bigot), *Mycteromyia etcheverryae* Coscarón and Philip and *Mycteromyia obscuripennis* (Philippi). However, in Coscarón and Philip revision [68], *Mycteromyia obscuripennis* does not appear to be included in the genus, as well as in the tabanid list of

Chile of Coscarón and González, in 1991 [46] or in the catalog of Fairchild and Burger [34]. Coscarón and Papavero [20] reported Kröber [31, 69] as the reason to keep the species in the genus. *Mycteromyia* sp. was captured in a survey performed in summer of 1971, in province of Coquimbo, Chile (Andean region, Central Chilean subregion) [70]. González conducted morphological studies of the mouthparts of *Mycteromyia conica* Bigot using scanning electron microscopy [71].

3.2.1.3. Genus Promycteromyia (Coscarón and Philip, 1979)

This brings together nine species, endemic to Chile, mainly in Andean region, Central Chile subregion [20, 34, 46]. *Promycteromyia cinerascens* (Bigot), a Chilean species, is the most well studied in the genus (as *Mycteromyia*) [68, 72].

3.2.1.4. Genus Sivestriellus (Brèthes, 1910)

This genus is with four more specialized species, which are distributed to Chile and Argentina in provinces of Pampa (Chacoan subregion) and Central Patagonia (Patagonian subregion) [20, 68].

3.2.2. The tribe Pangoniini (Rondani, 1856)

This has 130 Neotropical species, more or less restricted in habitats and is considered the most ancestral species between tabanids [19, 72]. They have naked eyes, prominent appendix in fork R4 vein, face not produced conically and proboscis rarely exceeding the length of the head, as frequent in Scionini [19]. With the exception of the species of *Esenbeckia*, specimens of this tribe are rarely collected and do not seem to be very active bloodsuckers [73]. The tribe has 14 genera in which the majority is in South America.

3.2.2.1. Genus Esenbeckia (Rondani, 1863)

This brings most tribe species, occurring throughout South America and is considered the most specialized among Pangoniini [19]. They are medium to large, slender and robust specimens, usually narrow frons, with or without a slender to clavate callus, bare eyes, usually with long proboscis and small compact labella, body pilosity short to sparse, often pattern wings [19]. The genus *Esenbeckia* is divided in five subgenera [20].

3.2.2.1.1. Subgenus Esenbeckia (Rondani, 1863)

It was reviewed by Wilkerson and Fairchild in 1985 [74], and brings together 51 species in the tropical region: *Esenbeckia (Esenbeckia) rafaeli* Limeira-de-Oliveira does not appear in Coscarón and Papavero catalog [20, 36, 75]. From neotropical species, only *Esenbeckia (Esenbeckia) illota (Williston)* do not occurs in the South America. In Amazonian subregion, *Esenbeckia (Esenbeckia) prasiniventris* (Macquart) was collected in forest in Roraima province [50], *Esenbeckia (Esenbeckia) matogrossensis* Lutz, in eastern Amazonian forest, Para province [53], and *Esenbeckia (Esenbeckia) clari* Lutz (as *lemniscata* Enderlein) in Pantanal province [76]. *Esenbeckia (Esenbeckia) osornoi* Fairchild was recorded in Cerrado province of Chacoan subregion, state of Tocantins, Brazil [77]. *Esenbeckia (Esenbeckia) lugubris* (Macquart) a large, glossy and dark-colored fly, with a powerful flight, and painful sting was by the first time reported in Atlantic Forest, from specimens collected on Marambaia Island, Rio de Janeiro state, Brazil [65].

3.2.2.1.2. The subgenus Ricardoa (Enderlein, 1922)

This comprises 38 species distributed in Central America and Mexico and will not be treated here [20].

3.2.2.1.3. The subgenus Proboscoides (Philip, 1943)

This comprises 11 species, all occurring in South America and ranging from Panama to Paraguay. Fairchild and Wilkerson [78] provided a key to females of 11 species of *Proboscoides*. They are not common in collections and there is much interspecific variation. The species mentioned in more recent studies are *Esenbeckia (Proboscides) farraginis* Fairchild and Wilkerson, collected in Central Brazil, Chacoan subregion, Cerrado province [77]; and *Esenbeckia (Proboscoides) suturalis* (Rondani), which occur in northern and eastern Amazon forest [50, 53].

3.2.2.1.4. The subgenera Astomyia (Burger, 1999) and Palassomyia (Fairchild, 1969)

They are monotypic with both endemic species of Chile. *Esenbeckia* (*Astomyia*) *media* Burger was proposed from a specimen deposited in Arthropod Collection of the State of Florida, with label written by Fairchild reading "*Esenbeckia (Astomyia) media* n. sp., n. subg.", leading to the conclusion that the author planned to propose these new taxa. However, Fairchild did not accomplish his purpose and Burger, describing the species, preserved the names suggested [79]. *Esenbeckia (Palassomyia) fascipennis* (Macquart) seems to be the least specialized among the genus *Esenbeckia* [19].

3.2.2.2. Genus Protosilvius (Enderlein, 1922)

This comprises five species, all occurring in Brazil and is a part of a more primitive group among *Pangoniini* [19]. They are species smaller in size, with slender, long wings, narrow forehead without callus, short proboscis, naked eyes, and has a third antennal segment with a variable number of segments [19]. The species occur only in Brazil, mainly in southeast Cerrado, Atlantic forest and Parana subregion, with one species in Amazon Basin [38]. There is record of a misidentified specimen of *Protosilvius termitiformis* Enderlein, in Paraná state [11, 38].

3.2.2.3. Genus Veprius (Rondani, 1863)

This has five species present in Central Chile and Midwest Argentina. They are flies with head almost twice broader than height, black eyes with no band and abundant pilosity, eyespots present, broad forehead with an inconspicuous or absent callus, third antennal segment basal plate and set aside 4-style, big and little labela sclerotized [19, 71]. These species occur in Central Chile and West Argentina [80]. Gonzáles described the male and re-described the female of *Veprius fulvus* Coscarón, Philip and Fairchild [80].

3.2.2.4. *Genus Protodasyapha (Enderlein, 1922)*

This comprises species similar to those in genus *Veprius*, but with 8-annulate antennae, subulate, the basal plate consolidated, and pilose eyes, sometimes pilosity is sparse [19]. The genus has been reviewed by Coscarón [72] and meets four species in two subgenera.

3.2.2.4.1. *Subgenus Curumyia (Coscarón 1976)*

It has only one species *Protodasyapha (Curumyia) lugens* (Philippi) occurring in Chile and Argentina [20].

3.2.2.4.2. *Subgenus Protodasyapha (Enderlein, 1922)*

This contains three endemic species from Central Chile [72]. González described the larvae and pupae of *Protodasyapha (Protodasyapha) hirtuosa* (Philippi) and compared with others Pangoniini species from Australia and North America; the larvae were found 3–5 cm beneath the soil surface of a *Lithraea* forest, on a steep and humid hillside [81].

3.2.2.5. *Genus Fairchildimyia (Philip and Coscarón, 1971)*

This comprises only two species that occur in Midwest Argentina. Coscarón considered that this genus and *Chaetopalpus* Philippi, 1865, form a monophyletic branch in Pangoniini [72]. The species have dark eyes, subulate antennae, frons with a circular-shaped callus, palpi with a short apical segment, sternite 8 of female very wide basally [19]. Chainey and Hall provided a picture of the front view of male head of *Fairchildimyia penai* Philip and Coscarón, comparing it with *Boliviamyia fairchildi* Chainey and Hall, the species described in that paper [73].

The other genera of Pangoniini are all monotypic: *Archeomyotes* Philip and Coscarón, 1971, *Austromyans* Philip and Coscarón, 1971 *Chaetopalpus* Philippi, 1865, each with an endemic species of Chile [20]. The recent genus *Boliviamyia* Chainey and Hall has only one endemic species from Bolivia, *Boliviamyia fairchildi* Chainey and Hall [73].

3.2.3. *Tribe Scepsini (Bequaert, 1930)*

It has only the genus *Scepsis* Walker, 1850 with a single Neotropical species, *Scepsis appendiculata* Macquart, which appears in catalogs as *Scepsis nivalis* Walker [20, 31, 34, 82]. This is a slender-body fly, with whitish milky wings, atrophied mouthparts, and wide frons without callus in both sexes [19]. The species is found on sand beaches, from coast of Rio de Janeiro state (Brazil) to probably northern Argentina [19]. The species has nonhematophagous habits and can be considered autogenous. Turcatel have reports with specimen coming from Guarapuava, inside the Paraná state, Brazil, plateau region of mixed rain forest, putting in doubt the information of collection [11]. The species was observed on Marambaia Island, Rio de Janeiro, southeastern Brazil, on the white sand beaches, it has a short and low flight, not reaching more than 25 cm height and 1.5 cm away [64, 65].

3.2.4. Tribe Scionini (Enderlein, 1922)

This species have robust bodies, well-developed ocelli, no frontal callus, pilose eyes, long proboscis, and short palpi [19]. The tribe comprises over 280 species in 17 genera, austral in distribution, occurring in Australia, New Guinea, New Zealand, and South America [66].

3.2.4.1. Genus Pseudomelpia (Enderlein, 1922)

First recognized as subgenus of *Scaptia* [66], it has only the species *Pseudomelpia horrens* (Enderlein), with little body densely hairy, with robust and cylindrical palp, basal antennal ring, partial and irregularly fused with the basal plate [66]. Both male and female are nectar feeding [83]. The species is endemic to Chile, from Santiago to Maule province, Central Chilean and Subantartic regions [66].

3.2.4.2. Genus Osca (Walker, 1850)

This comprises 11 species previously placed in subgenus *Scaptia*, all from southeastern South American, in temperate regions and at high altitudes in Ecuador, Peru, Bolivia, and Chile [20, 66]. They are moderately size flies similar to *Tabanus*, with short proboscis, palpi over half length of proboscis, and antennae with eight annuli [19]. González and Sanhueza using scanning electron microscopy, conducted detailed studies of the morphology and structure of the oral armor of *Osca varia* Walker (who does not have mandibles), *Osca lata* (Guérin-Meville), and rufa *Osca rufa* (Macquart) (all as *Scaptia* (*Scaptia*) (the latter two with mandibles with marginal teeth, suggesting bloodsucking habits) [84]. Larvae of *Osca lata* (Guérin-Méneville) (as *Scaptia* (*Scaptia*) were found under fallen logs in Puyehue forest area in the Patagonian, Subantartic subregion, Valdivian Forest province, Chile [85].

3.2.4.3. Genus Lepmia Fairchild, 1969

This species have moderate-size body and thick proboscis, small and reduced labella, face bulging and few projected, palpi usually short, broad, with extensive bare area [19]. The genus currently comprises six species [66]: two original species and four transferred from subgenus *Scaptia* (*Pseudoscione*) [66]. Pino and colleagues performed a survey during summer of 1971 in province of Coquimbo, Chile (Andean region, Central Chilean subregion) and found *Lepmia atra* (Philippi), as *Scaptia* (*Pseudoscione*), one of the most abundant species [69]. *Lepmia seminigra* (Ricardo), as *Scaptia (Lepmia)*, was collected in sandbank, during early afternoon, in Parana subregion, southeast Brazilian Atlantic Forest province; the species has a powerful flight that produces loud [65].

3.2.4.4. Genus Parosca (Enderlein 1922)

This comprises medium-size species, face conspicuously projected, diverging frons, proboscis long and slender with thick labella, broad palpi extensively flattened triangular dorsally rotate [19]. Three species are included in genus, all transferred from *Scaptia* (*Pseudoscione*) [66]. The terrestrial larva of austral horsefly *Parosca latipalpis* (Macquart), as *Scaptia* (*Pseudoscione*), was identified by molecular techniques and described from specimens found 2–3 cm below of the soil surface and associated with larvae of Coleoptera, Lepidoptera, and Diptera in southern Chile, Osorno, and subantartic subregion [86].

3.2.4.5. Genus Pseudoscione (Lutz, 1918)

This comprises nine species from the former subgenus *Scaptia* (*Pseudoscione*) [66]. They are from small to medium size stout species, with pale markings along sutures of scutum (like *Scione*), with wing cell M_3 open [66]. They are predominantly Patagonian, occurring in Chile, Argentina, also Brazil [87]. Coscarón reviewed the former subgenus *Pseudoscione*, and offered a key to 15 species (not for the sixteenth species), with its redescriptions and figures [66, 88]. The genus *Pseudoscione* includes the species of former subgenus *Scaptia* (*Pseudoscine*), excluding those transferred for *Lepmia* and *Parosca* [66]. Pino and collaborators performed a survey during summer of 1971 in province of Coquimbo, Chile (Andean region, Central Chilean subregion) and found *Pseudoscione dorsoguttata*, as *Scaptia (Pseudoscione)*, one of the most abundant species [70].

3.2.4.6. Genus Scione (Walker, 1850)

This comprises 41 recognized species uniformly mottled, small and slender body, and well-projected face, undeveloped labella, slender legs, cloud marks on wing veins, closed R_3 and M_3 cells [19, 66]. They are considered typically Andean species [88], occurring mainly in the mountainous regions of northwest South America, Venezuela, and Bolivia. Coscarón reviewed the genus and re-described *Scione claripennis* Ricardo and *Scione flavohirta* Ricardo, both the Andean region Argentina, also describing the male of the second species [88]. *Scione aurulans* (Fairchild), *Scione ablusus* Fairchild, and *Scione flavohirta* Ricardo, all were recorded to feed on man, with the latter also recorded to feed on cattle [89, 90]. There is no current formally review of the genus over 80 years, and lacks descriptive and uniform descriptive characters [66].

3.2.4.7. The genus Fidena (Walker, 1850)

This comprises currently 99 species, characterized by medium to large stout body, face shining and snout-like, proboscis extremely long and slender, reduced labella and widely open wing cell M_3 [19, 66]. The species of this genus are considered difficult to study, by the large number of taxa, few males have been described, great variability of characteristics and lack studies of immature stages [91]. *Fidena* species are widely distributed in South America, mainly mountains of southeastern Brazil and just off the subandean region [66]. They are separated into four subgenera.

3.2.4.7.1. Subgenus Fidena (Walker, 1850)

This comprises 94 species, with flattened palpi without subcallus, scuttum without strong vittae, femora, and tibiae without long hairs, wing cell R5 usually closed without long petiole; they occur in South America from Colombia to Argentina, predominantly in Brazil [66]. The

Brazilian species *Fidena (Fidena) rufohirta* (Walker) has a proboscis as long as the body length. Ilde performed morphological studies on *Fidena (Fidena) nigripes* (Röder), *Fidena (Fidena) brachycephala* Kröber, *Fidena (Fidena) florisuga* Lutz, *Fidena (Fidena) rufibasis* Kröber, and *Fidena (Fidena) fusca* (Thunberg), that are very useful to study external anatomy of the group [25, 26, 28, 37]. Coscarón redescribed females of *Fidena (Fidena) abominata* Philip, *Fidena (Fidena) atripes* (Röder), *Fidena (Fidena) latifrons* Kröber, *Fidena (Fidena) longipalpis* Enderlein, *Fidena (Fidena) neglecta* Kröber, *Fidena (Fidena) nigripes* (Röder), *Fidena (Fidena) ochrapogon* Wilkerson, *Fidena (Fidena) opaque* (Brèthes), and *Fidena (Fidena) sorbens* Wiedemann, so as males of these species, except *atripes* and *ochrapogon* [91]. *Fidena (Fidena) fusca* Thunberg is reported from mountainous regions of Parana subregion, southeastern Brazil [92, 93]. *Fidena (Fidena) auripes* (Ricardo), *Fidena (Fidena) eriomeroides* Lutz, and *Fidena (Fidena) aurulenta* Gorayeb, were captured in the western Amazon [53]. Rafael and collaborators report *Fidena (Fidena) schildi* (as *childi*) in Sierra de Pacaraima, northern Brazilian Amazon region [50]. Coscarón described the male *Fidena (Fidena) haywardi* Philip, a species from foothills of Argentine Andes, Puma subregion [88]. *Fidena (Fidena) pseudoaurimaculata* (Lutz) was collected in "campinarama" and canopy forest of Central Amazon Forest province [54, 94]. *Fidena (Fidena) freemani* Barretto, *Fidena (Fidena) analis* (Fabricius), and *Fidena (Fidena) loricornis* Kröber were also reported in the Central Amazon Forest [54, 56, 57, 95]. Records of *Fidena* species in the Amazonian subregion broaden the distribution of this group, in addition to the highlands of Parana and Subandean region [88]. Larvae and pupae of *Fidena (Laphriomyia) rufopilosa* (Ricardo) were found in phytotelmata of terrestrial bromeliads *Canistrum lindenii* (Regel) Mez, *Nidularium innocentii* Lemaire e *Vriesea friburgensis* (Mez), that grow on granitic rocks in secondary Atlantic forest, Brazilian southeast [96]. *Fidena (Fidena) longipalpis* Enderlein was captured in Planalto Serrano and coastal zone Parana subregion, and pampas of Chaco subregion, of southern and south-eastern Brazil [59, 60, 63]. Buestán collected *Fidena (Fidena) aureopygia* Kröber (as *aureopigia*) above 2000 m altitude in the Andean Cordillera in the transition zone between the humid forests of Chocó and the dry forests of southern Ecuador [97]. Lima reports the occurrence of *Fidena (Fidena) bistriga* Fairchild and Rafael, *Fidena (Fidena) castanea* (Perty), *Fidena (Fidena) fumifera* (Walker), and *Fidena (Fidena) lissorhina* Gorayeb and Fairchild in the state of Tocantins, Central Brazil, Chacoan subregion, Cerrado province [77]. Cárdenas and collaborators collected *Fidena (Fidena) rhinophora* (Bellardi) between 500 and 2000 m on both sides of the Andes, in misty rainforest in Ecuador [98–100]. Guimarães and collaborators collected *Fidena (Fidena) winthemi* (Wiedemann) in ecotone between Atlantic forest sandbank and rain forest in Parana subregion, in the southeast Brazilian Atlantic Forest Dominion [65].

3.2.4.7.2. Subgenus Laphriomyia (Lutz, 1911)

This comprises three species with femora and tibiae densely covered by long and conspicuous hairs [35]. There are three recognized species: *Fidena* (*Laphriomyia*) *kroeberi* Fairchild (previously in subgenus *Fidena*), *Fidena* (*Laphriomyia*) *mirabilis* Lutz, 1911 (subspecies of *rufopilosa* [20] or *rufopilosus* [11]), *Fidena* (*Laphriomyia*) *polidetarsis* Kröber (a synonym of *silvatica*) were elected as species of this subgenus [66]. They occur in Peru, Bolivia, and Brazil [66]. *Fidena* (*Laphriomyia*) *kroeberi* Fairchild was captured in both the ground level and canopy of western and central Amazonian Forest [53, 94].

3.2.4.7.3. Subgenus Leptofidena (Kröber 1930)

This also has only one species, with palpi thick, swollen and with a deep lateral concavity, frons with a protuberance callus-like, closure of wing cell R_3 with a long petiole [19, 66]. The species *Fidena (Leptofidena) morio* Wulp occur in Subandean region, western Argentina, and had the male described by Coscarón [20, 88].

3.2.4.7.4. Subgenus Neopangonia (Lutz, 1909)

This has only one species, with hairy face, with long and conspicuous hairs, scutum with a strong pattern, and wing cell R_5 broadly open [35]. *Fidena (Neopangonia) pusilla* Lutz occurs only in Brazilian Atlantic Forest province, Parana subregion, southeastern Brazil [20].

3.2.4.8. Genus Pityocera (Giglio-Tos, 1896)

This species has a body from small to medium-size, with antennal flagellum with tufts of hairs on one or more flagellomeres, face projected and shiny, proboscis equal to or longer than body's length [19]. They occur in northern South America [20]. Krolow and collaborators reviewed the genus in 2015, when five new species were also described [37]. The genus comprises 10 species in three subgenera.

3.2.4.8.1. Subgenus Elaphella (Bezzi, 1913)

This has only one species from Subcaribbean and north Amazonian subregions, *Pityocera (Elaphella) cervus* (Wiedemann) [20]. The species has first flagellomere long and finger-like projection, long projections on dorsal surfaces of the second to sixth flagellomeres, and wing with stump vein on M_1 [66].

3.2.4.8.2. Subgenus Pityocera (Giglio-Tos, 1896)

This species has pectinate antennae, with first six antennal flagellomeres with long projections on both dorsal and ventral surfaces, seventh and eighth fused, long and finger-like. The single species, *Pityocera (Pityocera) festai* Giglio-Tos (*festai* according to Coscarón and Papavero and Fairchild; and *festae* according to Lessard and Krolow [20, 37, 66, 101] occur from Panama to Ecuador, Caribbean subregion, and feeds on man [19, 101].

3.2.4.8.3. Subgenus Pseudelaphella (Kröber, 1930)

This currently has eight species after the review of Krolow and collaborators [37], but only three appear in Coscarón and Papavero's catalog [20]; in these species lack the dorsal projections on antennal segments, but there is a dense dorsal patch of hairs on enlarged first annulus of third segment [19]. They occur in Ecuador, Bolivia, and Brazil, in Amazon basin [66]. The new review of Lessard on Tribe Scionini [66] did not include the new species described by South-American authors, which are not mentioned by Coscarón and Papavero [20]: *Pityocera (Pseudelaphella) barrosi* Gorayeb and Krolow and *Pityocera (Pseudelaphella) gorayebi* Limeira-de-Oliveira and Krolow, both described from Brazilian Cerrado [37], *Pityocera (Pseudelaphella)*

http://dx.doi.org/10.5772/67108

pernaquila Gorayeb and Krolow, from Central and Oriental Brazilian Amazon [37] *Pityocera (Pseudelaphella) rhinolissa* Krolow and Henriques, from Central Brazilian Cerrado and Bolivian eastern plateau [37], and *Pityocera (Pseudelaphella) ecuadorensis* Buestán and Krolow, from coastal zone of Ecuador [37].

According to Lessard, the current genus *Scaptia* Walker, 1859 comprises only species occurring in Australia [66]. But in this text, the records of species of *Scaptia* are preserved as in major original references.

3.3. Subfamily Tabaninae

Neotropical species can be separated from the other subfamilies species by the absence of hind tibial spurs and functional ocelli, male with genitalia style truncate, ducts of spermathecal with cup-like extensions on caudal ends, eyes plain or with horizontal stripes [19]. Tabaninae are divided in two tribes: **Diachlorini** Lutz, 1909 and **Tabanini** Latreille, 1802. In neotropic species, the presence or absence of strong setae on basicosta to separate Tabanini from Diachlorini is often unreliable. In addition, others characters are used, as the sclerotized labella and vestiges of ocelli, which are common in Diachlorini but nearly unknown in Tabanini [19].

3.3.1. Tribe Diachlorini (Lutz, 1909)

This includes more than half of Neotropical Tabaninae, gathering nearly 600 species in 39 genera [20]. The reading of specialized literature to study this tribe, which has a large variety of species, both primitive and specialized, is strongly recommended. The more primitive species are dull colored, from small to medium size, occurring in colder areas and include species of genera *Dasybasis* e *Stenotabanus*; the remaining species are considered more specialized, mostly, are strictly tropical [19]. Trojan [102] published study of South-American Diachlorini distribution and considered that these species "are generally restricted to the northern part of the continent", occupying the Caribbean Archipelagos, limited by Andes in eastern border, and from Santa Catarina state in Brazil to Chaco and Salta in Argentina, in South border. Following some considerations about the main genera, which include the most common species recorded in surveys conducted by South-American researchers.

3.3.1.1. The genus Acanthocera (Macquart, 1834)

This comprises 28 species resembling wasps (Hymenoptera: Vespidae). They have slender and medium-sized body, antennae very long, first antennal segment at least 1,5 times the length of the second, and third always longer than the first and second together, vestigial or absent ocelli, partially sclerotized labella, palpi slender or swollen, slender abdomen with narrowed second tergite [103, 104]. Fairchild redefined the characteristics of the genus and provided a key to 16 species then known [103]. In catalog of Coscarón and Papavero [20] 20 species are listed, and it do not mention the study of Henriques and Rafael [104], where they described *Acanthocera* (*Nothocanthocera*) *distinta* Henriques and Rafael, transferred 11 species from the genus *Nothocanthocera* to genus *Acanthocera*, synonymized *Acanthocera* (*Acanthocera*) *lutzi* to *Acanthocera* (*Acanthocera*) *coarctata*, and the subgenus *Acanthocera* (*Mimodynerus*) Enderlein to *Acanthocera* (*Acanthocera*) Macquart. The species of *Acanthocera* are difficult to capture because they inhabit the canopy of forest [104]. The species of genus *Acanthocera* are currently distributed in four subgenera.

3.3.1.1.1. Subgenus Acanthocera (Macquart, 1834)

This subgenus has 16 species that have at least a tubercle or dorsal angle on antennal basal plate, usually a fairly long tooth or slender spine, frons rarely as wide as high, generally narrower [104]; the subgenus comprises 10 species in South America. *Acanthocera* (*Acanthocera*) *longicornis* (Fabricius), one of the most recorded species in studies, was captured in an ecotone area between rainforest and sandbanks on Marambaia Island, Rio de Janeiro, and in coastal zone in Parana subregion, Brazilian Atlantic Forest province [59, 64, 65].

3.3.1.1.2. Subgenus Nothocanthocera (Fairchild, 1969)

This comprises 12 species with short basal antennal segment, bare or partially bare frontoclypeus and gena, not wholly sclerotized labella, usually pale scutellum, without diagonal wing band, often resembling wasps [19, 104]; 11 species occurring in South America and one in Central America [104]. *Acanthocera (Nothocanthocera) distincta* Henriques and Rafael, was omitted by Coscarón and Papavero, from Amazonian forest, Amazonian subregion, Imeri province [36, 104].

3.3.1.1.3. Subgenus Polistimina (Fairchild 1969)

This has the single species *Acanthocera (Polistimina) politiformis* Fairchild, described from a male specimen from Amapa, northern Brazil [104]. The female was also described from Amapa: this red-yellowish species resembles wasps of the genus *Polistes* (Hymenoptera) [104]. The immature stages of *Acanthocera (Polistimina) vespiformis* Burger inhabit the tunnels opened by beetles in the trunks of guanandi *Callophyllum brasiliense* Cambess. The larva transforms sap that flows through these tunnels into a sticky mass with bad smell that attracts flies, which are trapped and are predated by the larvae, which are always found in tunnels less than 2 m above ground [105].

3.3.1.1.4. Subgenus Querbetia (Fairchild, 1964)

It is accepted by Fairchild [32], Moucha [82], Fairchild and Burger [34], Coscarón and Papavero [20] but is not mentioned by Henriques and Rafael in their revision of the genus [104]. These are species with bare eyes, frons less than twice as high as basal width, with basal callus as wide as frons, antennae with first segment very greatly inflated and shiny, labella extensively sclerotized and shiny, basicosta lacking strong setae [19]. There are two species *Acanthocera (Querbetia) chaineyi* Fairchild and Burger from Ecuador and Peru, and *Acanthocera (Querbetia) inopinata* (Fairchild), in Peru and Bolivia [20].

3.3.1.2. Genus Agelanius (Rondani, 1863)

This comprises 12 species, and it is considered as a part of the most primitive group within the tribe Diachlorini [106]. They are brown medium-size species, narrow frons, frontal callus not touching eyes, which are pilose and without bands, without dorsal prolongation on basal flagellomere, palpi slender and elongate, bare subcallus, and with abundant setae on basicosta, so that is difficult to use keys to separate the group [19]. They are similar to *Dasybasis* and differ from it by narrower frons, ridge-like or clavate callus and vestigial ocelli at vertex [19]. The genus is endemic to southern South America, Subandean Patagonia province, and occurs in Peru, southern Chile, and Argentina [34, 106]. During the last decade, González described *Agelanius verai* González [106], *Agelanius fuscus* González [107], *Agelanius burger* González [108], and *Agelanius chiloensis* González [109] all from Central Chile, Andean subregion. He also described the immature stages of *Agelanius fuscus*, which were found 5–10 cm beneath soil surface in forest of roble beech *Nothofagus obliqua* (Mirb.) Oerst [106] and the immature stages of *Agelanius cortesi* (González) collected beneath the soil surface near a small stream and with abundant *Gunnera chilensis* Lam. (giant-rhubarb) [110].

3.3.1.3. Genus Bolbodimyia (Bigot, 1892)

This comprises 13 species from which nine occurring in South America [20]. They are black or black and yellow, with subcallus and first antennal segment swollen and black shiny, wings wholly black, except the hyaline apex, vein R_4 strongly curved, swollen tibiae [19, 111]. Theses species are infrequently collected [112]. Stone reviewed the genus and provided a key to identification of 10 species then known [112]. Gómez and collaborators recorded *Bolbodimyia brunneipennis* Stone, *Bolbodimyia celeroides* Stone, *Bolbodimyia nigra* Stone and *Bolbodimyia philipi* Stone from Venezuela, and provides a key to the identification of five species reported in the country [111].

3.3.1.4. Genus Catachlorops (Lutz, 1913)

This comprises 66 species, characterized by small size body, frons narrow, frontal callus ridge-like or clavate [19]. Kröber [113] reviewed the genus and Barretto [23] provided a key to the females in Brazil and described the males of several species. Coscarón also reviewed the genus, re-described three species and described two new from Argentina [114]. The last reference to genus was made by Turcatel, when reviewed the records of species from Parana region, southeastern Brazil [11]. The species are distributed in six subgenera.

3.3.1.4.1. Subgenus Amphichlorops (Lutz, 1913)

This has seven species resembling to those in *Catachlorops*, from which are separated by yellow and fuses wings, often darker apical half [19]. All species occur in South America. *Catachlorops (Amphichlorops) flavus* Wiedemann was collected in area next to marsh and woodland in the evening, on Marambaia Island, Brazilian southeast, Parana subregion, Brazilian Atlantic Forest province [65]. They are well distributed in South America occurring in Colombia, Ecuador, Brazil, Peru, Bolivia, Paraguay, and Argentina [20].

3.3.1.4.2. Subgenus Catachlorops (Lutz, 1913)

This has 27 species occurring in South America [20]; they have small and medium-sized body, slender, callus usually clavate, brown to black tinted, black wings with a large rounded patch in discal cell, and hyaline apex [19]. *Catachlorops (Catachlorops) halteratus* Kröber and *Catachlorops (Catachlorops) rufescens* (Fabricius) inhabitat primary Amazonian Forest, Central Amazon subregion, Varzea province and the first was collected in February and from June to December, and the last, in April and from June to October [95]. *Catachlorops (Catachlorops) leptogaster* Barretto was collected in area next to marsh and woodland after 17:00 h, on Marambaia Island, Parana subregion, Brazilian Atlantic Forest province [65].

3.3.1.4.3. Subgenus Hadrochlorops (Fairchild, 1969)

This consists of six species characterized by large and stout body, hyaline wings faintly tinted, brownish or with dark cross veins margins [19]; they occur in Bolivia, Argentina and Brazil [20].

3.3.1.4.4. Subgenus Psalidia (Enderlein, 1922)

This has 13 species of which 7 occur in South America [20]. They are species with very slender palpi, very long antennal tooth, often curved in apex, first posterior cell closed, coarctate or slightly narrowed discal cell, wings always hyaline at base [19]. *Catachlorops (Psalidia) overali* Fairchild and Rafael was captured in canopy of "terra firme" Amazonian Forest, Central Amazon, subregion, Varzea province [94]. In the same region, *Catachlorops (Psalidia) rubiginosus* (Summers) occurs from June to November [95].

3.3.1.4.5. Subgenus Psarochlorops (Fairchild, 1969)

This has species related to *Psalidia,* but with the wing pattern reduced to an irregular small band bellow stigma and clouds around cross veins and fork of third vein [19]. This subgenus comprises nine species, and only one does not occur in South America [20]. *Catachlorops (Psarochlorops) difficilis* (Krober) inhabits the primary Amazonian Forest, Central Amazon subregion, Varzea province, and is collected from September to November [95].

3.3.1.4.6. Subgenus Rhamphidommia (Enderlein, 1922)

This species is characterized by clavate or ridge-like frontal callus, as wide as base frons, flagellum with hook-like projection, labella partially or wholly sclerotized, abdomen with median triangular spots most of tergites, wing with an irregular diagonal dark band [19, 115]. Four species occur in southeast South America, in Brazil and one of them, *Catachlorops (Rhamphidommia) potator* (Wiedemann), also in northern Argentina [20]. Henriques and Krolow described *Catachlorops (Rhamphidommia) dubius* Henriques and Krolow, the first species of the subgenus in Amazonian subregion, Madeira province, and provided a key to determine the species within subgenus [115].

3.3.1.5. Genus Chlorotabanus (Lutz, 1909)

This was created to *Tabanus mexicanus* Linnaeus, without providing a description or point type species, not meeting the rules of the International Code of Zoological Nomenclature [116]. The same paper was reprinted in 1911, keeping the faults [117]. In 1913, Lutz published a paper entitled "On the Systematics of horseflies, subfamily Tabaninae", republished in 1914 [118, 119]. This issue was currently discussed by Krolow and Henriques [120] and Guimarães et al. [121]. The date of 1913 was accepted for *Chlorotabanus* by Borgmeier [122] and Kröber [31], as well Fairchild [32]. Barretto was the first author to question the validity of the name [123]. Fairchild and Burger [34] also elected the year 1913 to designate the date of *Chlorotabanus*, in which were followed by Coscarón and Papavero [20, 33], but not in their last manual [35]. *Chlorotabanus* species are crepuscular and nocturnal, greenish pale color, without frontal callus, sclerotized labella, and unicolor eyes [19]. Coscarón completed the diagnosis of the genus adding features of gentitalia [72]. The genus appears in Coscarón and Papavero catalog comprising six species [20]; but Krolow and collaborators, in an excellent review, pointed 11 valid species, from which ten occur in South America, and one species in the United States, described three new species, and described the males of *Chlorotabanus leucochlorus* Fairchild and *Chlorotabanus flagellatus* Krolow and Henriques [57]. *Chlorotabanus inanis* (Fabricius) and *Chlorotabanus mexicanus* Linnaeus occur in savannah in French Guyana [17]. In Central Amazonian, Varzea dominion, *Chlorotabanus inanis* was observed in two periods of activity: in the morning, between 05:20 and 05:50 h, and at afternoon, between 17:45 and 18:20 h [124]. *Chlorotabanus inanis* was also captured on Maraca Island, Amazonian subregion, Guyana province, and in the state of Tocantins, Brazilian Chacoan subregion, Cerrado province [50, 77]. Guimarães and collaborators reported *Chorotabanus inanis* on Marambaia Island, Parana subregion, southeast Brazilian Atlantic Forest province; they observed that females prefer to feed on legs of horses, and when feeding, they become seemingly indifferent to the environment and are easily captured [65]. The species seems to be bivoltine, and appears from April to May and from October to December [64].

3.3.1.6. Genus Cryptotylus (Lutz, 1913)

This consists of five species with one subspecies, greenish color, with reduced or absent frontal callus, antennae with strong dorsal angle or tooth, labella wholly sclerotized and clear wings; they seldom attack man and are crepuscular and nocturnal species [19]. They are present in northern Amazonian subregion and one species in Chacoan subregion, in Paraguay and Argentina [20]. Fairchild provided a good key for species of the genus [125]. Philip and Fairchild reviewed the genus as a subgenus of *Chlorotabanus* [126]. Coscarón elected the key features for the diagnosis of genus adding feature of genitalia [72]. Gorayeb and Fairchild provided a new key for the genus and described *Cryptotylus firkin* Gorayeb and Fairchild, from Amazonian subregion, province of Para [127]. Coscarón and collaborators collected larvae of *Cryptotylus unicolor* (Coscarón and Poi of Neif) on *Pistia stratiotes* Linnaeus, in ponds in a region of dry forest, northeastern Argentine, province of Formosa, Argentine Chaco [128].

3.3.1.7. Genus Dasybasis (Macquart, 184)

This is one of the most numerous in tropical fauna, with 70 valid taxa, all present in South America, and also is well represented in of southeastern Subantartic subregion, Chile and Argentina, along Andean region [19, 129]. The genus comprises species that represent part of the most primitive group among Diachlorini [19]. They are species with callus filling the generally broad frons, or rarely reduced or absent, no tubercle at vertex, or at least, without vestigial ocelli, antennae without tooth, clear wings or clouded crossveins, and pollinose body [19]. The genus was reviewed by Coscarón and Philip in 1967, when the authors re-described the female *Dasybasis mendozana* (Macquart) that occurs in the Andean pre-cordillera region in Argentina [130]; the male was described by Coscarón [131]. Coscarón also described the larva and pupa of *Dasybasis nigra* (Enderlein), collected at dry season, found in small pits, remaining a dry creek, in Patagonian Subregion, Central Patagonian province [132]. An unidentified species of *Daybasis* was found during a survey performed during summer of 1971 in province of Coquimbo, Andean region, Central Chilean subregion [70]. *Dasybasis fairchildi* Coscarón and Philip had described immature stages from specimens collected in cold water streams in the Peruvian Andean highlands, at 1 cm deep in the sand or among the roots of the vegetation [133]. González described the immature stages of *Dasybasis (Dasybasis) nigrifrons* (Philippi), and *Dasybasis bruchi* (Brèthes) from moss of wetlands in Central and northern Chile [134]. The same author also described the immature stages of *Dasybasis pruinivitta* (Kröber) and *Agelanius cortesi* (González) from the same region [110, 135].

3.3.1.8. Genus Dasychela (Enderlein, 1922)

This consists of nine brown species, with a protuberant face and very long proboscis, tri- or biramous antennae with one or two long and slender dorsal spines with erect hairs and bare eyes [19]; they occur in southeastern South America [20]. Two subgenera are recognized.

3.3.1.8.1. Subgenus Dasychela (Enderlein, 1922)

This has six species, five of which occur in South America in Colombia, Ecuador and Brazil [20].

3.3.1.8.2. Subgenus Triceratomyia (Bequaert, 1937)

This has two species occurring in Ecuador, Peru e Bolivia.

3.3.1.9. Genus Diachlorus (Osten Sacken, 1876)

This comprises flies usually yellow and black colored, wings with a dark pattern, dark band in apex, colored eyes with patches and bands similar to *Chrysops*, variable frons, frontoclypeus bare and shiny [19]. These small flies occur in all South America, except in Chile [20]. *Diachlorus* has 29 species of which only *Diachlorus ferrugatus* (Fabricius) does not occur in South America (Central and North America). The first key to the genus was Kröber's [136], and currently the most elaborated study of the genus is Fairchild's, in which the author provided a key to identification of 23 species [137]. Coscarón added genitalia characteristics to the key characters for diagnosis of species [72]. Wilkerson and Fairchild provided a revised key and described five new species from South America [138]. *Diachlorus jobbinsi* Fairchild, *Diachlorus bicinctus* (Fabricius), and *Diachlorus curvipes* (Fabricius) are well distributed in northern and central Brazilian Amazonian region, in primary forest or varzea, level ground or canopy forest [50, 95]. *Diachlorus bivittatus* (Wiedemann) is a very aggressive species, and was the most abundant species in survey performed on Marambaia Island, Parana subregion, southeast Brazilian Atlantic Forest province. The species presented two generations per year (bivoltine) [64, 65]. *Diachlorus distinctus* Lutz was also found in that survey and has morphological and ethological similarities with *Diachlorus bivittatus* [65].

3.3.1.10. Genus Dichelacera (Macquart, 1838)

This comprises small to medium size species, with slender body, callus almost always as broad as frons, eyes usually with a single band, and labella wholly sclerotized [19]. According to Coscarón and Papavero [20] there are 80 valid species; but further studies increased this number to 83, with the descriptions of *Dichelacera matogrossensis* Henriques and Krolow, 2015 [139], *Dichelacera (Dichelacera) gemmae* Limeira-de-Oliveira and Gorayeb [140], and *Dichelacera (Dichelacera) walteri* Guimarães, Gorayeb and Rodrigues-Guimarães [41]. They are divided into four subgenera.

3.3.1.10.1. Subgenus Desmatochelacera (Fairchild, 1969)

This has only two species of which one occur from Costa Rica to Ecuador and one in Colombia and Peru [20].

3.3.1.10.2. Subgenus Dichelacera (Macquart, 1838)

This currently has 65 valid species after the revision of the genus *Acanthocera* by Henriques and Raphael and description of new species [41, 104, 139]. A total of 48 species occurs in South America. This subgenus is the largest in number of species, and characterized by labella wholly sclerotized, eyes nearly always with bands, callus more or less square, as wide as frons; all species are small to medium size [19]. *Dichelacera amazonenses* Henriques, *Dichelacera cervicornis* (Fabricius), *Dichelacera damicornis* Fabricius), *Dichelacera marginata* Macquart, *Dichelacera paraensis* Henriques, *Dichelacera trisuca* Fairchild and Philip, *Dichelacera villavoensis* Fairchild and Philip, are well distributed species in several environments in northern, central and eastern Amazonian subregion [54–57, 94, 95]. The type species of *Dichelacera* (*Dichelacera*) *matogrossensis* were collected in Chacoan subregion, Brazilian, Cerrado province [139]. Barros recorded the occurrence of *Dichelacera scutellata* Williston in Brazilian Pantanal, Amazonian subregion [76]. *Dichelacera alcicornis* (Fabricius) was the most abundant species collected in highlands of southeast Parana subregion [62, 63]. *Dichelacera (Dichelacera) walteri* Guimarães, Gorayeb and Rodrigues-Guimarães was described from specimens collected from August to September, on forest sandbanks from Marambaia Island, Parana subregion, Brazilian Atlantic Forest province, Rio de Janeiro [41]. *Dichelacera (Dichelacera) alcicornis* was also collected in the same place [65]. This last species was also recorded from Chacoan subregion, Pampa province, southern Brazil [60].

3.3.1.10.3. Subgenus Idiochelacera (Fairchild, 1969)

It has only one species, *Dichelacera (Idiochelacera) subcallosa* Fairchild and Philip that occurs from Costa Rica to Peru [20].

3.3.1.10.4. Subgenus Orthostyloceras (Lutz, 1933)

This comprises three species: *Desmatochelacera* (*Orthostyloceras*) *ambigua* (Lutz and Neiva) and *Desmatochelacera* (*Orthostyloceras*) *nubiapex* Fairchild and Philip occurring in Brazil, and *Desmatochelacera* (*Orthostyloceras*) *aurata* Wilkerson, occurring in Colombia [20].

3.3.1.11. Genus Dicladocera (Lutz, 1913)

This comprises 32 Andean species [19], although the Coscarón and Papavero catalog [20] also suggests Brazilian species are included in the genus; most species is distributed between Colombia and Peru [20]. This genus includes species with long antennal tooth, short proboscis, soft and pollinose labella, some setae on basicosta, eyes often pilose and wings with a dark band with a fenestrae on discal cell [19]. In Ecuador, *Dicladocera macula* (Macquart) was recorded in both side of Andean cordillera between 1600–3400 m, in montane forest, paramo and Andean shrubs [98]. Coscarón re-described the female e described the male of *Dicladocera nubipennis* (Rondani), a species from Argentine Subandean subregion [141].

3.3.1.12. Genus Lepiselaga (Macquart, 1838)

This has four small and robust species of black color, glossy palps, wings with black pattern with contracted discal cell, in two subgenera [19].

3.3.1.12.1. Subgenus Lepiselaga (Macquart, 1838)

This has the single *Lepiselaga (Lepiselaga) crassipes* (Fabricius), which occurs from Mexico to northern Argentina [20]. This is a very well-studied species. Lutz observed larvae of *Lepiselaga crassipes* (Fabricius) on moorhen lettuce, *Pistia stratiotes* Linnaeus in southeast Atlantic Brazilian Forest [142]. Later, Fairchild suspected that the larvae of the species also found on *Pistia* in mangroves in Panama Canal Zone, would be dependent on a more complex environment, formed by a maze composed of floating debris, mats of filamentous algae, *Salvinia* (water fern) and small specimens of *Pistia* [143]. *Lepiselaga crassipes* was also found in Central Amazon Subregion, Varzea province [55, 95] In a survey conducted in Pantanal, Brazilian Chaco subregion, the species was the fourth most abundant, occurring throughout the year, but more often in September and October [76]. The species also occurs in the transition zone between Cerrado and Pantanal, Brazil [144].

3.3.1.12.2. Subgenus Conopesalaga (Barretto, 1949)

This comprises three species with forehead as wide as high, or wider, inflated notopleurals lobes [19], which are distributed from Western Colombia to Argentina [20].

3.3.1.13. The genus Leucotabanus (Lutz, 1913)

This comprises 15 small- to medium-sized species, which have frons narrow to moderate, vertex with prominent tubercle, nearly always with vestiges of ocelli, callus clavate or ridge-like, basicosta sparsely or abundantly setose, usually black and shiny [19]. Eleven species occur in South America [20]. The genus has been well studied by Fairchild: in 1941 [145] he reviewed the genus and provided a key to 11 species with figures of eight; in 1953 [146], he reviewed the genus again and updated the key to 15 species; and, in 1985 [147] he updated the studies with a discussion of genus taxonomic position and offered a key to females of 18 species. Godoi and Rafael [148] described the immature stages of *Leucotabanus albovarius* (Walker) from specimens collected in rotten wood of the palm *Bactris gasipaes* Kunth (Arecaceae); they observed the adults active throughout the year in open areas and in primary Amazonian Forest, Central Amazon subregion, Varzea province [95]. *Leucotabanus exaestuans* (Linnaeus), a widely distributed species, is collected in the same environment all year long; it attacks horses and other animals on the head, near base of the ear [53, 95]. *Leucotabanus janinae* Fairchild is another species collected in the same environment from July to December, as well as *Leucotabanus palculus* Fairchild, but collected from July to December [95]. Specimens of *Leucotabanus sebastianus* Fairchild were captured next to marsh and rain forest area, on Marambaia Island, Parana subregion, Brazilian Atlantic Forest province [65].

3.3.1.14. *The genus Myiotabanus (Lutz, 1928)*

This comprises four species, three occurring in South America [20]; they are small species, with unusually long proboscis, small and partly sclerotized labella, inflated and short palpi [19]. They are similar to sarcophagids flies [149]. In 2004, Rafael and Ferreira reviewed the genus and provided a key to known species [148]. Coscarón and collaborators found larvae of *Myiotabanus barrettoi* Fairchild on *Pistia stratiotes* Linnaeus, in northeastern Argentine, province of Formosa, Argentine Chaco, region of dry forest [150].

3.3.1.15. *Genus Phaeotabanus (Lutz, 1913)*

It has 15 medium to large flies, greenish when alive or recently dead, unicolor eyes, narrow frons, slender callus, labella sclerotized, antennal plate with an obtuse dorsal angle, wings with dark markings [19]. The majority of species occurs in Brazil [20]. *Phaeotabanus cajennensis* (Fabricius), a large widely distributed species in South America, was captured in Trinidad using traps baited with white mice [151]. The same species was captured in canopy of "terra firme" Amazonian Forest, Central Amazon subregion, Varzea province [94]. *Phaeotabanus cajennensis* was also captured in ecotone between sandbank and rainforest and *Phaeotabanus limpidapex* (Wiedemann), and *Phaeotabanus litigiosus* (Walker) (more abundant from 17:00 to 19:00 h) were captured next to marsh and rain forest area on Marambaia Island, Parana subregion, Brazilian Atlantic Forest province [65]. This last species was also captured on Mel Island, coastal zone of Parana subregion, southeast Brazilian [152]. *Phaeotabanus limpidapex* was also captured at coastal zone of Parana subregion [59]. *Phaeotabanus fervens* (Linnaeus) feeds on caiman, in Pantanal and Central Amazon [95, 153]. The species occur in primary forest in areas of "campinas" and "campinarana", as well as in open areas near rivers and small stream banks, in Central Amazonian subregion, Varzea province [55, 95]; but according to Ferreira-Keppler and collaborators, it is active preferentially in "clareira" than forest [56]. The species was found also on Maraca Island, Amazonian subregion, Guyana province [50]. Other species captured in primary Amazonian Forest are *Phaeotabanus innotescens* (Walker) and *Phaeotabanus nigroflavus* (Kröber), both commonly collected near surface of water during drier months [95].

3.3.1.16. Genus Philipotabanus (Fairchild, 1943)

It comprises 29 species that are small to medium size flies, slender, narrow to very narrow frons, with clavate to threadlike callus, tubercle at vertex, unicolor eyes and palpi nearly always slender [19]. An excellent review with dichotomous key for the three subgenera and eleven species of genus *Philipotabanus* from records in Amazon was provided by Henriques [154].

3.3.1.16.1. Subgenus Melasmatabanus (Fairchild, 1964)

This has four species and one subspecies, similar to *Philipotabanus,* with a solid wing pattern, without fenestrae around cross veins, with the species all black [19]. All species occur in South America: they are largely distributed and can be seen from Panama to Midwest Brazil, in Andean areas, Amazonian Forest, and Cerrado [20]. Gorayeb and Rafael provided a key to females of species and subspecies of the genus and described *Philipotabanus* (*Melasmatabanus*) *pictus* Gorayeb and Rafael, from specimens collected in Pantanal subregion, Rondonia state, Brazil [155].

3.3.1.16.2. Subgenus Mimotabanus (Fairchild, 1964)

This comprises nine species with eight occurring in South America, Colombia, Ecuador and Peru; similar to foregoing group, they have solid wing pattern or a reduced shade below stigma, broader frons, clavate callus, and stouter palpi [19, 20]. The subgenus was first characterized by Fairchild in 1964 in a key for four species, and in 1975 the author reviewed the genus and provided a key to eight species than known [156]. The last species described for the genus was *Philipotabanus (Mimotabanus) tanypterus* Wilkerson, 1979. Lima reports the occurrence of *Philipotabanus* (*Mimotabanus*) *henriquesi* Limeira-de-Oliveira, Gorayeb and Rafael, from Brazil, in Chacoan subregion, Cerrado province [77].

3.3.1.16.3. Subgenus Philipotabanus (Fairchild, 1943)

This comprises 16 species with frons always narrow to very narrow, palpi slender, eyes bronzy in life, dark wing pattern, with hyaline fenestrae around crossveins and fork of third vein [19]. The genus is represented in Central America, but there are eight South-American species seen from Colombia to Bolivia and northern Amazon region [20]. *Philipotabanus* (*Philipotabanus*) *stigmaticalis* Kröber is a widely distributed species in Amazon Basin, more frequently captured in canopy of primary Amazonian Central Forest, Varzea province and is active throughout the year [94, 95]. Henriques described *Philipotabanus (Philipotabanus) obidensis Henriques, 2006* from eastern Peru and Bolivia, Puna subregion [154].

3.3.1.17. Genus Stenotabanus (Lutz, 1913)

This comprises 74 very small to medium size species, difficult to characterize, bare eyes with at least two transverse bands, moderate to broad frons, and callus as wide as frons [19]. Seven subgenera are currently recognized [20, 157]. Fairchild provided a key to the genera [158].

3.3.1.17.1. Subgenus Aegialomyia (Philip, 1941)

This has 25 species, but only four in South America [20]. *Stenotabanus (Aegialomyia) tobagensis* Fairchild occurs in Trinidad, Antillean dominion, and was observed attacking man on beach and caiman [159], *Stenotabanus (Aegialomyia) aberrans* Philip, described from Ecuador (northwestern South American dominion, Magdalena province), *Stenotabanus (Aegialomyia) geijskesi* Fairchild, from Suriname (Humid Guyana province) and Brazil (Para province, Amazonian subregion), and *Stenotabanus (Aegialomyia) ixyostactes* (Wiedemann), from Brazil (Chacoan subregion, Cerrado province) [20, 160].

3.3.1.17.2. Subgenus Brachytabanus (Fairchild, 1942)

This has three South-American species occurring in Colombia, Venezuela, Bolivia and Argentina, with one of them also occurring in Costa Rica and Panama [20]. *Stenotabanus (Brachytabanus) longipennis* Kröber, attracted by light in Colombia, *Stenotabanus (Brachytabanus) platyfrons* Fairchild, from Argentina, and *Stenotabanus (Brachytabanus) sphaeriscapus* Wilkerson, from Bolivia [20, 89, 156].

3.3.1.17.3. Subgenus Cretotabanus (Fairchild, 1969)

This has only one species, *Stenotabanus (Cretotatabanus) cretatus* Fairchild, recorded from eastern and central Amazonia, that appears before the rainy season and is collected preferably near the ground [55, 95].

3.3.1.17.4. Subgenus Melanotabanus (Lutz e Neiva, 1914)

This subgenus comprises two species from southeast Brazil: *Stenotabanus (Melanotabanus) brunnipes Kröber*, 1929 and *Stenotabanus (Melanotabans) fuliginosus* (Lutz & Neiva), 1914.

3.3.1.17.5. Subgenus Stenochlorops (Fairchild, 1969)

There are four Brazilian species: two from Amazon and two from Cerrado. *Stenotabanus (Stenotabanus) bequarti* Rafael, Fairchild and Gorayeb, occurs only in Amazon, along Rio Negro and its black water tributaries, flying during drier months [95].

3.3.1.17.6. Subgenus Stenotabanus (Lutz, 1913)

This has 59 species seen from Mexico to Argentina, Antilean and some in USA; 26 species are recorded from South America [156, 160]. They are small to very small species, with parallel-sided frons, round or square callus as wide as frons, middle frons with dark hair patch, eyes with two bands; it is the largest subgenera in South America [19, 20]. *Stenotabanus (Stenotabanus) obscurus* Kröber is a widely distributed species, occurring from Costa Rica to Argentina; Coscarón redescribed the female and described the male [161].

3.3.1.17.7. Subgenus Wilkersonia (Fairchild and Burger, 1994)

This has a single species, *Stenotabanus* (*Stenotabanus*) *roxannae* Wilkerson, from Caribbean subregion, Chocó, Colombia [20].

3.3.1.18. The genus Stibasoma (Schiner, 1867)

This comprises 19 species that are similar to bees of the genera *Centris* Fabricius, *Bombus* Latreille, *Xylocopa* Latreille and *Euglossa* Latreille [11]. They have robust bodies, with variable colors, very pilose legs, short antenna with a long dorsal spine, inflated palpi, sclerotized labella and fringed tibiae [19]. The genus was early studied by Ricardo [162] when it comprised six species, and by Knab, in 1913, who provided a key to 10 Neotropical species [163]. Coscarón completed the diagnosis of the genus adding genitalia features [72]. All species of the genus are recorded only from South America [20]. More recently, the genus was reviewed and two new species were described from southeast Brazilian Atlantic Forest: *Stibasoma manauensis* Turcatel, Rafael and Carvalho, and *Stibasoma ruthae* Turcatel, Rafael and Carvalho [164]. The larvae of the species in this genus are usually found in water of phytotelmata of bromeliads (Bromeliaceae). *Stibasoma flaviventre* Macquart and *Stibasoma venenata* Osten Sacken develop in arboreal bromeliads and *Stibasoma fulvohirtum* (Wiedemann) in terrestrial bromeliads, as found in Panama [165]. *Stibasoma fulvohirtum* (Wiedemann), *Stibasoma festifestivum* (Wiedemann), *Stibasoma flaviventre* (Macquart), and *Stibasoma currani* Kröber are the most common species appearing in surveys performed in Amazon [54–56, 94, 95].

3.3.1.19. Genus Rhabdotylus (Lutz, 1913)

This appears in Coscarón and Papavero's catalog [20] as a subgenus of *Stibasoma,* but Trojan, in 1998, revalidated the genus, based on characteristics of body pilosity and leg structure [166]. The genus comprises four species from which three are recorded from South America, and one has unknown distribution [20]. *Rhabdotylus planiventris* (Wiedemann) and *Rhabdotylus viridiventris* (Macquart) were captured in an ecotone area between sandbank and rain forest, on Marambaia Island, Parana subregion, southeast Brazilian Atlantic Forest province [65].

3.3.1.20. Genus Stigmatophthalmus (Lutz, 1913)

It has only one species, *Stigmatophthalmus altivagus* Lutz, which occurs in southeast Brazil, and was collected in mountainous region (800–2150 m above sea level), and in an ecotone area between the sandbank and rain forest on Marambaia Island, coastal zone in Parana subregion, and southeast Brazilian Atlantic Forest province [65, 167].

3.3.1.21. Genus Stypommisa (Enderlein, 1914)

This comprises 34 species, mostly small and slender, frons near always narrow, callus drop-shaped, marked tubercle at vertex, short proboscis, soft labella, palpi somewhat slender, clouds on crossveins, or anterior or posterior infuscation, and third appendiculate vein forked [19, 20]. The species can be seen from Nicaragua to Argentina, and only two species do not occur in South America [20]. Fairchild and Wilkerson reviewed the genus in 1986, including 28 species then known and provided a key to 26 of those species [168]. Coscarón elected genitalia features as key characters for define the genus [72]. *Stypommisa grandicolor* (Lutz) was the third most abundant species in the Central Amazon in a study conducted using Malaise trap in tropical forest [54]. *Stypommisa captiroptera* (Kröber), *Stypommisa glandicolor* (Lutz), and

Stypommisa modica (Hane) are widely distributed and fairly common species captured in surveys carried out in several environments in Amazon [54–57, 94, 95].

Recently, Brazilian researchers proposed the following two new genera within tribe Diachlorini.

3.3.1.22. Genus Muscotabanus (Henriques and Krolow, 2013)

This has the species *Muscotabanus rafaeli,* which was proposed from unidentified specimens from Entomological Collection of the National Research Institute Amazon, all collected in the Central Amazon [169].

3.3.1.23. Genus Elephantotus (Gorayeb, 2014)

The species *Elephantotus tracuateuensis* Gorayeb is described from specimens collected on the edge of a mangrove forest, coastal area of Brazil's western Amazon, near nests of *Eudocimus ruber* (Linnaeus) (scarlet ibis), *Nycticorax nycticorax* (Linnaeus) (socó sleeper) and *Ardea alba* (Linnaeus) (great egret) [170].

Others genera in the tribe Diachlorini are under-represented in more current surveys, and have no economic or sanitary importance.

3.3.2. Tribe Tabanini (Latreille, 1802)

With 207 Neotropical species characterized by setose basicosta, labella wholly pollinose, without ocelli [19, 36] There are some groups of species of *Leucotabanus, Stypommisa* e *Tabanus* (those with long antennal spine) difficult to place because the setose basicosta. These problems can be solved by the study of Fairchild [19]. In Neotropics, Tabanini comprises five genera, but only three with South-American species.

3.3.2.1. Genus Phorcotabanus (Fairchild, 1961)

With two South-American species, this can be seen from Colombia to Argentina [20]. *Phorcotabanus cinereus* (Wiedemann) occurs in Central Amazon, and was captured in "clareira," being the most abundant among the tabanid species captured in canopy of the forest [56].

3.3.2.2. The genus Poeciloderas (Lutz, 1921)

This comprises nine species, all endemic of South America; they are mainly observed in south temperate or Andean region, but not in Chile, although *Poeciloderas quandripunctatus* (Fabricius) is well distributed from Mexico to Argentina [20]. It is a fairly homogeneous group, with closely related species and very similar to those of genus *Tabanus*. This group is understudied and lacks higher setting to characterize the species in the genus [19]. Coscarón and Fairchild reviewed the genus in Argentina and provided a key to the four species in that country [171]. The more common species captured in current surveys is *Poeciloderas quandripunctatus* on Central Amazon and on Maraca Island, Amazonian subregion, Guyana province [50, 56]. The species was also captured in Tocantins, Brazil, Chacoan subregion, Cerrado province [77], and in southeastern Brazilian coastal zone and plateau, Parana subregion [59, 63]. The species was also collected in open meadows, from 10:00 h until ca. 16:00 h, during the sunniest and hottest hours of the day, on Marambaia Island, Parana subregion, southeast Brazilian Atlantic Forest province [65]. This species is recognized having a wide distribution for all Neotropics.

3.3.2.3. Genus Tabanus (Latreille, 1802)

This comprises world-wide distributed species with bare eyes, no tubercle at vertex, short proboscis, soft labella, setose basicosta, basal plate of third antennal segment with an acute or obtuse angle, rarely with a tooth or spine; in tropical species the wings can be tinted, spotted on crossveins, margined brown veins, entirely dark or black, but never banded [19]. Coscarón and Papavero [20] listed 191 species in tropical region, of which 110 occur in South America: in their catalog lacks *Tabanus bibanda* considered *nomen nudum* [37], recorded in the southeastern Brazilian plateau, Parana Forest province, Parana subregion [61]. Fairchild [172] reported *Tabanus nereus* Fairchild and *Tabanus eldridgei* Fairchild in mangrove areas of Colombia and Ecuador, and introduced the measure of "frontal index," as a morphological key feature to tabanid identification. Coscarón reviewed the genus and provided good illustrations to identify 16 Argentine species [173]. In 1983, Fairchild published an excellent study of the *Tabanus lineola* complex, providing keys to males and females of South America species *Tabanus campestris* Brèthes, *Tabanus colombensis* Macquart, *Tabanus commixtus* Walker, *Tabanus curtus* Hine, *Tabanus eldridgei* Fairchild, *Tabanus guapiensis* Wilkerson, *Tabanus nereus* Fairchild, *Tabanus occidentalis* Linnaeus *Tabanus penai* Philip, *Tabanus secundus* Walker (as *stenocephalus* Hine), *Tabanus Triangulum* Wiedemann, *Tabanus vittiger* Thomson, *Tabanus wilkersoni* Fairchild, and *Tabanus wokei* Fairchild [174]. Fairchild also offered a very relevant study of the larger species of *Tabanus* of eastern South America [175]. *Tabanus importunus* Macquart, *Tabanus occidentalis* Linnaeus and *Tabanus pungens* Wiedemann were reported as occurring in French Guiana, in pasture area, transition zone between savannah and eastern Brazilian Amazon forest [176]. *Tabanus aaptus* Fairchild, Tabanus *augustifrons* Macquart, *Tabanus antarticus* Linnaeus, *Tabanus callosus* Macquart, *Tabanus claripennis* Wiedemann, *Tabanus discus* Wiedemann, *Tabanus importunus* Wiedemann, *Tabanus lineifrons* Lutz, *Tabanus nebulosus* DeGeer, *Tabanus nematocallus* Fairchild, *Tabanus occidentalis* Linnaeus, *Tabanus piceiventris* Rondani, *Tabanus sannio* Fairchild, *Tabanus trivittatus* Fabricius, and *Tabanus unimacula* (Kröber) were recorded on Maraca Island, Amazonian subregion, Guyana province [50]. Several surveys conducted since 1999 till 2010, revealed that *Tabanus amapaensis* Fairchild, *Tabanus amanuensis* (Barretto), *Tabanus angustifrons*, Macquart, *Tabanus antarticus* Linnaeus, *Tabanus callosus* Macquart, *Tabanus claripennis* (Bigot), *Tabanus crassicornis* Wiedemann, *Tabanus discus* Wiedemann, *Tabanus importunus* Wiedemann, *Tabanus lineifrons* Lutz, *Tabanus nematocallus* Fairchild, *Tabanus occidentalis* Linnaeus, *Tabanus piceiventris* Rondani, *Tabanus pungens* Wiedemann, *Tabanus sannio* Fairchild, *Tabanus sextriangulus* Gorayeb and Rafael, *Tabanus trivittatus* Fabricius, and *Tabanus xuthopogon* Fairchild, are the most common species in central Amazon [54, 55, 57, 95]. *Tabanus importunus* Wiedemann is reputed as the most important vector in the Pantanal, Chaco subregion, Midwestern Brazil, being most abundant in November. It is the most common species in the region, followed by *Tabanus occidentalis* Linnaeus, a common species found in September and December, by *Tabanus claripennis* Bigot, more abundant during July to October [76]. A survey performed in Areguá, Paraguay, Chacoan subregion, Pampa province, found *Tabanus triangulum* Wiedemann, *Tabanus secundus* Walker (as *stenocephalus* Hine), *Tabanus occidentalis* Linnaeus, and *Tabanus pungens* (Wiedemann) among the most abundant species [49]. A survey performed in Tocantins, Brazil, Chacoan subregion, Cerrado province found *Tabanus antarcticus* Linnaeus, *Tabanus* cf. *cicur* Fairchild, *Tabanus fuscofasciatus* Macquart, *Tabanus glaucus* Wiedemann, *Tabanus importunus* Wiedemann, *Tabanus mucronatus* Fairchild, *Tabanus occidentalis* var. *dorsovittatus* Macquart, *Tabanus occidentalis* var. *modestus* Wiedemann, *Tabanus palpalis* Brèthes, and *Tabanus xuthopogon* Fairchild [77]. The pupae of *Tabanus triangulum* Wiedemann and *Tabanus platensis* Brèthes, and larvae of *Tabanus nebulosus ornativentris* Kröber on *Pistia stratiotes* Linnaeus were respectively collected in Santa Fé, Buenos Aires and Formosa, Argentina, Chacoan subregion, Pampa province [132]. *Tabanus angustus* Macquart takes place in the hills of Argentine Pampean subregion [131]. In southeastern Brazilian plateau, Parana Forest province, occur *Tabanus fuscus* Wiedemann, *Tabanus colombensis* Macquart, *Tabanus eldridgei* Fairchild, *Tabanus nebulosus ornativentris* Kröber, and *Tabanus wokei* Fairchild [63]. *Tabanus augustus* Macquart, *Tabanus claripennis* (Bigot), *Tabanus fuscus* Wiedemann, and *Tabanus triangulum* Wiedemann were captured in coastal zone of Parana state, Brazilian Atlantic Forest [59]. *Tabanus claripennis* (Bigot), *Tabanus discus* Wiedemann, *Tabanus fuscus* Wiedemann, *Tabanus importunus* Wiedemann, *Tabanus obsoletus* Wiedemann, *Tabanus occidentalis* Linnaeus, *Tabanus pungens* Wiedemann, and *Tabanus triangulum* Wiedemann were collected in several environments on Marambaia Island, Parana subregion, southeast Brazilian Atlantic Forest province [65].

4. Tabanids and diseases

There are few studies concerning transmission of diseases caused by tabanids in South America. Most of the researchers have the scope of knowing the species found in different environments, seasonal fluctuation, and biotic and abiotic factors that affect the behavior of tabanid populations. The first author to relate tabanids with animal disease in South America was Lutz. He pointed the tabanids as the main mechanical vector of *Trypanosoma evansi*, the etiological agent of "mal-de-caderas" or "surra" of equines [176]. Tabanids have been recorded as an important mechanical vector of *Trypanosoma vivax* in South America [177]. Raymond found *Trypanosoma vivax* was transmitted by *Tabanus importunus* between zebu bulls by interrupted blood meal, in French Guiana [17]. In Colombia, three specimens of tabanid (without identification) were found infected with flagellates morphologically compatible with *Trypanosoma vivax* [52], and in a livestock region, was found a strong positive correlation between incidence of *Trypanosoma vivax* in cattle and tabanid population [14]. An experimental essay demonstrated that *Tabanus nebulosus* is able to transmit *Trypanosoma vivax* between cattle, when interrupted blodmeal is resumed within 10 minutes or less [14]. *Cryptotylus unicolor* was able to transmit experimentally *Trypanosoma vivax* between the livestock [178]. Monzón and collaborators recorded equine trypanosomiases transmitted by *Tabanus* sp. in Argentina [180]. Tabanids have been recorded as an important mechanical vector of *Trypanosoma evansi* in Brazilian and Bolivian Pantanal [179]. Outbreaks of the cattle disease caused by *Trypanosoma evansi* have been associated with the rainy season when tabanids are more abundant [15]. The most important vector of trypanosomiasis of cattle in Pantanal is *Tabanus importunus* and it is more abundant during the rainy season, from September/October to January [181–185].

Due to major environmental changes imposed by human productive activity, new interactions between agents, vectors, and hosts have occurred. A specimen of *Tabanus importunus* was found infected with a *Leishmania* sp.: the diagnosis was performed using DNA amplification technique, in São Paulo, Brazil [186]. *Borrelia burgdorferi* was found naturally infecting tabanids on Marambaia Island, Rio de Janeiro, southeast Atlantic Brazilian Forest (no published data from Guimarães et al). *Tabanus* sp. was reported as vector of human botfly, *Dermatobia hominis* Linnaeus Jr., in Rio Grande do Sul, Brazil [187]. In Ecuador it was reported the transmission of the botfly by *Chrysops varians* [188].

The following is offered as an Appendix, in which the main studies performed in South-American countries are presented.

Appendix

In this study of South American tabanids, the authors included Trinidad, because of its geographic proximity to South America mainland and by the affinity of its tabanofauna with that of South America, it is considered to be in Caribbean subregion [189]. For this aspect, Panama should also be included, but studies on tabanids in that country are very extensive and we chose to omit them. Still, many species that occur in Panama also occur in South America Caribbean subregion.

The first studies on the tabanofauna **Trinidad** were from Bequaert, which listed 23 species belonging to the genera *Chrysops* (five species), *Esenbeckia, Selasoma, Stibasoma* (two species), *Dichelacera, Tabanus* (12 species) and *Acanthocera* [189]. In 1944, Bequaert brought the number of species as 31, 29 in Trinidad and two in Tobago, adding a species of the genus *Lepiselaga,* four in genus *Tabanus* and *Diachlorus* [190]. Callan increased the number of species to 34, registering a species for each genus of *Tabanus* (*Chlorotabanus*), *Stibasoma,* and *Fidena* [191]. Fairchild and Aitken added 11 species, describing *Acanthocera trinidadensis* and *Stibasoma flaviventre,* and corrected previous errors in identification, bringing to 45 species in Trinidad [158]. The record of *Phaeotabanus cajennensis* on the island rose to 46 species of tabanids in Trinidad [151].

About horseflies in **Guyanas**, Desquesnes and Rocque published an important compilation of knowledge on biology, sanitary importance, and control of tabanids [192]. In **French Guiana**, a survey performed by Raymond found some differences in relative abundances of tabanids as the environment: in savannah, the most abundant species were *Tabanus importunus, Tabanus occidentalis* var. *dorsovittatus, Tabanus wilkersoni,* and *Chlorotabanus mexicanus;* in the rain forest the most abundant species were *Phaeotabanus cajennensis, Chlorotabanus inanis, Stenotabanus cinereus,* and *Tabanus occidentalis.* Raymond also noted that in ecotone environments, there is a greater variety of species and greater chance of finding rare species [17]. In **Surinam** the available information

http://dx.doi.org/10.5772/67108

on tabanid is the Coscarón and Papavero catalog which list *Tabanus antarcticus* and *Tabanus nebulosus;* and the site InsectoidInfo points *Chlorotabanus inanis* as occurring in that country [20, 193]. In **Guyana** the tabanofauna is very diverse, with species in several genera, but information is restricted to catalogs [30, 32, 34, 82].

Studies on tabanids from **Colombia** were very profitable after 1940s. In 1946, Bequaert published a catalog of Colombian tabanids listing 129 species with their respective localities, and provided a key to the genera [194]. A list of 39 species of tabanids collected from Valle del Cauca, Pacific Ocean coast, was published in 1969 [195]. In 1979, Wilkerson published a list of 158 species of tabanids from Western Colombian in which are described 31 species, one subspecies and one subgenus; he also provided a checklist of 226 Colombian species [196]. During years 2000–2001 Orozco collected tabanids in eastern plains of southeast Colombia, Amazonian subregion, and recorded 64 species in 14 genera (*Catachlorops, Chlorotabanus, Cryptotylus, Chrysops, Dasychela, Diachlorus, Dichelacera, Fidena, Leucotabanus, Phaeotabanus, Pityocera, Poeciloderas, Stypommisa,* and *Tabanus*) [197]. In studies in Antioquia, Colombia, western region of Andes, Caribbean subregion, were recorded *Lepiselaga crassipes,* the most abundant species, followed by *Chrysops variegatus* (as *variegata*), *Tabanus occidentalis, Tabanus claripennis, Tabanus importunus, Tabanus albocirculus,* and *Tabanus nebulosus*. The authors also found three specimens of tabanid (without identification) infected with flagellates morphologically compatible with *Trypanosoma vivax* [52]. In 2016, Wolf and Miranda-Esquivel published the more currently catalogue in which are listed 256 species of Colombian tabanids [198].

In **Venezuela** the first list containing 31 species of tabanid was prepared by Pechuman [199]. The list was supplemented by Stone in 1944 which reported 52 species in 18 genera, and provided a key to the species of the genus *Cryptotylus* [200]. In savannah of the Venezuelan Caribbean subregion it was found the most abundant species *Tabanus pungens, Tabanus claripennis, Tabanus antarticus, Tabanus nebulosus, Chrysops venezuelensis* and *Esenbeckia prasiniventris;* the authors also emphasized the importance of tabanids in the transmission of bovine trypanosomiasis in the region [201]. Gorayeb and collaborators recorded 16 species to Venezuela, among the genera *Fidena* (four species), *Catachlorops, Diachlorus* (three species), *Dichelacera, Dicladocera, Leucotabanus, Philipotabanus Stenotabanus, Stypommisa* (three species) from deposited material the Museum of Agricultural Zoology, Institute Francisco Fernández Yépez of the Central University of Venezuela [202]. In 2005, it was published a list of 20 species of tabanids from Caribbean subregion, northern Venezuela [203]. In 2010, Gómez and collaborators reviewed some species of genus *Bolbodimyia* and provided a key to the recorded five species of this genus in Venezuela [111].

In 1968, Patrick and Hays published a tabanids list of eastern **Ecuador**, with 27 species among genera *Esenbeckia, Elaphela, Fidena, Chrysops, Diachlorus, Dichelacera, Lepiselaga, Chlorotabanus* (two species), *Stibasoma, Stenotabanus, Stypommisa* (three species), *Philipotabanus, Leucotabanus, Phaeotabanus* (two species), and *Tabanus* (nine species) collected in 1965–1966 in light areas of tropical forest next Limoncocha, Napo province, Amazonian subregion [204]. Buestán conducted a survey in coastal zone of Ecuador, Western and Arid provinces, Caribbean subregion of Ecuador, and found the most abundant species *Tabanus pungens,* followed by *Tabanus colombensis, Tabanus occidentalis, Tabanus albocirculus,* and *Lepiselaga crassipes* [61]. Fairchild and

León in 1986 published a revisional list of 81 species of tabanids of Ecuador, providing references of original descriptions, geographical data and a key to determine the genera [205]. In 2005, Cárdenas and Vieira published a list of 42 new records, updating the knowledge of Ecuadorian tabanofauna, with 181 species [206]. In 2007, Buestán and colleagues published a list rising to 204 tabanids species of Ecuador, providing species morphological data and information about the collections environments [207]. In 2009, Cárdenas, Buestán and Dangles published studies on the diversity and distribution of tabanofauna of Ecuador and a catalog listing 198 species; in this study, the authors discussed the distribution of *Chrysops varians tardus, Dicladocera macula* and *Fidena rhinophora* using georeferenced localities and niche modeling analyses [99]. More recently, Cárdenas studied the distribution of tabanids according to altitudes and climatic factors, finding that most specimens have their activities limited by extremes temperature and humidity [98, 100].

In **Peru**, the first specific publication on Peruvian tabanofauna was Soukoup's in 1945, which listed 81 species in the country [208], most of them were already mentioned in Kröber's catalog of 1934 [31]. In 1951, Kröber reported the result of the expedition held in southern Peru, adding 14 species to those known [209]. Philip published two lists of species, mostly from Peru, from specimens collected during an expedition of California Academy of Science to west coast of South America [210, 211]. Carrasco in 1972, published a list of 163 tabanid species of which 29 were collected from southern Peru, distributed mainly in the genera *Esenbeckia* (seven species), *Fidena* (five species), *Phaeotabanus* (five species), *Scaptia* (five species), *Scione* (20 species), *Chrysops* (12 species), *Stenotabanus* (12 species), *Dasybasis* (19 species), *Dicladocera* (12 species), *Stypommisa* (eight species), and *Tabanus* (31 species) [212]. Wilkerson and Fairchild in 1985 provided a checklist of 228 species known from Peru, with a key to subfamilies, tribes and genera. The list includes 73 tabanid species which were collected in lowland forest, southeast Peru, Amazonian subregion, Pantanal province, considered by authors as a site of great diversity in tabanids [74].

In **Brazil**, last years of 1900, a new generation of researchers in tabanids brought new knowledge of tabanofauna almost in the entire country. Gorayeb, in 1985, conducted a survey in the western Amazon and recorded 15 species of *Tabanus*, four *Chrysops*, four *Fidena*, three *Catachlorops*, seven *Diachlorus*, four *Dichelacera*, three *Phaeotabanus*, and four species of *Stypommisa* [53]. In surveys conducted in Central Amazon, between 1982 and 2007, using Malaise traps at ground level and hanging traps on water depth and 25 and 40 m high, 60 species belonging to the following genera of tabanids were collected: *Tabanus* (29 species), *Chrysops* (seven species), *Diachlorus* (six species), *Leucotabanus* (five species), *Fidena* (four species), *Phaeotabanus* (four species), *Stibasoma* (four species), *Dichelacera* (two species), and *Stypommisa* (three species) [54–57, 94, 95, 213–215]. One of the most abundant collected species in these surveys was *Phorcotabanus cinereus* (Wiedemann), which also occurs in Brazil's eastern savannah, and in the Argentine Chaco [54, 56, 57, 94, 95, 213–215]. In northern Amazon border with Colombia were recorded 20 species of *Tabanus*, two *Esenbeckia*, one *Fidena*, two *Chrysops*, and 13 *Diachlorini* (in genera *Catachlorops, Diachlorus, Dichelacera, Leucotabanus, Phaeotabanus* and *Stypommisa* [50]. In a study conducted in Central Amazonian subregion, Varzea province, were caught *Stenotabanus cretatus, Stenotabanus bequarti* Rafael, Fairchild and Gorayeb, *Phaeotabanus nigriflavus*, and *Tabanus occidentalis* feeding on caiman

Caiman crocodilus (L.) and anaconda *Eunectes murinus* (Linnaeus) [7]. In Pantanal, Brazilian Amazonian subregion, Barros recorded 10 species of *Tabanus,* and others in several general; most abundant species were *Tabanus importunus, Tabanus claripennis, Tabanus occidentalis, Lepiselaga crassipes,* and *Chrysops* sp. [76]. In Chaco subregion, in Cerrado of the Brazilian Midwest, two species were recorded in *Esenbeckia,* four *Fidena,* six *Catachlorops,* two *Dichelacera,* three *Stypommisa,* and 11 species of *Tabanus* [77]. *Tabanus occidentalis, Lepiselaga crassipes, Tabanus importunus, Tabanus claripennis,* and *Tabanus sorbilans* were the most abundant species recorded in the transition zone between Cerrado and Pantanal, Amazonian subregion [144]. In Parana subregion, a survey performed in Atlantic Forest, Marambaia Island, and southeastern Brazil, recorded 31 species of tabanids of genera *Chrysops* (two species), *Esenbeckia, Scepsis, Fidena, Scaptia, Acanthocera, Catachlorops* (two species), *Chlorotabanus, Diachlorus* (three species), *Dichelacera* (two species), *Leucotabanus, Phaeotabanus* (three species), *Rhabdotylus* (two species), *Stigmatophtalmus, Poeciloderas, Tabanus* (eight species); the most abundant species was *Diachlorus bivittatus* [64, 65]. Also in Parana subregion, southeastern Brazilian plateau, were recorded species of *Acanthocera* (three), *Chrysops* (six), *Dichelacera* (three), *Diachlorus, Fidena* (two), *Lepiselaga, Phaeotabanus, Poeciloderas,* and *Tabanus* (eight) [59, 63]. Dutra and Marinoni captured *Catachlorops furcatus* Wiedemann, *Catachlorops fuscinevris* (Macquart), *Chlorotabanus inanis, Chrysops* sp., *Diachlorus bivitattus, Dichelacera alcicornis, Phaeotabanus litigiosus, Poeciloderas quadripunctatus, Pseudacanthocera sylverii* (Macquart), *Stenotabanus* sp., and *Tabanus occidentalis* on Mel Island, coastal zone of Parana subregion, Brazilian southeast: the most abundant species was *Dichelacera alcicornis* [152]. Survey conducted between 1995–1997, using Malaise traps, southern Chacoan subregion, Pampa province, Brazilian far south, resulted in a list of 30 species, mainly representatives of *Chrysops* genera (seven species), *Dichelacera* (three species) and *Tabanus* (seven species) [60]. The first published study on seasonal fluctuation in Brazil, is Bouvier's, in 1952, which studied 52 species of tabanids and its seasonal variation in the northwest region of São Paulo, Brazil, Parana Subregion, Brazilian Atlantic Forest province [216].

In **Bolivia**, Chainey and colleagues published a preliminary list and a key to 32 genera and 167 species of tabanids, they also report tabanid collections in lowlands and mountainous areas, including grasslands, forests and wetlands, Chacoan subregion, eastern Bolivia; in this region, during August and October, cattle must be gathering in tight groups to reduce tabanid hematophagism impact on beef and milk yields [90]. Chainey and Hall proposed new genus and species, *Boliviamyia fairchildi* from specimens collected in the forest of the Bolivian southwest, Amazon subregion of Bolivia [73]. Coscarón re-described *Dichelacera boliviensis* (Brèthes) and *Dichelacera micracantha* Lutz from Bolivia [217]. Wilkerson provided a key of species of *Stenotabanus (Brachytabanus)* and described *Stenotabanus (Brachytabanus) sphaeriscapus,* from Bolivia [218]. Gutiérrez and Rumiz in 2002 conducted a great study to assess the degree of specialization relative to habitat of four groups of insects: among tabanids, *Pseudacanthocera brevicorne* (cerrado), *Tabanus sorbilans* (fields), *Phaeotabanus fervens* (riparian forest), and *Tabanus nebulosus* (woods gallery) were more specialized as the habitat; *Tabanus occidentalis* (cerrado, fields, riparian galleries and semi deciduous forests), and *Lepiselaga crassipes* (cerrado, fields, riparian forests and semi deciduous) were less demanding species as the habitat use [219].

About tabanids from **Chile**, Macquart, Rondani and Walker were the first researchers of Chilean tabanofauna [46]. They also highlights the work of Philippi, with the publication of a list of Chilean tabanids in 1865 [220]; the catalog "Insects Diptera of Chile," by Reed in 1888, which lists 61 species of tabanids in Chile [221]; and the works of Enderlein and Kröber who reviewed and published studies on the Chilean species [32, 69, 222]. It should be noted Pine work that offers an excellent review of studies in Chile about tabanids [223]. In 1968, Philip reviewed the types of 28 Chilean species previously described by Philippi in 1865 [221, 224]. In 1973, the first study on tabanid ecology in Chile was published, from collections made using Manitoba trap in north coast of Center-Chilean subregion, recording species of the genera *Scaptia* (three species), *Mycteromyia,* and *Dasybasis* [70]. And in 1991, Coscarón and González published a list of 110 species of Chilean tabanids in 16 genera, with geographical distribution, bibliography, and a key to subfamilies, genera and subgenera [46]. The most recent studies on Chilean tabanids concern the description of *Dasybasis elquiensis* [134]; redescription of male and female of *Scaptia (Pseudoscione) varies* (Walker) and the male's description of *Scaptia (Pseudoscione) atra* (Philippi) [91]; description of female and male redescription of *Dicladocera hoppi* Enderlein [225]; descriptions of adults and immature stages of *Agelanius fuscus* [107]; adult female of *Agelanius verai* and *Agelanius chiloensis* [106, 109]; description of immature stages of *Scaptia (Scaptia) lata* [85], *Dasybasis pruinivitta* [135], and *Agelanius cortesi* [110]; and description of *Dasybasis antillanca* and *Dasybasis collagua* [40]. Studies in wetlands ("humedales") in Chilean Andes in Centro-Chilean subregion, determined that the order Diptera is the most abundant among insects collected by Malaise trap, and in this Order, Tabanidae is the second group most abundant, after Tipulidae [226].

The study of tabanids in **Argentina** have been very fruitful, especially the works of Coscarón that since the beginning of 1960s has been studying the morphology, taxonomy and biology of the group. In addition to describing various species of several genera, Coscarón revised *genera Dicladocera* Lutz in Argentina [141] and *Lepiselaga* Macquart [227], the subgenus *Mesomyia (Coracella)* Philip [47], genera *Dichelacera* Macquart [214], *Stenotabanus* Lutz and *Myiotabanus* Lutz [161], *Catachlorops* Lutz [114], *Leucotabanus* Lutz, *Pseudacanthocera* Lutz, *Bolbodimyia* Bigot and *Pachyschelomyia* Barretto [228], *Diachlorus* Osten Sacken, *Stibasoma* Schiner, *Stypommisa* Enderlein, *Cryptotylus* Lutz and *Chlorotabanus* Lutz [229], *Tabanus* Linnaeus [173], *Chrysops* Meigen [230], *Phaeotabanus* Lutz and *Acanthocera* Macquart [231], *Scione* Walker [88], *Fidena* Walker [91], *Poeciloderas* Lutz [171], the subgenus *Scaptia* (*Lepmia*) Fairchild [232], the genus *Dasybasis* Macquart [130], the tribe Mycteromyiini [68], and the subgenus *Scaptia* (*Scaptia*) Walker [233]. He also studied the immature stages of several species in several genera [85, 128, 132, 133, 149, 234–237]. In 2002, Coscarón compiled an illustrate key to larvae and pupae of Argentine tabanids [238]; and in 2014, Coscarón and Papavero published another more elaborated key to tabanids immature stages of Neotropical tabanids [239]. In 2016, Dufek and colleagues published a study about the tabanids from west Argentina, Chaquean subregion, reporting species of *Chrysops* (four), *Fidena, Lepiselaga, Diachlorus, Dichelacera, Phaeotabanus, Phorcotabanus, Poeciloderas* (two species) and *Tabanus* (five species) [240]. In 1951 and 1953, Hack provided good morphological studies of Argentine tabanids [241, 242]. There is only one list of tabanids from Argentina published by Coscarón (1998) in which he listed 350 species [243].

Tabanids studies in **Paraguay** are still incipient. The first study of Paraguay tabanofauna was performed by Kröber, who described three species of *Poeciloderas*, all currently synonymized, and two species transferred to the genus *Tabanus* [244]. Tabanids were implicated by Russo in transmission of *Trypanosoma equinun* to equine population in Paraguay since 1954 [245]. In 1977, Coscarón summarized available literature on tabanids in Paraguay [246]. Strickman [48] collected tabanids during 1978–1980, in south-central Paraguayan region, Chaquean subregion: nine species of *Tabanus*, five species of *Chrysops*, two *Dichelacera*, two *Diachlorus*, two *Stypommisa*, and one for each genus *Acanthocera*, *Lepiselaga* and *Poeciloderas*. The same author studied the seasonal variation and climatic factors on the bionomics of *Chrysops variegatus*, in central Paraguay, in an area adjacent to a lake, consisted of a sandy and woody strip, bordered by the lake and a grassy swamp [49]. The author also points the species as a possible vector of the equine disease 'mal de caderas,' caused by *Trypanosoma evansi* [48]. No other record of Paraguayan tabanids was found in literature.

Bibliographic information about the tabanids from **Uruguay** is only available in the catalogs of Kröber [31], Fairchild [32], Moucha [82], Fairchild and Burger [34], and Coscarón and Papavero [20]. A list of 21 species of tabanids within the genera *Chrysops* (three species), *Esenbeckia*, *Fidena*, *Acanthocera*, *Catachlorops* (two species), *Dasybasis* (two species) *Lepiselaga*, *Poeciloderas*, *Stenotabanus*, *Stibasoma*, and *Tabanus* (six species) is available at the site "InsectoidInfo" [247].

Acknowledgements

The authors thank Dr. Inocêncio de Souza Gorayeb, Departamento de Zoologia, Museu Paraense Emilio Goeldi; Dr. Juan J. Morrone, Museu de Zoologia "Alfonso L. Herrera", Departamento de Biología Evolutiva, Facultad de Ciencias, Universidad Nacional Autónoma de México (UNAM); Dr. Rafael E. Cárdenas, Museo de Zoologıa QCAZ, Laboratorio de Entomologıa, Escuela de Ciencias Biologicas, Pontificia Universidad Católica del Ecuador, Quito, Ecuador; Dr. James Buestán, Subproceso de Entomología-Salud Animal del Instituto de Higiene y Medicina Tropical Leopoldo Izquieta Pérez, Guayaquil, Ecuador, for support and permission of images used.

Author details

Ronald Rodrigues Guimarães[1,2*], Harlan Ronald Storti Rodrigues[2] and Ronald Rodrigues Guimarães Júnior[2]

*Address all correspondence to: ronaldrguimaraes@gmail.com

1 UNESA – Estácio de Sá University, Rio de Janeiro, Brazil

2 CEMA – Center for Education and Research in Environmental Medicine, Rio de Janeiro, Brazil

References

[1] Henriques AL. Tabanidae. In: Vasconcelos CRF, Magalhães C, Rafael JÁ, Franklin E, editors. A fauna de Artrópodes da Reserva Ducke: estado atual do conhecimento taxonômico e biológico. Manaus: Editora INPA; 2009. pp. 169–178.

[2] Catalogue of Life [Internet]. 2013. Available from: http://www.catalogueoflife.org/col/browse/tree/id/15849896 2012 [Accessed: 2013-02-04]

[3] Ferreira RLM, Rafael JA. Criação de Imaturos de Mutuca (Tabanidae: Diptera) Utilizando Briófitas e Areia como Substrato. Neotropical Entomology 2006;**35(1)**:141–144. DOI: 10.1590/S1519-566X2006000100020

[4] Fairchild GB. Tabanidae. In: Hulbert SH, Rodríguez e Santos, ND, editors. Aquatic Biota of Tropical South America, Part 1: Arthropoda. San Diego: San Diego State University; 1981. pp. 290–230.

[5] Philip CB. A Collection of four species of Tabanid Flies taken from an Anaconda Snake in Peru in May, 1984. Pan-Pacific Entomologist. 1986;**62(1)**:23.

[6] Henriques AL, Ferreira RLM, Vidal JF, Rafael JA. *Betrequia ocellata* Oldroyd (Diptera: Tabanidae: Rhinomyzini) blood feeding on *Caiman crocodilus* (Linnaeus) (Crocodylia: Alligatoridae) in Manaus, Brazil. Revista Brasileira de Zoologia. 2000;**17(3)**:609–613.

[7] Ferreira RLM, Henriques AL, Rafael JA. Activity of Tabanids (Insecta: Diptera: Tabanidae) attacking the reptiles *Caiman crocodilus* (L.) (Alligatoridae) and *Eunectes murinus* (L.) (Boidae), in the Central Amazon, Brazil. Memórias do Instituto Oswaldo Cruz. 2002;**97(1)**:133–136. DOI: 10.1590/S0074-02762002000100024

[8] Limeira-de-Oliveira F, Rafael JA, Henriques AL. *Phorcotabanus cinereus* (Wiedemann, 1821) (Diptera, Tabanidae), an Ornithophilic species of Tabanid in Central Amazon, Brazil. Memórias do Instituto. Oswaldo Cruz. 2002;**97(6)**:839–842. DOI: 10.1590/S0074-02762002000600015

[9] Krinsky WL. Animal disease agents transmitted by horse flies and deer flies (Diptera: Tabanidae). Journal of Medical Entomology. 1976;**13(3)**:225–275. DOI: 10.1093/jmedent/13.3.225

[10] Foil LD. Tabanids as vectors of disease agents. Parasitology Today. 1989;**5(3)**:88–96 + addendum: 19–21.

[11] Turcatel M, Carvalho CJB, Rafael JA. Mutucas (Diptera: Tabanidae) do estado do Paraná, Brasil: chave de identificação pictórica para subfamílias, tribos e gêneros. Biota Neotropical. 2007;**7(2)**:2. Available from: http://www.scielo.br/scielo.php?script=sci_arttext&pid=S1676-06032007000200029&lng=en. DOI: 10.1590/S1676-060320070002 0029.

[12] Davies CR. Interrupted feeding of blood-sucking insects: causes and effects. Parasitology Today. 1990;**6(1)**:19–22. DOI: 10.1016/0169-4758(90)90387-J

[13] Guimarães RR, Guimarães-Junior RR, Buys S, Harlan-Ronald RS, Rodrigues-Guimarães R, Wermelinger ED, Carvalho, RW. Remarks on behavior of horse guard wasps (Hymenoptera, Crabronidae, Bembicinae) and interrupted hematophagism of horse flies (Diptera, Tabanidae). International Journal of Biology, 2015;**8**:27. DOI: 10.5539/ijb.v8n1p27

[14] Otte MJ, Abuabara JY. Transmission of South American *Trypanosoma vivax* by the Neotropical horsefly *Tabanus nebulosus*. Acta Tropica 1991;**49(1991)**:73–76.

[15] Silva RAMS, Herrera HM, Barros ATM. Trypanosomosis outbreak due to *Trypanosoma evansi* in the Pantanal, Brazil. A preliminary approach on risk factors. Revue D'élevage et de Médecine Vétérinaire des Pays Tropicaux. 1996;**48**:315–319.

[16] Chvála M, Stary P. Notes on the horse-flies of Cuba. Acta Entomologica Bohemoslovaca. 1967;**64**:392–401.

[17] Raymond HL. Répartition des principales espèces de taons (Diptera, Tabanidae) de la zone côtière de Guyane françase. Cahiers Orstom. Série Entomologie Médicale et Parasitologie. 1986;**24(3)**:219–224.

[18] Mackerras IM. The classification and distribution of Tabanidae. I. General review. Australian Journal of Zoology. 1954;**2(3)**:431–454. DOI: 10.1071/ZO9540431

[19] Fairchild GB. Notes on Neotropical Tabanidae. XII. Classification and distribution, with keys to genera and subgenera. Arquivos de Zoologia. 1969;**17**:199–255. DOI: 10.11606/issn.2176-7793.v17i4p199-255

[20] Coscarón S, Papavero N. Catalogue of Neotropica Diptera. Tabanidae. Neotropical Diptera. 2009;**16**:1–199. Available from: http://revistas.ffclrp.usp.br/Neotropical_Diptera/article/viewFile/210/165. [Accessed: 2014-03-02].

[21] Borror DJ, Delong DM, editors. Introdução ao Estudo dos Insetos. Rio de Janeiro: USAID and Editora Eddgar Blücher, 1969. 655 p.

[22] Brown BV, Borkent A, Cumming JM, Wood DM, Woodley NE, Zumbado MA, editors. Manual of Central America Diptera. Ottawa: NRC Research Press; 2008, Vol. 1, 714 p.

[23] Barretto MP. Estudos sobre tabanidas brasileiros. I. Morfologia externa do *Tabanus (Poeciloderas) quadripunctatus* Fabr. (Diptera, Tabanidae). Anais da Faculdade de Medicina da Universidade de São Paulo. 1946;**22**:113–149.

[24] Iide P. Estudos sobre *Tabanus (Lophotabanus) importunus* Wiedemann, 1828 (Diptera: Tabanidae). Revista Brasileira de Biologia. 1969;**29(2)**:207–223.

[25] Iide P. Estudo sobre o morfologia, a terminologia e o valor taxonômico das estruturas de genitália masculina dos Tabanidae, com base em *Fidena* (*Fidena*) *sorbens* (Wied., 1828) (Diptera, Tabanidae, Pangoniinae). Revista Brasileira de Biologia. 1982;**42(1)**:197–211.

[26] Iide P. Estudos morfológicos e taxonômicos sobre os Scionini brasileiros. O gênero *Fidena*. Parte I. (Diptera, Tabanidae, Pangoniinae). Revista Brasileira de Biologia. 1982;**42(2)**:347–358.

[27] Iide P. Estudos morfológicos e taxonômicos sobre os Scionini brasileiros. O gênero *Fidena*. Parte II. (Diptera, Tabanidae, Pangoniinae). Revista Brasileira de Biologia. 1988;**48(2)**:139–154, 45.

[28] Iide P. Estudos sobre os Scionini Neotropicais do gênero *Fidena* – o subgênero *Leptofidena*. Memórias do Instituto Oswaldo Cruz. 1989;**84(Suppl. 4)**:267–273. DOI: 10.1590/S0074-02761989000800047

[29] Martins AV. Os tabanídeos do Estado de Minas Gerais [thesis]. Belo Horizonte, Faculdade de Medicina, Universidade Federal de Minas Gerais; 1940.

[30] Hunter WD. A catalogue of the Diptera of South America. Part II, Homodactyla and Mydiadae. Translations of American Entomological Society. 1900/1901;**27**:121–155.

[31] Kröber O. Catalogo dos Tabanidae da America do Sul e Central, incluindo o Mexico e as Antilhas. Revista de Entomologia. 1934;**4(2-3)**:222–276, 291–333.

[32] Fairchild GB. Family Tabanidae 28. In: Papavero, N, editor. A Catalogue of the Diptera of the Americas South of the United States. Museo de Zoologia: São Paulo; 1971. pp. 1–163.

[33] Coscarón S, Papavero N, editors. An illustrated manual for the identification of the Neotropical genera and subgenera of Tabanidae (Diptera). Museu Paraense Emilio Goeldi. Coleção Emile Snethlage: Belém; 1993. 150 p.

[34] Fairchild GB, Burger JF. A Catalog of the Tabanidae (Diptera) of the Americas South of the United States. Memoirs of the American Entomological Institute. 1994;**55**:1–249.

[35] Coscarón S, Papavero N. Manual of Neotropical Diptera. Tabanidae. Neotropical Diptera. 2009;**6**:1–137. Available from: http://revistas.ffclrp.usp.br/Neotropical_Diptera/article/viewFile/229/179 [Accessed: 2014-03-02].

[36] Henriques AL, Krolow TK, Rafael JA. Corrections and additions to catalogue of Neotropical Diptera (Tabanidae) of Coscarón e Papavero (2009). Revista Brasileira de Entomologia. 2012;**56(3)**:277–280. DOI: 10.1590/S0085-56262012005000042

[37] Krolow TK, Henriques AL, Gorayeb IS, Limeira-De-Oliveira F, Buestán J. Taxonomic revision of the Neotropical genus *Pityocera* Giglio-Tos, 1896 (Diptera: Tabanidae: Scionini). Zootaxa. 2015;**3904(3)**:301–333. DOI: 10.11646/zootaxa.3904.3.1

[38] Rafael JA, Marques DWA, Limeira-de-Oliveira F. Tabanidae (Diptera) of Maranhão state, Brazil. V. Description of *Protosilvius gurupi* sp. n. (Pangoniinae, Pangoniini) and key to *Protosilvius* species. ZooKeys. 2012;**235**:41–50. DOI: 10.3897/zookeys.235.3786

[39] González CR. Two new species of the genus *Dasybasis* Macquart, 1847, from Chile (Diptera: Tabanidae: Diachlorini). Zootaxa. 2014;**3893**:293–300. DOI: dx.doi.org/10.11646/zootaxa.3893.2.8.

[40] Gorayeb IS, Gómez ZT, Rios MV. Description of a new species of *Stenotabanus* from Venezuela (Diptera: Tabanidae). Boletim do Museu Paraense Emílio Goeldi. Ciências Naturais. 2013;**8(1)**:41–47.

[41] Guimarães RG, Gorayeb IS, Rodrigues-Guimarães R, Seppa GS, Carvalho RW. Description of *Dichelacera (Dichelacera) walteri* n. sp. (Diptera: Tabanidae) with a key to related species in the subgenus *Dichelacera* Macquart. Neotropical Entomology. 2015;**44(5)**:474–480. DOI: 10.1007/s13744-015-0311-8.

[42] Fairchild GB. Arboreal Tabanidae in Panama. Proceedings of the Entomological Society of Washington. 1953;**55(5)**:241–275.

[43] Morrone JJ. Biogeografía de América Latina y el Caribe. Sociedad Entomológica Aragonesa: Zaragoza; 2001, Vol. 3, 148 p.

[44] Morrone JJ. Biogeographic areas and transition zones of Latin America and the Caribbean Islands based on panbiogeographic and cladistic analyses of the entomofauna. Annual Review of Entomology. 2006;**51**:467–494. DOI: 10.1146/annurev.ento.50.071803.130447.

[45] Martins Neto RG, Santos JCK. Um novo gênero e uma nova espécie de mutuca (Insecta, Diptera, Tabanidae) da Formação Santana (Cretáceo Inferior), bacia do Araripe, Nordeste do Brasil. Acta Geologica Leopoldensia. 1994;**59**:289–297.

[46] Coscarón S, González CR. Tabanidae de Chile: Lista de especies y clave para los géneros conocidos de Chile (Diptera-Tabanidae). Acta Entomologica Chilena. 1991;**16**:125–150.

[47] Coscarón S. Contribución al conocimiento de los Tabanidae Neotropicales I. El subgénero *Mesomyia* (*Coracella*) Philip. Idesia. 1972;**2**:87–97.

[48] Strickman D. Notes on Tabanidae from Paraguay. Journal of Medical Entomology. 1982;**19(4)**:399–402. DOI: 10.1093/jmedent/19.4.399

[49] Strickman D, Hagan DV. Seasonal and meteorological effects on activity of *Chrysops variegatus* in Paraguay. Journal of American Mosquito Control Association. 1986;**2(2)**:212–216.

[50] Rafael JA, Gorayeb IS, Rosa MSS, Henriques AL. Tabanidae (Diptera) da Ilha de Maracá e Serra Pacaraima, Roraima, Brasil, com descrição de duas espécies novas. Acta Amazonica. 1991;**21**:351–367. 10.1590/1809-43921991211367

[51] Bermúdez LV, Bermúdez EM. Immature stages of some Tabanidae (Diptera) species in Mexico. Memoirs on Entomology International. 1999;**14**:257–270.

[52] Parra-Henao G, Alarcón-Pineda EP, López-Valencia G. Ecology and parasitological analysis of horse flies (Diptera: Tabanidae) in Antioquia, Colombia. Caldasia. 2008; **30(1)**:-188.

[53] Gorayeb IS. Tabanidae (Diptera) da Amazônia Oriental: sazonalidade, ataque e estratificação arbórea [thesis]. Manaus: Instituto Nacional de Pesquisas da Amazônia, Universidade Federal do Amazonas; 1985.

[54] Barbosa MGV, Henriques AL, Fonseca CRV, Rafael JA. Aspectos da riqueza e da similaridade entre habitats em relação às espécies de Tabanidae (Insecta: Diptera) da Reserva Ducke, Amazônia Central, Brasil. Amazoniana. 2005;**18**:251–266.

[55] Henriques AL. Tabanidae (Insecta: Diptera) from Parque Nacional do Jaú. II. Chapter 9. In: Borges SH et al., editors. Windows for Biodiversity in Parque Nacional do Jaú: a strategy for the study of biodiversity in the Amazon. Fundação Vitória Amazônica: Manaus; 2004. pp. 143–152.

[56] Ferreira-Keppler RL, Rafael JA, Guerrero JCH. Sazonalidade e uso de ambientes por espécies de Tabanidae (Diptera) na Amazônia Central, Brasil. Neotropical Entomology. 2010;**39(4)**:645–654. DOI: 10.1590/S1519-566X2010000400028

[57] Krolow TK, Henriques AL, Rafael JA. Tabanidae (Diptera) no dossel da floresta amazônica atraídos por luz e descrição de machos de três espécies. Acta Amazonica. 2010;**40**:605–612. DOI: 10.1590/S0044-59672010000300022

[58] Limeira-De-Oliveira F. Taxonomia de adultos de mutucas (Diptera: Tabanidae) do Estado do Maranhão, Brasil [thesis] Manaus: Instituto Nacional de Pesquisas da Amazônia (INPA), Universidade Federal do Amazonas; 2003.

[59] Bittencourt-Vaz D, Pédra LG, Krüger RF, Krolow TK. Levantamento das espécies costeiras de Tabanidae (Diptera) nas unidades de conservação da Planície Costeira do Rio Grande do Sul. Proceddings of the 21° Congresso de Iniciação Científica /4ª Mostra Científica; Universidade Federal de Pelotas. 2012; Pelotas, Rio Grande do Sul. Avaiable from: http://www2.ufpel.edu.br/cic/2012/anais/pdf/CB/CB_01334.pdf [Accessed: 2014-03-16]

[60] Krolow TK, Krüger RF, Ribeiro PB. Chave pictórica para os gêneros de Tabanidae (Insecta: Diptera) do bioma Campos Sulinos, Rio Grande do Sul, Brasil. Biota Neotropica. 2007;**7(2)**:253–264. DOI: 10.1590/S0044-59672010000300022

[61] Buestán J. Identificación y distribución estacional de Tabanidae en las Provincias de Guayas, Los Ríos y Cañar. Revista Ecuatoriana de Higiene y Medicina Tropical. 1980;**33**:101–116.

[62] Bassi RMAA, Cunha MCI, Coscarón S. Estudo do comportamento de Tabanídeos (Diptera, Tabanidae) do Brasil. Acta Biológica Paranaense. 2000;**29(1–4)**:101–115.

[63] Miletti LC, Colombo BR, Cardoso CP, Stalliviere FM, Tavares KCS, Komati LKO, Vieira LL, Christen SE, Ramos CJR. Prevalence, seasonality and behavior of Tabanidae (Diptera) captured on a horse in the Planalto Serrano of Santa Catarina State, Brazil. International Journal of Tropical Insect Science. 2011;**31(1–2)**:122–126. DOI: 10.1017/S1742758411000130.

[64] Guimarães RR. Tabanidae (Insecta: Diptera): Caracterização, Ecologia e Interação com a População Quilombola da Ilha da Marambaia, Rio de Janeiro, Brasil [thesis]. Rio de Janeiro: Escola Nacional de Saúde Pública Sérgio Arouca, Fundação Oswaldo Cruz; 2015.

[65] Guimarães RR, Guimarães-Júnior RR, Harlan-Rodrigues RS, Rodrigues-Guimarães R, Carvalho RR. 2016. Checklist and Notes on Behavior of Horse Flies (Diptera: Tabanidae) from Marambaia Island, Rio de Janeiro, Brazil, with New Records for the State. EntomoBrasilis. 2016;**9(2)**:73–80. DOI: 10.12741/ebrasilis.v9i2.585.

[66] Lessard BD. Revision of the austral horse fly tribe Scionini (Diptera: Tabanidae). Austral Entomology. 2014;**53(2)**:203–239. DOI: 10.1111/aen.12076

[67] Krolow TK, Henriques AL, González CR. Taxonomic revision of the Neotropical genus *Caenopangonia* Kröber, 1930 (Diptera: Tabanidae). Zootaxa. 2016;**4154(5)**:526–540. 10.11646/zootaxa.4154.5.3

[68] Coscarón S, Philip C. A revision of the Mycteromyiini ('genus *Mycteromyia*' of authors) a new tribe of Neotropical horse flies. Proceedings of the California Academy of Science. 1979;**41(19)**:427–452.

[69] Kröber O. Tabanidae. Diptera of Patagonia and South Chile In: British Museum, editor. Natural History. London: British Museum; 1930. pp. 106–161.

[70] Pino G, Candia V, Letelier S, Ostoic A, Sánchez J, Sánchez V. Los tábanos de Chile. 4. Estudios ecológicos preliminares en Fray Jorge. Revista Chilena de Entomologia. 1973;**7**:207–216.

[71] González CR. Morfología de la armadura bucal de la hembra de *Mycteromyia conica* (Bigot) (Diptera: Tabanidae: Mycteromyiini). Acta Entomologica Chilena. 2006;**30(2)**:21–24.

[72] Coscarón S. Contribución al conocimiento de los Tabanidae Neotropicales. II. Los Pangoniini del sur de Sudamérica y datos sobre la tribu Scepsidini. Revista del Museu de La Plata (n. s.) 12 (Zool.) 1976;**114**:75–116.

[73] Chainey JE, Hall MJR. A new genus and three new species of Pangoniini (Diptera: Tabanidae) from Bolivia. Memórias do Instituto Oswaldo Cruz. 1996;**91(3)**:307–312. DOI: 10.1590/S0074-02761996000300010

[74] Wilkerson RC, Fairchild GB. A checklist and generic key to the Tabanidae (Diptera) of Peru with special reference to the Tambopata Reserved Zone, Madre de Dios. Revista Peruana de Entomologia. 1985;**27**:37–53.

[75] Limeira-de-Oliveira F. Tabanidae (Diptera) do Estado de Maranhão, Brasil II. Descrição de *Esenbeckia* (*Esenbeckia*) *rafaeli*, sp. nov. Neotropical Entomology. 2008;**37(4)**:426–428. DOI: 10.1590/S1519-566X2008000400011

[76] Barros ATM. Seasonality and relative abundance of Tabanidae (Diptera) captured on horses in the Pantanal, Brazil. Memórias do Instituto Oswaldo Cruz, 2002;**96**:917–923. DOI: 10.1590/S0074-02762001000700006

[77] Lima HIL, Krolow TK, Henriques AL. Checklist of horse flies (Diptera: Tabanidae) from Taquaruçu, Tocantins, Brazil, with new records for the state. CheckList. 2015;**11**:1596. DOI: 10.15560/11.2.1596 DOI: http%3A%2F%2Fdx.doi.org%2F10.15560%2F11.2.1596.

[78] Fairchild GB, Wilkerson RC. New species of *Esenbeckia* (*Proboscoides*) (Diptera, Tabanidae) with a key to the subgenus. Florida Entomologist. 1981;**64(1)**:159–175. DOI: 10.2307/3494607

[79] Burger JF. Notes on and descriptions of new and little-known species of Neotropical Tabanidae (Diptera). Memoirs on Entomology International. 1999;**14**:51–74.

[80] González CR. Descripción del macho y redescripción de la hembra de *Veprius fulvus* Coscarón, Philip & Fairchild, 1979 (Diptera: Tabanidae: Pangoniini). Acta Entomologica Chilena. 2006;**30(1)**:39–42.

[81] González CR. The Immature Stages of *Protodasyapha (Protodasyapha) hirtuosa* (Philippi) and their Comparison with the Immature Stages of other Pangoniini (Diptera: Tabanidae: Pangoniinae). Memórias do Instituto Oswaldo Cruz. 1998;**93(4)**:465–470. DOI: 10.1590/S0074-02761998000400008

[82] Moucha J. Horseflies (Diptera: Tabanidae) of the world. Synoptic catalogue. Acta Entomologica Musei Nationalis Pragae. Supplementum. 1976;**7**:1–319.

[83] Coscarón S, González CR. Systematic position and redescription of *Scaptia* (*Pseudomelpia*) *horrens* Enderlein, 1925 (Diptera: Tabanidae: Pangoniinae). Acta Entomologica Chilena. 2001;**25**:31–40.

[84] González CR, Sanhueza Y. Estudio comparativo de la armadura bucal de hembras em 3 especies de *Scaptia* (*Scaptia*) Walker de distribuición chilena (Diptera: Tabanidae). Acta Entomologica Chilena. 2004;**28(1)**:23–31.

[85] Coscarón S, González C. Los estados preimaginales del "Colihuacho" *Scaptia (Scaptia) lata* (Guérin-Méneville) (Tabanidae, Diptera). Acta Entomologica Chilena 1989;**15**:249–256.

[86] González CR, Llanos L, Saldarriaga-Córdoba M. Description of the Terrestrial Larva of *Parosca latipalpis* (Macquart) (Diptera: Tabanidae) from southern Chile. Neotropical Entomology. 2016;**45(5):**548–553. DOI: 10.1007/s13744-016-0390-1

[87] Wilkerson RC, Coscarón S. A review of South American *Scaptia* (*Pseudoscione*) (Diptera: Tabanidae). Journal of Medical Entomology. 1984;**21**:213–236. DOI: 10.1590/S0074-02761998000400008

[88] Coscarón S. El género *Scione* em la Argentina (Diptera: Tabanidae). Revista de la Sociedad Entomologica Argentina. 2000;**59(1-4)**:85–90.

[89] Fairchild GB. Notes on Neotropical Tabanidae (Diptera). IV. Further new species and new records for Panama. Journal of Medical Entomology. 1964;**1(2)**:169–185. DOI: 10.1093/jmedent/1.2.169

[90] Chainey JE, MJR Hall, Amayo JL, Bettella P. A preliminary checklist and key to the genera and subgenera of Tabanidae (Diptera) of Bolivia with particular reference to Santa Cruz Department. Memórias do Inst. Oswaldo Cruz. 1994;**89**:321–345. DOI: 10.1590/S0074-02761994000300007

[91] Coscarón S. El género *Fidena* Walker 1850 em Argentina y Chile (Diptera: Tabanidae). Acta Entomologica Chilena. 2001;**25**:51–72.

[92] Lutz A. Tabaniden Brasiliens und einiger Nachbarstaaten. Zoologische Jahrbücher. 1909;**4**:619–692, pls 1–3. [Fac-simile reproduction in Benchimol & Sá, 2005;117–196. Portuguese translation in Benchimol & Sá, 2005;197–258].

[93] Lutz A, Neiva A. Contribuição para o conhecimento da fauna indígena de tabânidas. Memórias do Instituto Oswaldo Cruz, 1909;**1(1)**:28–32. DOI: 10.1590/S0074-02761909000100004

[94] Oliveira AF, Ferreira RLM, Rafael, JA. 2 Sazonalidade e atividade diurna de Tabanidae (Diptera: Insecta) de dossel na Reserva Florestal Adolpho Ducke, Manaus, AM. Neotropical Entomology. 2007;**36(5)**:790–797. DOI: 10.1590/S1519-566X2007000500022

[95] Henriques AL, Rafael JA. Tabanidae (Diptera) from Parque Nacional do Jaú, Amazonas, Brazil, with description of two new species of *Diachlorus* Osten Saken. In: A collection of articles on Diptera commemorative the life and work of Graham B. Fairchild. Memoirs on Entomology. International, Gainsville: Associated Publishers; 1999, Vol. 14, pp. 195–222.

[96] Zillikens A, Gorayeb IS, Steiner J, Marcondes CB. Aquatic Larvae and Pupae of *Fidena (Laphriomyia) rufopilosa* (Ricardo) (Diptera: Tabanidae) Developing in Bromeliad Phytotelmata in the Atlantic Forest of southern Brazil. Journal of the Kansas Entomological Society. 2005;**78(4)**:381–386. DOI: 10.2317/0410.16.1

[97] Buestán J. *Eristalotabanus violaceus* Kröber (Diptera: Tabanidae) una espécie muy poco conocida. Revista de la Universidad de Guayquil. 2013;**115**:13–18.

[98] Cárdenas R. Fine-scale climatic variation drives altitudinal niche partitioning of tabanid flies in a tropical montane cloud forest, Ecuadorian Chocó. Insect Conservation and Diversity. 2016;**9**:87–96. DOI: 10.1111/icad.12146.

[99] Cárdenas RE, Buestán J, Dangles O. Diversity and distribution models of horse flies (Diptera: Tabanidae) from Ecuador. Annales de la Société Entomologique de France. 2009;**45(4)**:511–528. DOI: 10.1080/00379271.2009.10697633

[100] Cárdenas RE, Hernández LN, Barragá AR, Dangles O. Differences in morphometry and activity among Tabanid fly assemblages in an Andean Tropical Montane Cloud Forest: indication of altitudinal migration? Biotropica. 2013;**45(1)**:63–72. DOI: 201310.1111/j.1744-7429.2012.00885.x

[101] Fairchild GB. Notes on Tabanidae (Dipt.) from Panama. VIII. The genera *Pityocera, Scione* and *Esenbeckia*. Annals of the Entomological Society of America. 1942:**35(2)**:183–199. DOI: 10.1093/aesa/35.2.183

[102] Trojan P. Distribution pattern- of Diachlorini s. str. (Diptera: Tabanidae). Fragmenta faunística. 1997;**40(6)**:5971.

[103] Fairchild GB. Notes on the genus *Acanthocera* Macquart (Dipt. Tabanidae) Revista de Entomologia. 1939;**10(1)**:14–27.

[104] Henriques AL, Rafael JA. Revisão do gênero Neotropical *Acanthocera* Macquart, 1834 (Diptera: Tabanidae). Acta Amazonica. 1995;**23(4)**:404–440.

[105] Burger JF. Description of five new species of Tabanidae (Diptera) from Costa Rica and revised key to species for the genera *Fidena* Walker, *Scione* Walker, and *Chrysops*

Meigen in Costa Rica. Proceedings of the Entomological Society of Washington. 2002;**104(4)**:928–940.

[106] González CR. *Agelanius verai*, a new species of horse fly from Chile (Diptera: Tabanidae). Zootaxa. 2004;**571**:1–5. DOI: 10.11646/zootaxa.571.1.1

[107] González CR. Descriptions of adults and immature stages of *Agelanius fuscus*, a new horse fly species fron central Chile (Diptera: Tabanidae: Diachlorini) Studia dipterologica. 2004;**11**:211–217.

[108] González CR. *Agelanius burgeri*, a new species of horse fly from south Chile (Diptera: Tabanidae). Zootaxa. 2006;**1364**:59–64.

[109] González CR. *Agelanius chiloensis*, una nueva especie de tábano del sur de Chile (Diptera: Tabanidae). Gayana. 2009;**73(1)**:12–16. DOI: 10.4067/S0717-65382009000100002

[110] González CR. Description of immature stages of *Agelanius cortesi* (Gonzállez) from central Chile (Diptera: Tabanidae). Studies on Neotropical Fauna and Environment. 2007;**42(1)**:5–9. DOI: 10.1080/01650520600875247

[111] Gómez ZT, Velásquez de Ríos M, Gorayeb I. Nuevos registros de *Bolbodimyia* Bigot, 1892 (Diptera: Tabanidae), con clave para las especies de Venezuela. Entomotropica. 2010;**25(1)**:7–10.

[112] Stone A. The genus *Bolbodimyia*. Annals of the Entomological Society of America. 1954;**47(2)**:248–254. DOI: 10.1093/aesa/47.2.248

[113] Kröber O. Das Tabanidengenus *Catachlorops* Lutz (Dipt.). Verhöff Deutsch Kolon. Museum Bremen. 1939;**2(3)**:211–232.

[114] Coscarón S. Notas sobre tabánidos argentinos. XIII. El género *Catachlorops* Lutz en la Argentina. Physis. 1975;**34(88)**:23–31.

[115] Henriques AL, Krolow TK. Description of a new species of Tabanidae (Diptera) from the Amazon region: *Catachlorops* (*Rhamphidommia*) *dubius* sp. nov. and a key to species of the subgenus. Zootaxa. 2009;**2295**:64–68.

[116] Lutz A. Collecção de Tabânidas. In: Instituto Oswaldo Cruz, editor. Instituto Oswaldo Cruz, em Manguinhos, Rio de Janeiro. Instituto Oswaldo Cruz: Rio de Janeiro; 1909. pp. 28–30.

[117] Lutz A. List of Tabanidae. In: Instituto Oswaldo Cruz, editor. Internationale Hygiene-Ausstellung Dresden. Institut Oswaldo Cruz, Rio de Janeiro; 1911. pp. 33–35.

[118] Lutz A. Sobre a systematica dos tabanideos, subfamilia Tabaninae. Brazil Medico. 1913;**27(45)**:486–490.

[119] Lutz A. Sobre a systematica dos tabanideos, subfamilia Tabaninae. Memórias do Instituto Oswaldo Cruz. 1914;**6(3)**:163–168. (Reprint of the previous article, with translation into German). DOI: 10.1590/S0074-02761914000300003

[120] Krolow TK, Henriques AL. Taxonomic revision of the New World genus *Chlorotabanus* Lutz, 1913 (Diptera: Tabanidae). Zootaxa. 2010;**2656**:1–40.

[121] Guimarães RR, Guimarães-Júnior RR, Harlan-Rodrigues RS, Guimarães RR, de Xerez R, Carvalho RW. Remarks on record of *Chlorotabanus inanis* (Fabricius, 1787) (Diptera, Tabanidae) and its occurrence in Rio de Janeiro Province, Brazil. Entomology, Ornithology and Herpetology. 2014;**4**:139. DOI: 10.4172/2161-0983.1000139.

[122] Borgmeier T. A propósito da nomenclatura dos Tabanidae da região neotropica. Revista de Entomologia. 1933;**3(3)**:286–303.

[123] Barretto MP. Estudos sobre tabânidas brasileiros. XI. Sôbre a validade nomenclatural dos nomes genéricos publicados em "Colleção de Tabânidas, Instituto Oswaldo Cruz em Manguinhos, Rio de Janeiro, 1909" (Diptera, Tabanidae). Papéis Avulsos de Zoologia. 1950;**9(5)**:61–68.

[124] Rafael JA, Charlwood JD. Atividade crepuscular de *Chlorotabanus inanis* (Fab.) and *Cryptotylus unicolor* (Wied.) (Diptera: Tabanidae). Acta Amazonica. 1981;**11(2)**:411–413.

[125] Fairchild GB. Notes on Tabanidae from Panama. I. The genera *Chlorotabanus* and *Cryptotylus*. Revista de Entomologia. 1940;**11(3)**:713–722.

[126] Philip CB, Fairchild GB. American biting flies of the genera *Chlorotabanus* Lutz and *Cryptotylus* Lutz (Diptera, Tabanidae). Annals of the Entomological Society of America. 1956;**49(4)**:313–324. DOI: 10.1093/aesa/49.4.313

[127] Gorayeb IS, Fairchild GB. Tabanidae (Dipt.) da Amazônia. VI. Descrição de uma espécie e chave para o género *Cryptotylus* Lutz. Publicações Avulsas do Museu Paraense Emilio Goeldi. 1985;**40**:101–111.

[128] Coscarón S, Mancebo OA, Coscarón-Arias CL. The preimaginal stages of *Cryptotylus unicolor* (Wiedemann) and *Tabanus nebulosus ornativentris* Kroeber (Tabanidae-Diptera-Insecta). Memórias do Instituto. Oswaldo Cruz. 1998;**93(1)**:91–97. DOI: 10.1590/S0074-02761998000100016

[129] Coscarón S. A new species of *Dasybasis* Macquart from NW of Argentina (Tabanidae, Diptera), Insecta). Memórias do Instituto Oswaldo Cruz. 1989;**84(Suppl. 4)**:125–128. DOI: 10.1590/S0074-02761989000800026

[130] Coscarón S, Philip CB. Revisión del género "*Dasybasis*" Macquart en la región Neotropical. Revista del Museu Argentino de Ciências Naturais Benardino Rivadavia, Entomologia. 1967;**2(2)**:15–266.

[131] Coscarón S. Male description [sic] of four species of Argentina horse flies (Diptera; Tabanidae). Memoirs on Entomology International. 1999;**14**:239–247.

[132] Coscarón S. Datos sobre estados preimaginales de tábanos neotropicales. Revista de la Sociedad Entomologica Argentina. 1969;**31(1-4)**:19–22.

[133] Coscarón S, Philip CB. Notas sobre biologia del macho de *Dasybasis fairchildi* Coscarón e Philip. Revista de la Sociedad Entomologica Argentina. 1967;**29(1-4)**:43–51.

[134] González CR. The immature stages of two species of *Dasybasis* from the Southern Neotropical region (Diptera: Tabanidae: Diachlorini). Annales Zoologici. 2000;**52(2)**:271–277.

[135] González CR. The immature stages of *Dasybasis pruinivitta* (Kröber), from Central Chile (Diptera: Tabanidae). Zootaxa. 2004;**573**:1–7. DOI: 10.11646/zootaxa.573.1.1

[136] Kröber O. Die amerikanischen Arten der Tabaniden. Subfamilie Diachlorinae End. Archiv fu'f Schiffs und Tropen Hygiene. 1928;**32**:1–55.

[137] Fairchil GB. Notes on Neotropical Tabanidae (Diptera). XII. The genus *Diachlorus* O.S. Florida Entomologist. 1972;**55(4)**:219–229. DOI: 10.2307/3493370

[138] Wilkerson RC, Fairchild GB. Five new species of *Diachlorus* (Diptera: Tabanidae) from South America, with a revised key to species and new locality records. Proceedings of the Entomological Society of Washington. 1982;**84(3)**:636–650.

[139] Henriques AL, Krolow TK. Description of *Dichelacera* (*Dichelacera*) *matogrossensis* sp. nov. (Diptera, Tabanidae) of Brazilian Midwestern. Iheringia, Série Zoologia. 2015;**105(2)**:242–244. DOI: 10.1590/1678-476620151052242244

[140] Limeira-de-Oliveira F, Gorayeb IS, Henriques AL. Tabanidae (Diptera) do Estado do Maranhão, Brasil. IV. Descrição de *Dichelacera* (*Dichelacera*) *gemmae* sp. nov. Neotropical Entomology. 2009;**38**:104–107. DOI: 10.1590/S1519-566X2009000100011

[141] Coscarón S. Notas sobre tabánidos argentinos. V. Revisión del género *"Dicladocera"* Lutz en la Argentina. Revista del Museu Argentino de Ciências Naturais Benardino Rivadavia, Entomologia. 1967;**2(1)**:1–14.

[142] Lutz, A. Estudios de zoología y parasitología venezolanas. Rio de Janeiro: s.n., 1928. 133 p., 26 pls. [Tabanidae, pp. 51–64, figs.]. [Republ. com diferente paginação, Caracas: s.n., 1955. 137p.]

[143] Fairchild GB. A note on the early stages of *Lepiselaga crassipes* Fab. (Diptera-Tabanidae). Psyche. 1940;**47**:8–13. DOI: 10.1155/1940/57946

[144] Koller WW, Barros ATM, Gomes A, Madruga CR, Araújo CP, Umaki A, Ismael APK. Sazonalidade de tabanídeos (Diptera: Tabanidae) em área de transição entre o Cerrado e Pantanal, no Mato Grosso do Sul, Brasil. In: Proceedings of the Congresso Brasileiro de Parasitologia Veterinária. 2002. Rio de Janeiro: R376.pdf (CD-ROM).

[145] Fairchild GB. Notes on Tabanidae (Dipt.) from Panama. IV. The genus *Leucotabanus* Ad. Lutz. Annals of the Entomological Society of America. 1941;**34(3)**:629–638. DOI: 10.1093/aesa/34.3.629

[146] Fairchild GB. Notes on Neotropical Tabanidae (Diptera) with descriptions of new species. Annals of the Entomological Society of America. 1953;**46(2)**:259–280. DOI: 10.1093/aesa/46.2.259

[147] Fairchild GB. Notes on Neotropical Tabanidae (Dipt.). XVIII. The genus *Leucotabanus* Lutz. Myia. 1985;**3**:299–331.

[148] Godoi FSP, Rafael JA. Descrição da larva, exúvia pupal e macho de *Leucotabanus albovarius* (Walker) (Diptera, Tabanidae) da Amazônia Central. Revista Brasileira de Entomologia. 2007;**51(1)**:101–106. DOI: 10.1590/S0085-56262007000100017

[149] Rafael JA, Ferreira RLM. Revisão do gênero *Myiotabanus* Lutz (Diptera, Tabanidae) com descrição de uma espécie nova. Revista Brasileira de Zoologia. 2004;**21(2)**:325–331. DOI: 10.1590/S0085-56262007000100017

[150] Coscarón S, Coscarón-Arias LC, Mancebo OA. The Immature Stages of *Myiotabanus barrettoi* Fairchild (Tabanidae-Diptera-Insecta). Memórias do Instituto Oswaldo Cruz. 1996;**91(1)**:21–26. DOI: 10.1590/S0074-02761996000100004

[151] Chadee DD, Tikasingh ES. Occurrence of *Phaeotabanus cajennensis* (new record) (Diptera: Tabanidae) in Trinidad. Journal of Medical Entomology. 1982;**19(2)**:211. DOI: 10.1093/jmedent/19.2.211

[152] Dutra RRC, Marinoni RC. Insetos capturados em armadilhas Malaise na ilha do Mel, baía de Paranaguá, Paraná, Brasil. 2. Tabanidae (Diptera). Revista Brasileira de Zoologia. 1995;**11(2)**:247–256. DOI: 10.1590/S0101-81751994000200007

[153] Barros AT. Seasonality of *Phaeotabanus fervens* (Diptera: Tabanidae) in the Pantanal Region, Brazil. Memórias do Instituto Oswaldo Cruz. 1996;**91(2)**:159. DOOI: 10.1590/S0074-02761996000200005

[154] Henriques AL. The genus *Philipotabanus* Fairchild (Insecta: Diptera: Tabanidae) in the Amazon Region, with a key to the female species and description of *P. obidensis* (sic) sp. nov. Acta Amazonica. 2006;**36(4)**:549–556.

[155] Gorayeb IS, Rafael JA. Tabanidae (Diptera) da Amazônia. V. Descrição de duas espécies novas. Boletim do Museu Paraense Emilio Goeldi. Zoologica. 1984;**1(1)**:45–55.

[156] Fairchild GB. Notes on Neotropical Tabanidae. V. Further new species and new records for Panama. Journal of Medical Entomology. 1964;**1**:69–85. DOI: 10.1093/jmedent/1.2.169

[157] Chainey JE, Hall MJR, Gorayeb IS. *Stenotabanus* Lutz (Diptera: Tabanidae): An identification guide to the South American species. Memoirs on Entomology International. 1999;**14**:75–134.

[158] Fairchild GB. Notes on Tabanidae (Dipt.) from Panama. IX. The genera *Stenotabanus* Lutz, *Lepiselaga* Macquart and related genera. Annals of the Entomological Society of America. 1942;**35(1)**:289–309. DOI: 10.1093/aesa/35.3.289

[159] Fairchild GB, Aitken THG. Additions to the Tabanidae (Diptera) of Trinidad, B. W. I. Annals of the Entomological Society of America. 1960;**53(1)**:1–8. DOI: 10.1093/aesa/53.1.1

[160] Gorayeb IS. Tabanidae (Diptera) da Amazônia. XV. Ocorrência de *Stenotabanus (Aegialomyia) geijskesi* Fairchild, 1953, no litoral do estado do Pará, Brasil, redescrição

da fêmea e descrição do macho. Boletim do Museu Paraense Emílio Goeldi. Zoologica. 1999;**13(1)**:77–83.

[161] Coscarón S. Notas sobre tabánidos argentinos. IX. Los géneros *Stenotabanus* Lutz y *Myiotabanus* Lutz (Diptera-Insecta). Informes. 1975;**76**:1–39.

[162] Ricardo G. Notes on the smaller genera of the Tabaninae of the family Tabanidae in the British Museum collection. Annals and Magazine of Natural History. 1904;**7(14)**: 349–373. DOI; 10.1080/03745480409443020

[163] Knab F. Gad-flies (Tabanidae) of the genus *Stibasoma*. Proceedings of United State National Museum. 1913;**46(2033)**:407–412. DOI: 10.5479/si.00963801.2033.407

[164] Turcatel M, Carvalho CJB, Rafael JA. A taxonomic revision of *Stibasoma* Schiner, 1867 (Diptera: Tabanidae). Zootaxa. 2010;**2368**:1–39.

[165] Goodwin JT, Murdoch P. A study of some immature Neotropical Tabanidae (Diptera). Annals of the Entomological Society of America. 1974;**67**:85–133.DOI: 10.1093/aesa/67.1.85

[166] Trojan, P. Supraspecific taxa of Tabaninae (Diptera: Tabanidae). III. The tribe Diachlorini and its taxonomic division. Annals of the Upper Silesian Museum (Entomology). 1998;**8–9**:5–91.

[167] Lutz A. Tabanidas do Brazil e alguns Estados visinhos. Memórias do Instituto Oswaldo Cruz. 1913;**5**:142–191.

[168] Henriques AL, Krolow TK. Description of *Muscotabanus* gen. nov. and *Muscotabanus rafaeli* sp. nov. (Diptera: Tabanidae: Diachlorini) from Amazon Basin, Brazil. Memórias do Instituto Oswaldo Cruz 2013;**108(3)**:383–385. DOI: 10.1590/S0074-02762013000300018

[169] Fairchild GB, Wilkerson RC. A review of the Neotropical genus *Stypommisa* (Diptera: Tabanidae). Contributions of the American Entomological Institute. 1986;**22(5)**:1–61.

[170] Gorayeb IS. Tabanidae (Diptera) of Amazônia XXI. Descriptions of *Elephantotus* gen. n. and *E. tracuateuensis* sp. n. (Diachlorini) from the Brazilian coast. ZooKeys. 2014;**395**:23–31. DOI: 10.3897/zookeys.395.7056.

[171] Coscarón S, Fairchild GB. El género *Poeciloderas* Lutz en Argentina (Tabanidae, Diptera, Insecta). Physis. 1976;**35(91)**:293–301.

[172] Fairchild GB. Notes on Neotropical Tabanidae (Diptera). XIV. Two new species of *Tabanus* from Panama and Colombia. Proceedings of the Entomological Society of Washington. 1973;**75(3)**:319–329.

[173] Coscarón S. Notas sobre tabánidos Argentinos XV. El género *Tabanus* Linnaeus. Obra Centenaria del Museo de la Plata, 1979;**6**:251–278.

[174] Fairchild GB. Notes on Neotropical Tabanidae (Diptera). XIX. The *Tabanus lineola* complex. Miscellaneous Publications of the Entomological Society of America,1983;**57**:1–50.

[175] Fairchild GB. Notes on Neotropical Tabanidae (Diptera). XX. The larger species of *Tabanus* of eastern South America. Contributions of the American Entomological Institute. 1984;**21(3)**:1–50.

[176] Raymond HL. Abondance relative et dynamique saisonnière des Tabanidae (Diptera) d'une savane de Guyane Française. Naturaliste Canadien. 1988;**115**:251–259.

[177] Lutz A. Zoologia medica. Tabanidae ou motucas. Folha Medica. 1922;**3(190)**:146–148.

[178] Herrera HM, Rademaker V, Abreu UGP, D'Andrea PS, Jansen AM. Variables that modulate the spatial distribution of *Trypanosoma cruzi* and *Trypanosoma evansi* in the Brazilian Pantanal. Acta Tropica. 2007;**102**:55–62. DOI: 10.1016/j.actatropica.2007.03.001

[179] Ferenc S, Raymond HL, Lancelot R. Essai de transmission mecanique de *Trypanosoma vivax* Ziemann (Kinetoplastida: Trypanosomatidae) par le taon neotropical *Cryptotylus unicolor* Wiedemann (Diptera: Tabanidae). Proceedings of the 18th International Congress of Entomology; 1988. p. 295.

[180] Monzón CM, Hoyos CB, Jara GA. Outbreaks of equine trypanosomiasis caused by *Trypanosoma evansi* in Formosa Province, Argentina. Revue scientifique et technique, 1995;**14(3)**:747–752. DOI:

[181] Nunes VLB, Oshiro ET, Dorval MEC, Garcia LAM, da Silva AAP, Bogliolo AR. Investigação epidemiológica sobre *Trypanosoma (Trypanozoon) evansi* no Pantanal sul-matogrossense. Estudos de reservatórios. Brazilian Journal of Veterunary Parasitology. 1993;**2**:41–44.

[182] Barros T, Foil L. Seasonal occurrence and relative abundance of Tabanidae (Diptera) from the Pantanal region, Brazil. Memoirs Entomology International, 1999;**14**:87–396.

[183] Dávila AMR, Souza SS, Campos C, Silva RA. The seroprevalence of equine trypanosomosis in the Pantanal. Memórias do Instituto Oswaldo Cruz, 1999;**94**:199–202. DOI: 10.1590/S0074-02761999000200014

[184] Silva RAMS, Victório AM, Ramirez L, Dávila AMR, Trajano V, Jansen AM. Hematological and blood chemistry alterations in coatis (*Nasua nasua*) naturally infected by *Trypanosoma evansi* in the Pantanal, Brazil. Revue D'élevage et de Médecine Vétérinaire des Pays Tropicaux. 1999;**52**:119–122.

[185] Barros ATM, Foil LD. Influência da distância na transferência de tabanídeos (Diptera: Tabanidae) entre eqüinos. Proceedings of the Simpósio sobre recursos naturais e socio-econômicos do Pantanal, Corumbá: Embrapa Pantanal; 2000, Vol. 3, pp.155–156.

[186] Coelho WMD, Bresciani KDS. Molecular and parasitological detection of *Leishmania* spp. in a dipteran of the species *Tabanus importunus* Revista Brasileira de Parasitologia Veterinária. 2013;**22(4)**:605–607. DOI: 10.1590/S1984-29612013000400025

[187] Ribeiro PB, Oliveira CMB, Costa PRP. Foréticos da *Dermatobia hominins* (L. Jr, 1781) (Diptera Cuterebridae) no Rio Grande do Sul, Brasil. Arquivos Brasileiros de Medicina Veterinária. 1985;**37(5)**:507–509.

[188] Buestán J. *Chrysops varians* var. *tardus* (Diptera: Tabanidae) como vector de huevos de *Dermatobia hominis* en Ecuador. Memorias de las XXX Jornadas Nacionales de Biología: Quito-Ecuador; 2006, Vol. 1, p. 193.

[189] Bequaert J. Tabanidae of the Island of Trinidad, B. W. I. Bullettin of Entomological Research. 1940;**30(4)**:447–453. DOI: 10.1017/S0007485300040621

[190] Bequaert J. Further studies of the Tabanidae of Trinidad, B. W. I. Psyche. 1944;**51:**2–21. DOI: 10.1155/1944/71329

[191] Callan EMcC. Observations on the distribution of Tabanidae in the Caribbean area, with new records of species from Trinidad, B.W.I. (Diptera). Psyche. 1952;**59(1)**:37–40. DOI: 10.1155/1952/10243

[192] Desquesnes M, La Rocque S. Horseflies of the Guyanas. Biology, Veterinary Significance & Control Methods. IICA Office in Guyana: Georgetown; 1993. 33 p.

[193] InsectoidInfo [Internet]. 2016. Available from: http://insectoid.info/checklist/tabanidae/guyana/ [Accessed: 2016-09-10].

[194] Bequaert J, Renjifo-Salcedo S. The Tabanidae of Colombia (Diptera). Psyche. 1946;**53(3–4)**:52–86. DOI: 10.1155/1946/40437.

[195] Lee VH, Fairchild GB, Barretto P. Artropodos hematófagos del Rio Raposo, Valle, Colombia, III. Tabanidae. Caldasia, 1969;**10(49)**:441–458.

[196] Wilkerson RC. Horse flies (Diptera: Tabanidae) of the Colombian departments of Chocó, Valle and Cauca. Cespedesia. 1979;**8(31–35)**:87–433.

[197] Orozco ISP. Horse flies (Diptera: Tabanidae) from the "Parque Nacional Natural Chiribiquete", Caquetá, Colômbia [thesis]. Bucaramanga: Universidade Industrial de Santander; 2005.

[198] Wolff M, Miranda-Esquivel DR. Family Tabanidae. Zootaxa. 2016;**4122(1)**:249–301. DOI: 10.11646/zootaxa.4122.1.23.

[199] Pechuman L. Lista provisional de los tabánidos de Venezuela. Boletin de Entomologia Venezolana. 1942;**1(3)**:51–58.

[200] Stone A. Some Tabanidae from Venezuela. Boletin de Entomologia Venezolana. 1944;**3(3)**:125–138.

[201] Velásquez de Ríos M, Gómez ZT, Gorayeb IS, Tamasaukas R. Abundancia estacional de tabánidos (Diptera: Tabanidae) en el sector Las Lajas, Municipio Miranda, estado Guárico, Venezuela. Entomotropica, 2004;**19(3)**:149–152.

[202] Gorayeb IS, Velásquez de Ríos M, Tiape Z. Nuevos registros de distribuición de Tabanidos (Dipetra: Tabanidae) para Venezuela. Entomotropica. 2006;**21(3)**:195–198.

[203] Gómez ZT, Velásquez de Ríos M, Gorayeb I. Lista preliminar de Tabánidos (Diptera: Tabanidae) del Noroccidente de Guárico y Sur de Aragua, Venezuela. Entomotropica. 2005;**20(1)**:75–78.

[204] Patrick CR, Hays KL. Some Tabanidae (Diptera) from eastern Ecuador. The Florida Entomologist. 1968;**51(4)**:219–221. DOI: 10.2307/3493422

[205] Fairchild GB, León LA. Lista provisional de Tabanidae (Diptera) del Ecuador. Publicaciones del Museu Ecuatoriano de Ciencias Naturales, Ser. Revista. 1986;**5(7)**:97–122.

[206] Cárdena RE, Vieira JM. Nuevas citas de tabânidas (Diptera: Tabanidae) para Ecuador. Boletín da la Sociedae Entomológica Aragonesa. 2005;**36**:153–156.

[207] Buestán J, Navarrete R, Mejia M. Lista actualizada de tábanos (Diptera: Tabanidae) del Ecuador. Revista Ecuatoriana de Higiene y Medicina Tropical. 2007;**44(1)**:23–62.

[208] Soukoup J. Lista de los dípteros peruanos descritos em los últimos años. Boletín del Museo de Historia Natural Javier Prado. 1945;**9(32–33)**:119–142.

[209] Kröber O. Tabanidae. Therevidae und Conopidae (Dipt.). Beitraege zur Fauna Perus. Band II. 1951:183–192.

[210] Philip CB. Descriptions of new Neotropical Tabanidae mostly in the California Academy of Sciences. Pan-pacific Entomologist 1958;**34(2)**:63–76.

[211] Philip CB. Further records of Neotropical Tabanidae (Diptera) mostly from Peru. Proceedings of the California Academy of Sciences. 1960;**31(3)**:69–102.

[212] Carrasco F. Tábanos del Peru. Revista Peruana de Entomología. 1972;**15(2)**:317–323.

[213] Rafael JA. Ocorrência sazonal e abundância relativa de Tabanidae (Diptera) no Campus Universitário, Manaus, Amazonas. Acta Amazonica. 1982;**12(1)**:225–229.

[214] Rafael JA, Charlwood JD. Idade fisiológica, variação sazonal e periodicidade diurna de quatro populações de Tabanidae (Diptera) no Campus Universitário, Manaus, Brasil. Acta Amazonica. 1980**;10(4)**:907–927.

[215] Rafael JA, Gorayeb IS. Tabanidae da Amazônia. I. Uma nova armadilha suspensa e primeiros registros de mutucas de copas de árvores. Acta Amazonica. 1982;**12(1)**:232–236.

[216] Bouvier G. Notes sur les tabanidés de la région de Campinas (Estado S. Paulo) Brésil. Memórias Insituto Oswaldo Cruz. 1952;**50**:581–596. DOI: 10.1590/s0074-02761952000100017

[217] Coscarón S. Notas sobre tabánidos argentinos. X. El género *Dichelacera* Macq. Revista de la Sociedad Entomologica Argentina. 1974;**34(3–4)**:229–242.

[218] Wilkerson RC. A new species of *Stenotabanus (Brachytabanus)* (Diptera: Tabanidae) with a key to the species of the subgenus. Proceedings of the Entomological Society of Washington. 1985;**87(1)**:221–224.

[219] Gutiérrez T, Rumiz D. Patrones de diversidad de grupos selectos de insectos en el Bosque Chiquitano Y Pampas del Cerrado de Santiago y Tucavaca, Santa Cruz – Bolivia. Revista Bolivia Ecológica. 2002;**11**:37–46.

[220] Philippi RA. Aufzahlung der chilenischen Dipteren. Verhandlungen der Zoologisch-Botanischen Gesellschaft in Wien. 1865;**15**:707–727.

[221] Reed EC, editor. Catálogo de los insectos dípteros de Chile. Anales de la Universidad de Chile, LXXIII. Santiago: Imprenta Nacional; 1888. p. 46.

[222] Enderlein G. Zur Kenntniss einiger von Herr Oskar Schoenemann gesammelten chilenischen Melpiinen. Wiener Entomologische Zeitung.1929;**46(2)**:66–71.

[223] Pino G. Los Tábanos de Chile. l. Sinópsis históricobibliográfica de los estudios sobre tábanos chilenos. Revista Chilena de Entomologia. 1968;**6**:115–122.

[224] Philip CB. The types of Chilean species of Tabanidae (Díptera) described by Dr. R. A. Philippi. Revista Chilena de Entomología. 1968;**6**:7–16.

[225] González CR. Descripción de la hembra y redescripción del macho de *Dicladocera hoppi* Enderlein (Diptera, Tabanidae). Acta Entomológica Chilena. 2002;**26**:95–99.

[226] Cepeda JP, Pola M, Zuleta C, González C. Relaciones de abundancia y diversidade de la Entomofauna del humedal Tambo-Puquíos.In: Cepeda JP, editor. Geoecología de los Andes desérticos. La Alta Montaña del Valle del Elqui. La Serena: Ediciones Universidad de La Serena; 2006; pp. 475–521.

[227] Coscarón S. Notas sobre tabánidos argentinos – VI. El género *Lepiselaga* Mac. Revista de la Sociedad Entomologica Argentina. 1968;**30(1–4)**:51–59.

[228] Coscarón S. Notas sobre tabánidos argentinos. XI. Sobre los géneros *Leucotabanus* Lutz, *Pseudacanthocera* Lutz, *Bolbodimyia* Bigot y *Pachyschelomyia* Barretto. Revista del Museu Argentino de Ciências Naturais Benardino Rivadavia, Entomologia. 1976;**5**:89–103.

[229] Coscarón S. Notas sobre tabánidos argentinos. XIV. Sobre los géneros *Diachlorus* Osten Sacken, *Stibasoma* Schiner, *Stypommisa* Enderlein, *Cryptotylus* Lutz y *Chlorotabanus* Lutz (Diptera). Revista de la Sociedad Entomologica Argentina. 1976. **35(1–4)**:39–50.

[230] Coscarón S. Notas sobre tabánidos argentinos. XVI. El género *Chrysops* Meigen. Acta Zoológica Lilloana. 1979;**35(1979)**:365–392.

[231] Coscarón S. Notas sobre tabánidos argentinos XVII. Los géneros *Phaeotabanus* Lutz y *Acanthocera* Macquart en la Argentina (Tabanidae: Diptera). Revista de la Sociedad Entomologica Argentina. 1978;**37(1/4)**:27–32 (1978).

[232] Coscarón S, Iide P. The Subgenus *Scaptia (Lepmia)* Fairchild: Redescription of Females and Description of a Male (Diptera: Tabanidae: Pangoniinae). Memórias do Instituto Oswaldo Cruz. 2003;**98(6)**:757–760. DOI: 10.1590/S0074-02762003000600008

[233] Coscarón S, Wilkerson RC. South American *Scaptia* (*Scaptia*) (Dipt., Tabanidae) with a key to species of the subgenus. Myia. 1985;**3**:277–298.

[234] Coscarón S. Los estados inmaduros de siete espécies neotropicales del género *Dasybasis* Macquart (Tabanidae, Diptera, Insecta). Acta Entomologica Chilena. 1991;**16**:7–23.

[235] Coscarón S, Led JE. Contribución al conocimiento de *Tabanus claripennis* (Big.) com especial referencia a la morfología y biología de sus estadios preimaginales. Analecta Veterinária. 1968;**1(l)**:13–15.

[236] Coscarón S, Poi de Neiff A. The pupa of *Cryptotylus unicolor* (Wiedemann) (Diptera-Tabanidae). Revista de la Asociación de Ciencias Naturales del Litoral. 1996;**27(1)**:65–68.

[237] Coscarón S, Mancebo AO, Coscarón-Arias CL. Description of Male, Larva and Pupa of *Stibasoma theotaenia* (Wiedemann) (Diptera-Tabanidae). Memórias do Instituto Oswaldo Cruz. 1999;**94(5**):619–623. DOI: 10.1590/S0074-02761999000500010

[238] Coscarón S. Clave ilustrada de larvas y pupas de Tabanidae de Argentina (Diptera, Insecta) In: Salomón OD, editor. Actualizaciones en artropodología sanitaria argentina. Buenos Aires: Ed. Mundo Sano; 2002. pp. 11–21.

[239] Coscarón S, Papavero N. Key to the known immature stages of Neotropical Tabanidae. Neotropical Diptera. 2014;**24**:1–22

[240] Dufek MI, Oscherov EB, Damborsky MP. Actualización de la distribución de tábanos (Diptera: Tabanidae) en la reserva natural Esteros del Iberá, Argentina. Entomotropica. 2016;**31(4)**:36–39.

[241] Hack WS. Contribución al estudio de los tabánidos argentinos (Diptera). Annales Instituto de Medicina Regional Universidad Nacional de Tucumán. 1951;**3(2)**:157–184.

[242] Hack WH. Segunda contribuición al estudio de los tabánidos argentinos (Diptera). Annales Instituto de Medicina Regional Universidad Nacional de Tucumán. 1953;**3**:339–347.

[243] Coscarón S. Tabanidae (Diptera-Insecta). In: Morrone JJ, Coscarón S, editors. Biodiversidad de artrópodos de Argentina. Una perspectiva biotaxonómica. La Plata: Ediciones Sur; 1998. pp. 341–352.

[244] Kröber O. Drei neue *Poeciloderas*-Arten aus Paraguay. Arbeiten ueber Morphologische und Taxonomische Entomologie Berlin. 1936;**3**:221–223.

[245] Russo JV. El mal de caderas en el Chaco Paraguayo (Tripanosomiasis Equina). Asuncion: Registro de la propiedad intelectual no. 2361. 1954. 75 p.

[246] Coscarón S. Tabanidae. In: Hurlbert SH, editor. Aquatic biota of Southern South America. San Diego: San Diego State University; 1977. pp. 297–304.

[247] InsectoidInfo [Internet]. 2016. Available from: http://insectoid.info/checklist/tabanidae/uruguay/ [Accessed: 2016-09-10].

Section 2

Symbiotic Relationships

Chapter 5

The Symbiome of *Llaveia* Cochineals (Hemiptera: Coccoidea: Monophlebidae) Includes a Gammaproteobacterial Cosymbiont *Sodalis* TME1 and the Known *Candidatus* Walczuchella monophlebidarum

Tania Rosas-Pérez, Arturo Vera-Ponce de León, Mónica Rosenblueth, Shamayim T. Ramírez-Puebla, Reiner Rincón-Rosales, Julio Martínez-Romero, Michael F. Dunn, Éva Kondorosi and Esperanza Martínez-Romero

Additional information is available at the end of the chapter

http://dx.doi.org/10.5772/66442

Abstract

The genome and transcriptome of the endosymbiotic flavobacterium *Candidatus* Walczuchella monophlebidarum revealed its role in the synthesis of essential amino acids for its host, the wax cochineal *Llaveia axin axin*. There were, however, missing genes in the endosymbiont for some biosynthetic pathways. Here, we characterized TME1, another cochineal symbiont that may metabolically complement *Walczuchella*. TME1 was ascribed to the gammaproteobacterial genus *Sodalis* on a phylogenomic basis using gene sequences from 143 proteins core genome sequences and the core average nucleotide identity (ANI) confirmed its position. Additionally, we describe *Sodalis* as a coherent genus. TME1 genome is around 3.4 Mb and has complete gene sequences for the biosynthesis of 10 essential amino acids, for polyamines, flagella, nitrate respiration, and detoxification among many others. Transcripts from ovaries and bacteriomes allowed the identification of differentially transcribed genes from the endosymbionts and host. Highly transcribed genes were identified in TME1 and transcripts involved in amino acid biosynthesis were found. We review here that cosymbionts that derived from different bacterial classes and genera seem to be advantageous for insects that have Flavobacteria as the primary endosymbionts.

Keywords: endosymbionts, scale insect, Gammaproteobacteria, *Sodalis*-like, Alphaproteobacteria, fungi

1. Introduction

All organisms are inhabited by microbes that exert different effects on their hosts. In insects, there are many examples of beneficial associations with symbiotic microbes that have been linked to the insect ecological success. Symbionts that are vertically transmitted from mother to offspring and with an intrinsic interdependence with the insect host are considered as primary endosymbionts and they have reduced genomes [1, 2]; they do not grow on standard laboratory media. In theory, endosymbionts evolved from gut bacteria [3] that are largely more complex and may be determined by the diet and the environment. Primary endosymbionts may reside inside insect cells called bacteriocytes that may be found in specialized host structures called bacteriomes. Bacteriomes may be equivalent to plant-root nodules considering that they are host structures harboring particular bacterial species with specific roles [4]. But even in plants, cosymbionts have been encountered; for example, the slow-growing actinobacteria *Micromonospora* is found in nodules formed by *Bradyrhizobium, Rhizobium,* or *Frankia* in several legumes or actinorhizal roots, although *Micromonospora* is unable to form nodules [5]. *Micromonospora* has been reported to enhance nodulation and promote plant growth, may enhance plant defense responses, or inhibit pathogens [6].

In insects, cosymbiosis is not uncommon and there are cases in which two or more bacterial symbionts are found in the bacteriome [7, 8]. Additionally, other microbes including fungi may be found in the hemolymph or in different insect tissues [9–11]. Fungal symbionts may be found as well in specialized insect structures known as mycangia [12] or inside insect cells called mycetocytes [13].

In insects, primary bacterial endosymbionts synthesize essential amino acids or vitamins for their hosts and reside intracellularly in bacteriomes. In some cases, complementation of metabolic pathways seems to occur among different insect symbionts [14–17]. Additionally, cosymbionts may have different roles, and some have been implicated in defense [18–21], tolerance to stress [22], resistance to high temperatures [23–25], to virus [26–28], or may manipulate sex differentiation [29]. There is an example in which a secondary endosymbiont substituted a lost primary *Buchnera* symbiont in an aphid [30]. Among others, alpha, gamma and beta-proteobacteria have been found as cosymbionts; for example, the primary endosymbiont *Candidatus* Sulcia muelleri ("*Sulcia*" from here on) (phylum Bacteroidetes, class Flavobacteria) with a highly reduced genome has betaproteobacteria as cosymbionts found in green rice leafhoppers [7], stinkbugs [31], and spittlebugs [32, 33]. In leafhoppers, the symbionts occupy different types of bacteriocytes that constitute the outer or inner regions of the bacteriome [7]. The *Sulcia* cosymbionts are *Hodgkinia, Zinderia, Nasuia* [34, 35] with very small genomes, and the gammaproteobacteria *Baummania, Arsenophonus,* or *Sodalis,* the latter considered as a new acquisition. Surprisingly a gammaproteobacterium may be found inside *Sulcia* cells and be transmitted to the next generation [36].

Scale insects (Hemiptera: Coccoidea) feed on plant sap, which is a nutritionally poor diet that lacks most of the essential amino acids. Therefore, these insects have built up symbiotic associations with bacteria that can synthesize them. Most of the scale insect families that have been analyzed, such as Monophlebidae, Coelostomidiidae, Orthezidae, Phenacoccinae from Pseudococcidae,

Coccidae, Lecanodiaspididae, Diaspididae, and a clade of Eriococcidae, harbor flavobacteria as primary symbionts and enterobacteria as secondary symbionts. [37–39]. It has been reported that the families from scale insects Dactylopiidae, some Eriococcidae, and Pseudococcinae from Pseudococcidae harbor different endosymbionts, which could indicate that they lost their flavobacteria and enterobacteria and acquired other endosymbionts [39]. Flavobacteria seem to be very ancient symbionts, perhaps starting symbiosis before the divergence of scale insects [39] (150–250 mya [40]). Although it has been suggested that Flavobacteria have cospeciated only within Monophlebidae, Coelostomidiidae, Ortheziidae, and Diaspididae [38–41], and host switches seem to have occurred in the other families [39]. Otherwise, enterobacteria have undergone more evolutionary events (losses, duplications, and host switches). Some scale insects have enterobacteria closely related to *Sodalis* endosymbionts (*Sodalis*-like). But others may have symbionts closely related to *Pantoea* and *Klebsiella* [39].

Sodalis cosymbionts have been identified mainly by their 16S rRNA but also by other gene sequences. They have been found within various insect orders including Diptera, Coleoptera, Phthiraptera, and Hemiptera [42–45]. The first described was *S. glossinidius*, the secondary symbiont of tsetse flies [46]. Later, bacteria with related gene sequences were referred as *Sodalis*-like [47] or *Sodalis*-affiliated but more recently several "*Sodalis*-like" bacteria and SOPE [48] are classified as *Sodalis*, others have been assigned to different genera. Still, scientists are in the process of making correct adscriptions for some of these bacteria [49].

The flavobacteria endosymbiont *Candidatus* Walczuchella monophlebidarum ("*Walczuchella*" from here on) was sequenced from the giant wax cochineal *Llaveia axin axin* (Llave) (Coccoidea: Monophlebidae) [50]. This insect has been used to obtain a lacquer to coat traditional art crafts by native people in Mexico and Guatemala since pre-Hispanic times [51]. The flavobacterial genome revealed that the endosymbiont's major role is to synthesize and provide amino acids to the insect host [50]. The Flavobacteria genome was obtained from the analysis of a metagenome of *L. axin axin*. From this metagenome, we could also ensemble sequences from other microorganisms. Here, we present the draft genome of another cosymbiont of *Walczuchella*, a *Sodalis*-like bacteria that is designated here as *Sodalis* TME1. We also present a comparison to the genomes of five other *Sodalis*, as well as preliminary data of a metatranscriptome performed in the bacteriome of *L. axin axin* adults and in the ovaries of senescent adults.

2. Materials and methods

DNA, sequencing, and assembly were performed from bacteriomes (Illumina HiSeq 2000) and from the homogenized of female adults (pyrosequencing) of *L. axin axin* collected in the state of Chiapas, Mexico, as described [50]. A photograph from *L. axin axin* female adults is shown in **Figure 1**. RAST and GosthKOALA from KEGG [52] were used for genomic and metabolic pathway annotation of the metagenomic data that was previously reported when we obtained the *Walczuchella* genome [50]. *Sodalis* TME1 genome sequence has been deposited at DDBJ/ENA/GenBank under the accession MNBX00000000. The version described in this chapter is MNBX01000000.

Figure 1. *L. axin axin* adult females on a *Jatropha curcas* plant.

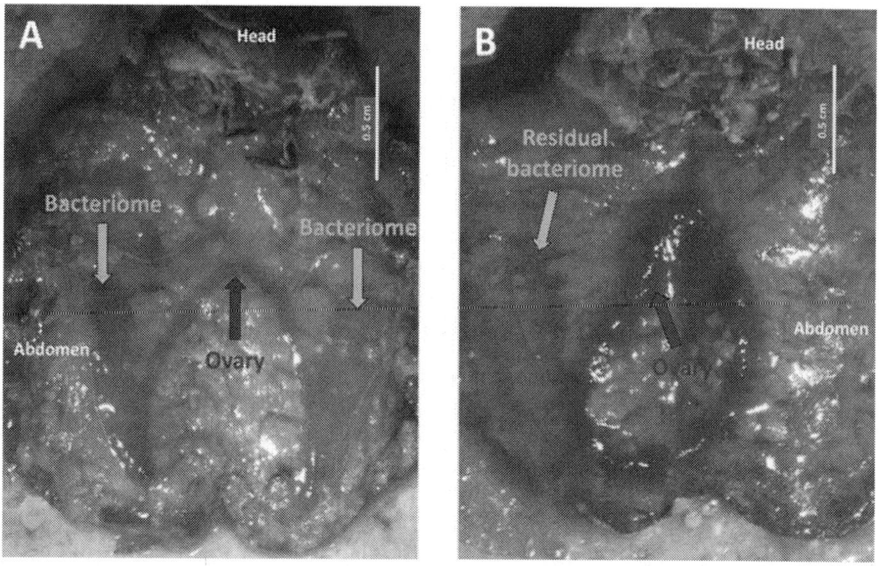

Figure 2. Dissected *L. axin axin* adult females used for the metatranscriptome analysis. (A) early stage and (B) late stage or senescent adults.

Comparative phylogenomic analysis was performed with 20 genomes of gammaproteobacteria from GeneBank. Gene calling of all genomes was performed using GeneMark version 2.5 [53]. The pangenome and core genome from orthologous genes of all strains were obtained by GET_HOMOLOGUES version 2.0 software [54] with -A -c -t 0 -M -n 35 and -A -c -t 0 -G -n 35 parameters. We selected a set of 143 unique single-copy orthologous genes from the core genome. Translated coding sequences of each gene were concatenated using BioEdit Version 7.2.5 and aligned with Clustal Omega version 1.2.1 [55]. Prottest3 version 3.4.2 [56] was used to select the best amino acid substitution model using the AICc correction. The edited alignment contained 47,803 amino acid positions. Maximum likelihood phylogeny was performed by PhyML software version 3.1 [57] using the CpREV model with the Shimodaira–Hasegawa-like procedure for internal branch support [58]. The genome of *Escherichia coli* K-12 MG1655 was used as outgroup.

Comparative genomics was carried out with the following *Sodalis* genomes: *S. glossinidius* morsitans from tsetse fly, *Sodalis*-like endosymbiont from the blood-feeding lice *Proechinophthirus fluctus* (an obligate ectoparasite of fur seals), *S. pierantonus* SOPE from rice weevils *Sitophilus oryzae*, the free-living *S. praecaptivus*, and *Sodalis*-like symbiont of the meadow spittlebug *Philaenus spumarius*. Orthologous genes and the core genomes were obtained by GET_HOMOLOGUES as described above. Core genome matrix was parsed from GET_HOMOLOGUES result, using the parsing_pangenome_matrix.pl script. Shared genes between *Sodalis*-like TME1 and all other strains were retrieved by parsing the core matrix using custom perl scripts. Annotation of each gene cluster was carried out by BLASTp 2.2.30+ [59] searches against Uniref100 database. Furthermore, average nucleotide identity (ANI) was determined for all *Sodalis* genomes described above using the ANIcalculator software described by Varghese et al. [60] with the default parameters.

RNA was extracted from the bacteriome of *L. axin axin* female adults and from the ovaries of senescent female adults that do not possess the structure of the bacteriomes (bacteriomes degrade in senescent adults) (**Figure 2**). Sequencing of cDNA was performed by SOLID technology. The sequences were mapped to the genomes of *Walczuchella, Sodalis*-like TME1, and two insect reference genomes, *Drosophila melanogaster* and to the aphid *Acyrthosiphon pisum.* Differentially expressed genes were identified by comparing expression values between samples and using Kal's *Z*-test of proportions [61]. Genes with a change in the expression more than twofold and a *p*-value of <0.01 in the Z-test were considered as differentially expressed genes.

To determine the uric acid and uricase activity, *L. axin axin* adult females were individually dissected under sterile conditions. Guts including the Malpighian tubules were extracted and metabolic activities were detected as described [62].

3. Results

We found gene sequences of an enterobacterium (gammaproteobacterium) related to *Sodalis* in the metagenome of the wax cochineal *L. axin axin* [50]. The phylogeny with a set of 143 conserved genes shows that the enterobacterium of *L. axin axin* is closely related to other *Sodalis*-like endo-

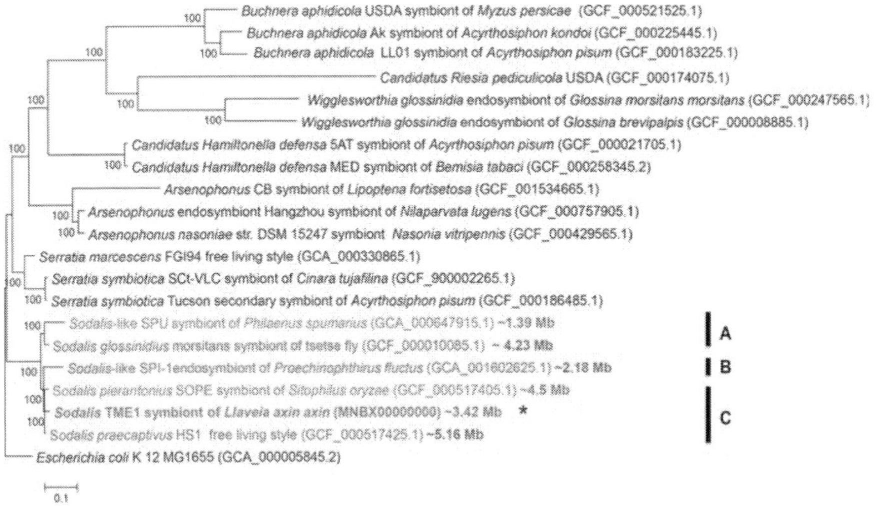

Figure 3. Maximum likelihood phylogeny of sequenced enterobacterial endosymbionts performed with 143 conserved genes. *Sodalis* endosymbionts of plant feeding host: green; blood feeding host: red; free-living style: blue. * : *Sodalis* TME1 used in this study. Scale bar indicates 1 % estimated sequence divergence. SH-aLRT values > 50 are indicated.

symbionts, especially close to the free-living *S. praecaptivus* [63] (**Figure 3**). The small branches in the *Sodalis* group may indicate that they have recently diverged while the large differences found in genome sizes among these endosymbionts indicate that evolution may be occurring mainly by genome reduction when compared to the larger genome of the free-living *Sodalis* (**Figure 3**).

TME1 was compared with the ANI (average nucleotide identity) metric to other *Sodalis* using the same core genome used in the phylogenomic analysis. TME1 showed ANIs well over 95% that is used to delineate species with *S. pierantonius* SOPE and *S. praecaptivus* HS1, but lower than 95% with *S. glossinidius* morsitans, *Sodalis*-like SPU, and *Sodalis*-like SPI-1 (**Table 1**). There was a good correlation of the ANI values obtained and phylogenetic positions that allowed the identification of three groups within *Sodalis* (**Figure 3** and **Table 1**).

The draft assembly of the enterobacterial endosymbiont *Sodalis* TME1 genome consisted of 679 scaffolds with an N50 of 7713 and an average G + C content of 55.6%. The scaffolds sum 3.4 Mb [50]. A total of 3067 genes were identified to which a functional annotation was assigned. The functional categories more represented by the annotated genes were catabolic and cellular process as well as carbohydrate, amino acid and transcription DNA dependent metabolism (**Figure 4**). Interestingly, many phage-related sequences were found as well as genes for different multidrug efflux pumps and type III and IV secretion systems. TME1 has genes for polyamine biosynthesis and excretion as well as Ankyrin repeat domains and for a lactoyl-glutathione lyase that is a detoxifying enzyme [64]. Among the conserved genes in the core genome of *Sodalis* TME1, *S. pierantonius* str. SOPE and *S. praecaptivus* str. HS1 are genes for the synthesis of flagella and for nitrate reduction (*nar*GHI) and nitrite reduction (*nfr*ABCD). Maybe nitrate serves in *Sodalis* as an electron acceptor in anaerobiosis as occurs in bacterial symbionts of marine bivalves *Lucinoma aequizonata* [65]. *Sodalis* TME1 genome has genes for uric acid utilization such as uricase (*ua*Z), allantoinase (*all*B), allantoate deiminase (*all*C), and urease (*ure*C and *ure*D). Comparative genomics with all *Sodalis* strains show that *all*C and the alpha subunit for urease gene (*ure*C) orthologous were only present in *Sodalis* TME1. Experimentally, uric acid and uricase activity were quantified in *L. axin axin* female adults. We detected 5.86 ± 0.77 ng of uric acid per tissue μg^{-1} and 32.87 ± 5. 25 mU of uricase per tissue μg^{-1} in female cochineals.

	Sodalis str. TME1	*Sodalis pierantonius* SOPE	*Sodalis praecaptivus* HS1	*Sodalis glossinidius* str. morsitans	*Sodalis*-like str. PSPU	*Sodalis*-like str. SPI-1
Sodalis str. TME1						
Sodalis pierantonius str. SOPE	**98.45**					
Sodalis praecaptivus str. HS1	**98.54**	**98.46**				
Sodalis glossinidius str. morsitans	91.24	91.04	91.27			
Sodalis-like str. PSPU	89.81	89.67	89.87	**95.48**		
Sodalis-like str. SPI-1	92.25	91.94	92.15	89.91	85.86	

Table 1. Average nucleotide identity (ANI) percentage among *Sodalis* strains. Values in bold are >95%. Colors correspond to green, plant-feeding host; red, blood-feeding host; blue, free-living style..

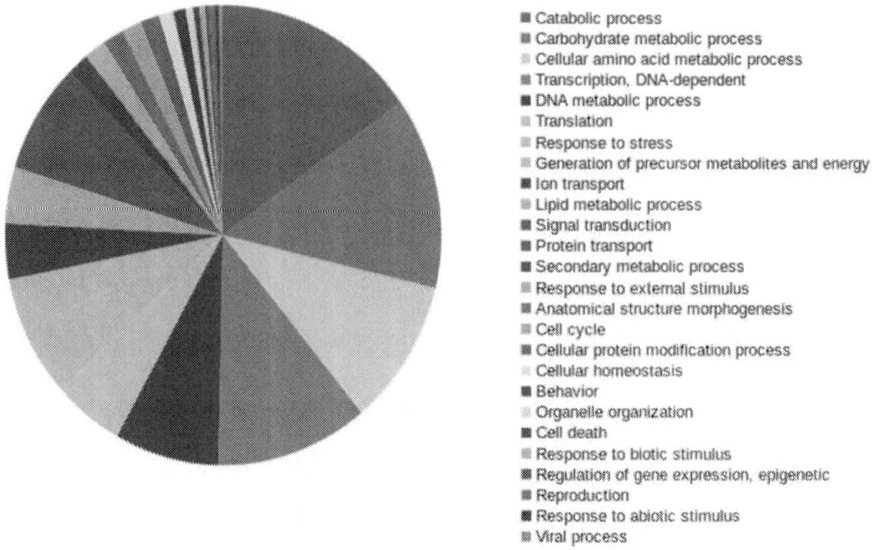

Figure 4. Gene functional categories of *Sodalis* TME1.

We obtained 11,042,037 and 11,042,428 reads from the cDNA sequence of the bacteriome and the ovaries, respectively. These two organs were selected for studying the differentially expressed genes because endosymbionts are transferred from bacteriomes to the ovaries for vertical transmission to their offspring. It was expected to find genes related to the migration of the endosymbionts from the bacteriome and the colonization of the ovaries. Reads mapped to the reference genomes are shown in **Table 2**. The number of genes that were statistically differentially expressed is shown in **Table 3**.

Walczuchella in the bacteriome tissue showed only two genes that exhibited differential expression, a putative hydrolase and the chaperone GroEL. Other genes showed a change in expression less than twofold compared to their expression in the ovary. The chaperonin GroES is almost at the limit for differential expression with 1.86-fold (**Table 4**).

From the ovary tissue, we found differential expression of *Walczuchella* genes that code for some ATP synthase subunits (some of them annotated previously as pseudogenes), cyto-

Reference genome	Bacteriome	Ovaries
Drosophila melanogaster (exons)	2,019,585	2,008,381
Acyrthosiphon pisum (mRNA refseq)	3,082,319	2,912,207
Walczuchella	1,052,077	87,502
Sodalis TME1	409,128	483,601

Table 2. Number of reads mapped to the reference genomes.

Reference genome	Bacteriome	Ovaries
Drosophila melanogaster (exons)	494	680
Acyrthosiphon pisum (mRNA refseq)	244	280
Walczuchella	2	89
Sodalis TME1	66	50

Table 3. Number of genes differentially expressed according to Z-test ($p < 0.01$).

	Walczuchella	*Sodalis* TME1	Insect
Bacteriome (differential expression) (high RPKM)	Putative hydrolase	T3SS-secreted effector	Chaperon Hsp70
	Chaperones GroEL, GroES	Allantoinase	ABC transporters
	Hypothetical proteins	Hypothetical proteins	Antiparasitic-like peptide
	ATP synthase B subunit	NAD biosynthesis	Asparaginase
	Amino acids biosynthesis genes	FtsE cell division gene	Unknown genes
		Transcriptional regulation	Extracellular glutamate receptor channel
		Flagellum synthesis	Phospholipids synthesis
			Transcriptional regulation
Ovary (differential expression) (high RPKM)	ATP synthase B and A subunits (pseudogenes)	Hypothetical proteins	ATPase subunit
	AhpC oxidative stress gene	NAD biosynthesis	Transmembrane transporters of sugars and amino acids
	Glycoprotease	Flagellum synthesis	Peptidoglycan-binding protein
	Amino acids biosynthesis genes	FtsE cell division gene	Lysozyme
	Cytochrome c oxydase	Glycolysis	Unknown genes
	SecY translocase	Phage lysozyme	Transcriptional regulation
	Hypothetical proteins	Transcriptional regulation	Phospholipids synthesis

Table 4. Highly expressed and differentially expressed genes in the bacteriome and the ovaries in the endosymbionts *Walczuchella* and *Sodalis* TME1 and the host *L. axin axin*.

chrome c oxidase, also some genes of protein translocation systems, tryptophan, histidine and chorismate biosynthesis, one gene related to oxidative stress, and a gene that encodes a possible component of an ABC transporter (**Table 4**).

In the bacteriome, the enterobacterium TME1 showed very strong overexpression of a gene that codes an effector protein possibly secreted by the type III secretion system (TTSS), expressed 66.8-fold compared to its expression in the ovaries. Also, a gene that codes an allantoinase that participates in uric acid metabolism is highly overexpressed in the bacteriome, showing a 50-fold change. Other genes with overexpression in the bacteriome are four ABC transporters, a peroxidase, the heme synthase, two genes related to nucleotides biosynthesis, two genes related to lipid A biosynthesis, and two genes of the type III secretion system (**Table 4**).

In the ovary, TME1-overexpressed genes were related to NAD synthesis, carbohydrate metabolism, stress response, and some transporters and transcriptional regulators (**Table 4**).

Among the insect differentially expressed genes in the bacteriome there were 19 putative transporters (for amino acids, carbohydrates, vitamins, drugs, or unknown substrates), five genes related to defense systems including an antiparasitic peptide with identity to Drosomycin, three from *D. melanogaster*, two genes related to heat-shock response, an oxidative stress response gene, seven genes related to amino acid metabolism, and some genes related to lipid, carbohydrate, and vitamin metabolism (**Table 4**).

On the other hand, we found that in the insect, in the ovaries there was overexpression of 15 transporters, 17 immune response genes, some genes related to heat shock, desiccation, oxidative stress, and hypoxia response, and genes related to lipids, vitamins, carbohydrates, nucleotides, amino acids, and chitin synthesis and metabolism (**Table 4**).

4. Discussion

Due to the annual cycle of the wax cochineal, we are only able to collect insects once a year during the rainy season. It is worth mentioning that in 2015 and 2016, we did not find cochineals in many of the places where we had collected previously. Considering the menace of mosquitoes transmitting Zika, or Chikungunya, extensive fumigations with chemical insecticides have been carried out in many places in Mexico, especially in Chiapas. The relation to the diminished populations of cochineals remains to be established.

A previous survey of symbiotic bacteria from scale insects in Mexico revealed the prevalence of Flavobacteria and Gammaproteobacteria [39]. Some of the Gammaproteobacteria had 16S ribosomal gene sequences closely related to those of TME1, and thus they may be considered as *Sodalis* as well. They were obtained from different scale insects such as *Insignorthezia* sp. and *I. insignia, Icerya purchasi, Cripticerya* sp., and *Pseudococcus longispinus* that together with *Llaveia* would be hosts for *Sodalis*.

While Flavobacteria and insects showed a co-divergent pattern of evolution, the phylogenetic relationships of the Gammaproteobacteria and insects were not parallel, indicating multiple enterobacterial transfers among the different hosts, and a more recent and less dependent symbiosis. In agreement, the genome size of the gammaproteobacterium TME1 is much larger than that from the primary endosymbiont from wax cochineals, the Flavobacteria *Walczuchella*, and also larger than those from other cosymbionts as the Betaproteobacteria that accompany the bacteroidete *Sulcia* found in some insects.

The genome from the gammaproteobacterium TME1 (3.4 Mb) is within the range of those from other *Sodalis* (1.4–4.7 Mb, **Figure 3**). There are very few genomes available from *Sodalis*, namely those from *Sodalis* found in blood-sucking insects as in lice [42] and tsetse flies [66], in plant-feeding insects as the rice weevils [44], in spittlebugs [45], and from a free-living bacterium [67]. The average nucleotide identity (ANI [68] being used for global genomic comparisons and considered now as a gold standard in prokaryote taxonomy [69]) was estimated for the *Sodalis* with available genomes. ANI values and the phylogenomic analysis performed showed *Sodalis* as a defined and coherent genus with three groups A–C. These groups could represent at least three different species according to the global standards [69]. Two of these groups were identified as different lineages by Lo et al. [49]. The phylogenetic groups that we described here have a 100 SH-like value support, group A contains *S. glossinidius* from tsetse flies and *Sodalis* from the meadow spittlebug *P. spumarius*, group B is constituted by *Sodalis* from the fur seal *P. fluctus*, and group C contains the closely related TME1, the free-living *S. praecaptivus* and *S. pierantonius* SOPE. The nucleotide sequence conservation among the group A symbiotic and free-living *Sodalis* may reflect that the former were recent acquisitions in insects without enough time for sequence divergence in their hosts. The presence of very similar *Sodalis* in distinct insect isolates reinforces the reports that indicate that they may frequently be transferred among hosts [39, 47].

TME1 has biosynthetic pathways for all essential amino acids and may supply the needs of the wax cochineal and of *Walczuchella* that does not have complete pathways for the biosynthesis of all essential amino acids. Since *Sodalis* TME1 has all enzymes for TCA it may complement this pathway in *Walczuchella*. It is worth noting that the flavobacterium *Candidatus* Uzinura diaspidicola, an endosymbiont from the armored scale insect *Aphytis melinus* that feed on parenchyma which may provide more nutrients than sap, supplies its host with all nutrients without the need of a cosymbiont [70]. Other armored scale insects have been reported to have a *Sodalis*-like endosymbiont [39].

In *S. glossinidius* that is a secondary symbiont of tsetse flies, a type III secretion system was found implicated in cell invasion and maybe required for colonizing the insect bacteriocytes [71]. Genes encoding for a similar system were found in TME1. Notably, genes that code for the type III secretion system (TTSS) as well as a gene coding for an effector protein that may be secreted by this system were among the most highly induced in the bacteriome of TME1. In *Salmonella enterica*, polyamines are required for full expression of TTSS and for some effector coding genes. Mutants in polyamine biosynthesis are affected in intracellular colonization and survival and may be complemented by adding polyamines to the medium [72]. Furthermore, the modulation of a TTSS by a spermidine transporter has been reported in *Pseudomonas aeruginosa*. Exogenous addition of spermidine to the wild *P. aeruginosa* strain increased the expression of genes that produce effector proteins [73]. TME1 has all genes for spermidine and putrescine biosynthesis as well as for the excretion of spermidine. Polyamines may regulate host defense responses as do some effectors secreted by TTSS. This remains to be tested.

Uric acid and uricase activity were detected in *L. axin axin* females. Uric acid is the final product of purine metabolism. Only few insects are capable of degrading uric acid into other products. In plant-feeding insects, bacterial and fungal symbionts are capable of recycling uric acid into other nitrogen sources [74–76]. *Sodalis* TME1 has uricase and allantoinase-codifying genes, and the latter was highly expressed in bacteriomes suggesting that *Sodalis* TME1 could participate in providing nitrogen to the host by uric acid recycling.

By reverse transcriptase-polymerase chain reaction (RT-PCR) using primers targeted to *Sodalis*, we found sequences from *Sodalis* in the bacteriome (our own unpublished results), thus we may suppose that *Sodalis* are localized in bacteriomes as *Walczuchella*. In *Llaveia*, in addition to *Walczuchella* and *Sodalis* we found sequences of alphaproteobacteria that are related to Rickettsiales and several fungi that are reported elsewhere (Vera Ponce de León, submitted). Coincidently, the seal lice with a *Sodalis* endosymbiont also harbor a *Rickettsia* that is very abundant. The role of the very little abundant *Rickettsia*-like bacterium in *Llaveia* is unknown. *Wolbachia* is found in members of the Coelostomidiidae family [37] that is closely related to Monophlebidae insect family that contains the Mexican wax cochineals.

Here, we used the term symbiome [27] to refer to the group of primary and secondary (cosymbionts) endosymbionts (and/or their genomes), residing in a host. We consider that the term symbiome is more adequate than the terms endosymbiotic community or consortium that are sometimes used instead.

The cosymbionts of different Flavobacteria in scale insects are diverse lineages of related Gammaproteobacteria [39]. Similarly, the cosymbionts of *Sulcia* (a flavobacterium as *Walczuchella*) are varied and may be different even in related hosts [7, 36]. *Sulcia* cosymbionts may belong to alpha, beta, or gammaproteobacteria, with alpha and betaproteobacteria looking like the oldest symbionts. It was reported that *Candidatus* Zinderia insecticola, the Betaproteobacteria of spittelbugs was probably substituted by a *Sodalis*-like symbiont in members of the Philaenini tribe of the spittelbugs [33, 45]. The displacement of betaproteobacterial cosymbionts by the gammaproteobacterium *Sodalis* seems recent and was described as an event *"in statu nascendi"* (in the stage of being born) in *Cicadella viridis* [77]. There are other examples where one endosymbiont may substitute another one or is on the way toward displacement of a highly reduced-genome endosymbiont [33, 77–79]. Distinct (apparently replaceable) cosymbionts may fulfill the different needs of insects that may change overtime and conditions specially if the insect changes habit [22], otherwise there may be cosymbiont redundancy, with different bacteria performing the same or very similar role (e.g., the synthesis of essential amino acids). The *Sodalis* cosymbiont in the wax cochineals seems to be recently acquired as in *C. viridis*. The insect symbiome seems plastic or dynamic with cosymbionts playing a key role in this plasticity. Here, we enlarged the list of putative functions of *Sodalis* that may include uric acid recycling, polyamine biosynthesis, or detoxification.

Acknowledgements

AVPL is a doctoral student from Programa de Doctorado en Ciencias Biomédicas, Universidad Nacional Autónoma de México (UNAM) and received the 331625 fellowship from Consejo Nacional de Ciencia y Tecnología (CONACyT).

Author details

Tania Rosas-Pérez[1,2], Arturo Vera-Ponce de León[2], Mónica Rosenblueth[2,*], Shamayim T. Ramírez-Puebla[2], Reiner Rincón-Rosales[3], Julio Martínez-Romero[2], Michael F. Dunn[2], Éva Kondorosi[4] and Esperanza Martínez-Romero[2]

*Address all correspondence to: mrosen@ccg.unam.mx

1 Cavanilles Institute of Biodiversity and Evolutionary Biology, University of Valencia, Spain

2 Center for Genomics Sciences, National Autonomous University of Mexico, Cuernavaca, Morelos, Mexico

3 Technological Institute of Tuxtla Gutiérrez, Tuxtla Gutiérrez, Chiapas, Mexico

4 Biological Research Centre, Hungarian Academy of Sciences, Szeged, Hungary

References

[1] McCutcheon JP. The bacterial essence of tiny symbiont genomes. Current Opinion in Microbiology. 2010;13:73–78. DOI: 10.1016/j.mib.2009.12.002.

[2] Martínez-Cano DJ, Reyes-Prieto M, Martínez-Romero E, Partida-Martínez LP, Latorre A, Moya A, Delaye L. Evolution of small prokaryotic genomes. Frontiers in Microbiology. 2015; **5**:742. DOI: 10.3389/fmicb.2014.00742.

[3] Husník F, Chrudimský T, Hypša V. Multiple origins of endosymbiosis within the Enterobacteriaceae (γ-Proteobacteria): convergence of complex phylogenetic approaches. BMC Biology. 2011;**9**:87. DOI: 10.1186/1741-7007-9-87.

[4] Ramírez-Puebla ST, Servín-Garcidueñas LE, Jiménez-Marín B, Bolaños LM, Rosenblueth M, Martínez J, Rogel MA, Ormeño-Orrillo E, Martínez-Romero E. Gut and root microbiota commonalities. Applied and Environmental Microbiology. 2013; **79**:2–9. DOI: 10.1128/AEM.02553-12.

[5] Trujillo ME, Alonso-Vega P, Rodríguez R, Carro L, Cerda E, Alonso P, Martínez-Molina E. The genus *Micromonospora* is widespread in legume root nodules: the example of *Lupinus angustifolius*. ISME Journal. 2010; **4**:1265–1281. DOI: 10.1038/ismej.2010.55.

[6] Trujillo ME, Riesco R, Benito P, Carro L. Endophytic Actinobacteria and the interaction of *Micromonospora* and nitrogen fixing plants. Frontiers in Microbiology. 2015; **6**:1341. DOI: 10.3389/fmicb.2015.01341.

[7] Noda H, Watanabe K, Kawai S, Yukuhiro F, Miyoshi T, Tomizawa M, Koizumi Y, Nikoh N, Fukatsu T. Bacteriome-associated endosymbionts of the green rice leafhopper *Nephotettix cincticeps* (Hemiptera: Cicadellidae). Applied Entomology and Zoology. 2012; **47**:217–225. DOI:10.1007/s13355-012-0110-1

[8] Bennett GM, Moran NA. Small, smaller, smallest: the origins and evolution of ancient dual symbioses in a Phloem-feeding insect. Genome Biology and Evolution. 2013; **5**:1675–1688. DOI: 10.1093/gbe/evt118.

[9] Tsuchida T, Koga R, Meng XY, Matsumoto T, Fukatsu T. Characterization of a facultative endosymbiotic bacterium of the pea aphid *Acyrthosiphon pisum*. Microbial Ecology. 2005; **49**:126–133. DOI: 10.1007/s00248-004-0216-2.

[10] Kaiwa N, Hosokawa T, Kikuchi Y, Nikoh N, Meng XY, Kimura N, Ito M, Fukatsu T. Primary gut symbiont and secondary, *Sodalis*-allied symbiont of the Scutellerid stinkbug *Cantao ocellatus*. Applied and Environmental Microbiology. 2010; **76**:3486–3494. DOI: 10.1128/AEM.00421-10.

[11] Chrudimský T, Husník F, Nováková E, Hypša V. *Candidatus* Sodalis melophagi *sp. nov.*: phylogenetically independent comparative model to the tsetse fly symbiont *Sodalis glossinidius*. PLoS One. 2012; **7**:e40354. DOI: 10.1371/journal.pone.0040354.

[12] Happ GM, Happ CM, Barras SJ. Fine structure of the prothoracic mycangium, a chamber for the culture of symbiotic fungi, in the southern pine beetle, *Dendroctonus frontalis*. Tissue and Cell. 1971; **3**:295–308. DOI: 10.1016/S0040-8166(71)80024-1

[13] Kuechler SM, Gibbs G, Burckhardt D, Dettner K, Hartung V. Diversity of bacterial endosymbionts and bacteria-host co-evolution in Gondwanan relict moss bugs (Hemiptera: Coleorrhyncha: Peloridiidae). Environmental Microbiology. 2013; **15**:2031–2042. DOI: 10.1111/1462-2920.12101.

[14] Wu D, Daugherty SC, Van Aken SE, Pai GH, Watkins KL, Khouri H, Tallon LJ, Zaborsky JM, Dunbar HE, Tran PL, Moran NA, Eisen JA. Metabolic complementarity and genomics of the dual bacterial symbiosis of sharpshooters. PLoS Biology. 2006; **4**:e188. DOI: 10.1371/journal.pbio.0040188.

[15] Cottret L, Milreu PV, Acuña V, Marchetti-Spaccamela A, Stougie L, Charles H, Sagot MF. Graph-based analysis of the metabolic exchanges between two co-resident intracellular symbionts, *Baumannia cicadellinicola* and *Sulcia muelleri*, with their insect host, *Homalodisca coagulata*. PLoS Computational Biology. 2010; **6**:e1000904.

[16] López-Madrigal S, Latorre A, Porcar M, Moya A, Gil R. Mealybugs nested endosymbiosis: going into the 'matryoshka' system in *Planococcus citri* in depth. BMC Microbiology 2013; **13**:74. DOI: 10.1186/1471-2180-13-74

[17] Rao Q, Rollat-Farnier PA, Zhu DT, Santos-Garcia D, Silva FJ, Moya A, Latorre A, Klein CC, Vavre F, Sagot MF, Liu SS, Mouton L, Wang XW. Genome reduction and potential metabolic complementation of the dual endosymbionts in the whitefly *Bemisia tabaci*. BMC Genomics. 2015; **16**:226. DOI: 10.1186/s12864-015-1379-6.

[18] Scarborough CL, Ferrari J, Godfray HC. Aphid protected from pathogen by endosymbiont. Science. 2005; **310**:1781–1781. DOI: 10.1126/science.1120180.

[19] Moran NA, Tran P, Gerardo NM. Symbiosis and insect diversification: an ancient symbiont of sap-feeding insects from the bacterial phylum Bacteroidetes. Applied Environmental Microbiology. 2005; **71**:8802–8810. DOI:10.1128/AEM.71.12.8802-8810.2005.

[20] Nakabachi A, Ueoka R, Oshima K, Teta R, Mangoni A, Gurgui M, Oldham NJ, van Echten-Deckert G, Okamura K, Yamamoto K, Inoue H, Ohkuma M, Hongoh Y, Miyagishima SY, Hattori M, Piel J, Fukatsu T. Defensive bacteriome symbiont with a drastically reduced genome. Current Biology. 2013; **23**:1478–1484. DOI: 10.1016/j.cub.2013.06.027.

[21] Leclair M, Pons I, Mahéo F, Morlière S, Simon J-C, Outreman Y. Diversity in symbiont consortia in the pea aphid complex is associated with large phenotypic variation in the insect host. Evolutionary Ecology. 2016; DOI:10.1007/s10682-016-9856-1 (published on line).

[22] Oliver KM, Degnan PH, Burke GR, Moran NA. Facultative symbionts in aphids and the horizontal transfer of ecologically important traits. Annual Review of Entomology. 2010; **55**:247–266. DOI 10.1146/annurev-ento-112408-085305.

[23] Chen DQ, Montllor CB, Purcell A H. Fitness effects of two facultative endosymbiotic bacteria on the pea aphid, *Acyrthosiphon pisum*, and the blue alfalfa aphid *A. kondoi*. Entomologia Experimentalis et Applicata. 2000; **95**:315–323. DOI: 10.1046/j.1570-7458.2000.00670.x

[24] Russell JA, Moran NA. Costs and benefits of symbiont infection in aphids: variation among symbionts and across temperatures. Proceedings of the Royal Society B: Biological Sciences. 2006; **273**:603–610. DOI:10.1098/rspb.2005.3348.

[25] Montllor C B, Maxmen A, Purcell AH. Facultative bacterial endosymbionts benefit pea aphids *Acyrthosiphon pisum* under heat stress. Ecological Entomology. 2002; **27**:189–195. DOI:10.1046/j.1365-2311.2002.00393.x

[26] Teixeira L, Ferreira A, Ashburner M. The bacterial symbiont *Wolbachia* induces resistance to RNA viral infections in *Drosophila melanogaster*. PLoS Biology. 2008; **6**:e2. DOI: 10.1371/journal.pbio.1000002

[27] Bouclas DG, Kariithi HM, Bourtzis K, Schneider DI, Kelley K, Miller WJ, Parker AG, Abd-Alla AM. Transgenerational transmission of the *Glossina pallidipes* hytrosavirus depends on the presence of a functional symbiome. PLoS One. 2013; **8**:e61150. DOI: 10.1371/journal.pone.0061150.

[28] Chrostek E, Marialva MS, Esteves SS, Weinert LA, Martinez J, Jiggins FM, Teixeira L. *Wolbachia* variants induce differential protection to viruses in *Drosophila melanogaster*: a phenotypic and phylogenomic analysis. PLoS Genetics. 2013; **9**:e1003896. DOI: 10.1371/journal.pgen.1003896

[29] Werren JH, Baldo L, Clark ME. *Wolbachia*: master manipulators of invertebrate biology. Nature Reviews Microbiology. 2008; **6**:741–751. DOI: 10.1038/nrmicro1969.

[30] Koga R, Tsuchida T, Fukatsu T. Changing partners in an obligate symbiosis: a facultative endosymbiont can compensate for loss of the essential endosymbiont Buchnera in an

aphid. Proceedings of the Royal Society B: Biological Sciences. 2003; **270**:2543–2550. DOI: 10.1098/rspb.2003.2537

[31] Hosokawa T, Kaiwa N, Matsuura Y, Kikuchi Y, Fukatsu T. Infection prevalence of *Sodalis* symbionts among stinkbugs. Zoological Letters. 2015; **1**:5. DOI: 10.1186/s40851-014-0009-5.

[32] McCutcheon JP, Moran NA. Functional convergence in reduced genomes of bacterial symbionts spanning 200 My of evolution. Genome Biology and Evolution. 2010: **2**:708–718. DOI:10.1093/gbe/evq055

[33] Koga R, Bennett GM, Cryan JR, Moran NA. Evolutionary replacement of obligate symbionts in an ancient and diverse insect lineage. Environmental Microbiology. 2013; **15**:2073–2081. DOI: 10.1111/1462-2920.12121.

[34] Van Leuven JT, Meister RC, Simon C, McCutcheon JP. Sympatric speciation in a bacterial endosymbiont results in two genomes with the functionality of one. Cell. 2014; **158**:1270–1280. DOI: 10.1016/j.cell.2014.07.047.

[35] Bennett GM, Abbà S, Kube M, Marzachì C. Complete genome sequences of the obligate symbionts "*Candidatus* Sulcia muelleri" and "*Ca*. Nasuia deltocephalinicola" from the Pestiferous Leafhopper *Macrosteles quadripunctulatus* (Hemiptera: Cicadellidae). Genome Announcements. 2016; **4** pii:e01604–15. DOI: 10.1128/genomeA.01604-15.

[36] Kobiałka M, Michalik A, Walczak M, Junkiert Ł, Szklarzewicz T. Sulcia symbiont of the leafhopper Macrosteles laevis (Ribaut, 1927) (Insecta, Hemiptera, Cicadellidae: Deltocephalinae) harbors Arsenophonus bacteria. Protoplasma. 2016; **253**:903–912. DOI: 10.1007/s00709-015-0854-x.

[37] Dhami MK, Turner AP, Deines P, Beggs JR, Taylor MW. Ultrastructural and molecular characterization of a bacterial symbiosis in the ecologically important scale insect family Coelostomidiidae. FEMS Microbiology Ecology. 2012; **81**:537–546. DOI: 10.1111/j.1574-6941.2012.01378.x

[38] Gruwell ME, Morse GE, Normark BB. Phylogenetic congruence of armored scale insects (Hemiptera: Diaspididae) and their primary endosymbionts from the phylum Bacteroidetes. Molecular Phylogenetics and Evolution. 2007; **44**:267–280. DOI: org/10.1016/j.ympev.2007.01.014

[39] Rosenblueth M, Sayavedra L, Sámano-Sánchez H, Roth A., Martínez-Romero E. Evolutionary relationships of flavobacterial and enterobacterial endosymbionts with their scale insect hosts (Hemiptera: Coccoidea). Journal of Evolutionary Biology. 2012; **25**:2357–2368. DOI: 10.1111/j.1420-9101.2012.02611.x

[40] Koteja J. Essay on the prehistory of the scale insects (Homoptera, Coccinea). In Annales Zoologici. Państwowe Wydawnictwo Naukowe: Warzawa, Wroclaw; 1985; **38**. p. 461–503.

[41] Dhami MK, Buckley TR, Beggs JR, Taylor MW. Primary symbiont of the ancient scale insect family Coelostomidiidae exhibits strict cophylogenetic patterns. Symbiosis. 2013; **61**:77–91. DOI: 10.1007/s13199-013-0257-8

[42] Boyd BM, Allen JM, Koga R, Fukatsu T, Sweet AD, Johnson KP, Reed DL. Two bacterial genera, *Sodalis* and *Rickettsia*, associated with the seal louse *Proechinophthirus fluctus* (Phthiraptera: Anoplura). Applied and Environmental Microbiology. 2016; **82**:3185–3197. DOI: 10.1128/AEM.00282-16.

[43] Toju H, Koga R, Nikoh N, Meng XY, Kimura N, Fukatsu T. "*Candidatus* Curculioniphilus buchneri" a novel clade of bacterial endocellular symbionts from weevels of the genus *Curculio*. Applied and Environmental Microbiology. 2010; **76**:275–282. DOI: 10.1128/AEM.02154-09

[44] Oakeson KF, Gil R, Clayton AL, Dunn DM, von Niederhausern AC, Hamil C, Aoyagi A, Duval B, Baca A, Silva FJ, Vallier A, Jackson DG, Latorre A, Weiss RB, Heddi A, Moya A, Dale C. Genome degeneration and adaptation in a nascent stage of symbiosis. Genome Biology and Evolution. 2014; **6**:76–93. DOI: 10.1093/gbe/evt210.

[45] Koga R, Moran NA. Swapping symbionts in spittlebugs: evolutionary replacement of a reduced genome symbiont. ISME Journal. 2014; **8**:1237–1246. DOI: 10.1038/ismej.2013.235.

[46] Dale C, Maudlin I. *Sodalis gen. nov.* and *Sodalis glossinidius sp. nov.*, a microaerophilic secondary endosymbiont of the tsetse fly *Glossina morsitans morsitans*. International Journal of Systematic Bacteriology. 1999; **49**:267–275. DOI: 10.1099/00207713-49-1-267.

[47] Smith WA, Oakeson KF, Johnson KP, Reed DL, Carter T, Smith KL, Koga R, Fukatsu T, Clayton DH, Dale C. Phylogenetic analysis of symbionts in feather-feeding lice of the genus *Columbicola*: evidence for repeated symbiont replacements. BMC Evolutionary Biology. 2013; **13**:109. DOI: 10.1186/1471-2148-13-109.

[48] Gil García R, Belda E, Gosalbes MJ, Delaye L, Vallier A, Vincent Monégat C, Heddi A, Silva GJ, Moya A, Latorre A. Massive presence of insertion sequences in the genome of SOPE, the primary endosymbiont of the rice weevil "*Sitophilus oryzae*". International Microbiology. 2008; **11**:4–48. DOI: 10.2436/20.1501.01.43.

[49] Lo WS, Huang YY, Kuo CH. Winding paths to simplicity: genome evolution in facultative insect symbionts. FEMS Microbiology Reviews. 2016; pii:fuw028. DOI: org/10.1093/femsre/fuw028

[50] Rosas-Pérez T, Rosenblueth M, Rincón-Rosales R, Mora J, Martínez-Romero E. Genome sequence of "*Candidatus* Walczuchella monophlebidarum" the flavobacterial endosymbiont of *Llaveia axin axin* (Hemiptera: Coccoidea: Monophlebidae). Genome Biology and Evolution. 2014; **6**:714–726. DOI: 10.1093/gbe/evu049

[51] Williams M, MacVean CM. Ethnococcidology: use of the giant margarodids, *Llaveia* spp. (Homoptera: Coccoidea: Margarodidae), by indigenous peoples of Mesoamerica in their culture, medicine and arts. Israel Journal of Entomology. 1995; **29**:147–148.

[52] Kanehisa M, Sato Y, Morishima K. BlastKOALA and GhostKOALA: KEGG tools for functional characterization of genome and metagenome sequences. Journal of Molecular Biology. 2016; **428**:726–731. DOI: 10.1016/j.jmb.2015.11.006

[53] Besemer J, Borodovsky M. GeneMark: Web software for gene finding in prokaryotes, eukaryotes and viruses. Nucleic Acids Research. 2005; **33**:451–454. DOI:10.1093/nar/gki487

[54] Contreras-Moreira B, Vinuesa P. GET_HOMOLOGUES, a versatile software package for scalable and robust microbial pangenome analysis. Applied and Environmental Microbiology. 2013; **79**:7696–7701. DOI:10.1128/AEM.02411-13

[55] Sievers F, Wilm A, Dineen D, Gibson TJ, Karplus K, Li W, Higgins DG. Fast, scalable generation of high-quality protein multiple sequence alignments using Clustal Omega. Molecular Systems Biology. 2011; **7**:539. DOI:10.1038/msb.2011.75

[56] Darriba D, Taboada GL, Doallo R, Posada D. ProtTest 3: fast selection of best-fit models of protein evolution. Bioinformatics. 2011; **27**:1164–1165. DOI:10.1093/bioinformatics/btr088

[57] Guindon S, Dufayard J-F, Lefort V, Anisimova M, Hordijk W, Gascuel O. New algorithms and methods to estimate maximum-likelihood phylogenies: assessing the performance of PhyML 3.0. Systematic Biology. 2010; **59**:307–321. DOI:10.1093/sysbio/syq010

[58] Shimodaira H, Hasegawa M. Letter to the editor multiple comparisons of log-likelihoods with applications to phylogenetic inference. Molecular Biology and Evolution. 1999; **16**:1114–1116. DOI:10.1177/0148607109348061

[59] Camacho C, Coulouris G, Avagyan V, Ma N, Papadopoulos J, Bealer K, Madden TL. BLAST+: architecture and applications. BMC Bioinformatics. 2009; **10**:421. DOI:10.1186/1471-2105-10-421

[60] Varghese NJ, Mukherjee S, Ivanova N, Konstantinidis KT, Mavrommatis K, Kyrpides NC, Pati A. Microbial species delineation using whole genome sequences. Nucleic Acids Research. 2015; **43**:6761–6771. DOI: 10.1093/nar/gkv657.

[61] Kal AJ, van Zonneveld AJ, Benes V, van den Berg M, Koerkamp MG, Albermann K, et al. Dynamics of gene expression revealed by comparison of serial analysis of gene expression transcript profiles from yeast grown on two different carbon sources. Molecular Biology of the Cell. 1999; **10**:1859–1872. DOI:10.1091/mbc.10.6.1859

[62] Vera-Ponce de León A, Sanchez-Flores A, Rosenblueth M, Martínez-Romero E. Fungal community associated with *Dactylopius* (Hemiptera: Coccoidea: Dactylopiidae) and its role in uric acid metabolism. Frontiers in Microbiology. 2016; **7**:954. DOI: 10.3389/fmicb.2016.00954

[63] Chari A, Oakeson KF, Enomoto S, Jackson DG, Fisher MA, Dale C. Phenotypic characterization of *Sodalis praecaptivus sp. nov.*, a close non-insect-associated member of the

Sodalis-allied lineage of insect endosymbionts. International Journal of Systematic and Evolutionary Microbiology. 2015; **65**:1400–1405. DOI: 10.1099/ijs.0.000091.

[64] Thornalley PJ. The glyoxalase system: new developments towards functional characterization of a metabolic pathway fundamental to biological life. The Biochemical Journal. 1990; **269**:1–11. DOI: 10.1042/bj2690001

[65] Hentschel U, Hand SC, Felbeck H. The contribution of nitrate respiration to the energy budget of the symbiont-containing clam *Lucinoma aequizonata*: A calorimetric study. The Journal of Experimental Biology. 1996; **199**:427–433.

[66] Toh H, Weiss BL, Perkin SA, Yamashita A, Oshima K, Hattori M, Aksoy S. Massive genome erosion and functional adaptations provide insights into the symbiotic lifestyle of *Sodalis glossinidius* in the tsetse host. Genome Research. 2006; **16**:149–156. DOI:10.1101/gr.4106106

[67] Clayton AL, Oakeson KF, Gutin M, Pontes A, Dunn DM, von Niederhausern AC, Weiss RB, Fisher M, Dale C. A novel human-infection-derived bacterium provides insights into the evolutionary origins of mutualistic insect–bacterial symbioses. PLoS Genetics 2012; **8**:e1002990. DOI: 10.1371/journal.pgen.1002990

[68] Konstantinidis KT, Tiedje JM. Genomic insights that advance the species definition for prokaryotes. Proceedings of the National Academy of Sciences of the United States of America. 2005; **102**:2567–2572. DOI:10.1073/pnas.0409727102.

[69] Richter M, Rosselló-Móra R. Shifting the genomic gold standard for the prokaryotic species definition. Proceedings of the National Academy of Sciences of the United States of America. 2009; **106:**19126–19131. DOI: 10.1073/pnas.0906412106.

[70] Sabree ZL, Huang CY, Okusu A, Moran NA, Normark BB. The nutrient supplying capabilities of *Uzinura,* an endosymbiont of armoured scale insects. Environmental Microbiology. 2013; **15**:1988–1999. DOI: 10.1111/1462-2920.12058.

[71] Dale C, Young SA, Haydon DT, Welburn SC. The insect endosymbiont *Sodalis glossinidius* utilizes a type III secretion system for cell invasion. Proceedings of the National Academy of Sciences of the United States of America. 2001; **98**:1883–1888. DOI:10.1073/pnas.98.4.1883

[72] Jelsbak L, Thomsen LE, Wallrodt I, Jensen PR, Olsen JE. Polyamines are required for virulence in *Salmonella enterica* serovar Typhimurium. PLoS One. 2012; **7**:e36149. DOI: 10.1371/journal.pone.0036149.

[73] Zhou L, Wang J, Zhang LH. Modulation of bacterial Type III secretion system by a spermidine transporter dependent signaling pathway. PLoS One. 2007;**2**:e1291. DOI: 10.1371/journal.pone.0001291.

[74] Potrikus CJ, Breznak JA. Anaerobic degradation of uric Acid by gut bacteria of termites. Applied and Environmental Microbiology. 1980; **40(1)**:125–132.

[75] Morales-Jiménez J, Vera-Ponce de León A, García-Domínguez A, Martínez-Romero E, Zúñiga G, Hernández-Rodríguez C. Nitrogen-fixing and uricolytic bacteria associated with the gut of *Dendroctonus rhizophagus* and *Dendroctonus valens* (Curculionidae: Scolytinae). Microbial Ecology. 2013; **66**:200–210. DOI:10.1007/s00248-013-0206-3

[76] Patiño-Navarrete R, Piulachs M-D, Belles X, Moya A, Latorre A, Peretó J. The cockroach Blattella germanica obtains nitrogen from uric acid through a metabolic pathway shared with its bacterial endosymbiont. Biology Letters. 2014; **10**:7–10. DOI:10.1098/rsbl.2014.0407.

[77] Michalik A, Jankowska W, Kot M, Gołas A, Szklarzewicz T. Symbiosis in the green leafhopper, Cicadella viridis (Hemiptera, Cicadellidae). Association in statu nascendi? Arthropod Structure and Development. 2014; **43**:579–587. DOI: 10.1016/j.asd.2014.07.005.

[78] Lamelas A, Gosalbes MJ, Manzano-Marín A, Peretó J, Moya A, Latorre A. *Serratia symbiotica* from the aphid *Cinara cedri*: a missing link from facultative to obligate insect endosymbiont. PLoS Genetics. 2011; **7**:e1002357. DOI: 10.1371/journal.pgen.1002357.

[79] Lefèvre C, Charles H, Vallier A, Delobel B, Farrell B, Heddi A. Endosymbiont phylogenesis in the dryophthoridae weevils: evidence for bacterial replacement. Molecular Biology and Evolution. 2004; **21**:965–973. DOI:10.1093/molbev/msh063

Chapter 6

The Role of Mites in Bark and Ambrosia Beetle-Fungal Interactions

Sneha Vissa and Richard William Hofstetter

Additional information is available at the end of the chapter

http://dx.doi.org/10.5772/67106

Abstract

Insects share complex interactions with mites and fungi that range from obligate mutualisms to antagonistic relationships. These multitrophic interactions often result in changes to the host environment and population dynamics of the insect. Here, we review Scolytidae and Platypodidae beetles (bark beetles and ambrosia beetles, Coleoptera: Curculionidae) and their micro-organismal interactions with mites and fungi. Many bark beetles and ambrosia beetles are closely associated with mutualistic fungi used as a food source. These fungi are carried by the beetles in specialized pockets called "mycangia." In addition to beetle-specific fungi, secondary fungi are often vectored by mite populations phoretic on the beetles. These secondary introductions create a complex fungal micro-biome within the host tree of the associated Scolytid beetles, with a myriad of consequences to beetle success and tree mortality. In this chapter, we provide a detailed review of specific beetle-fungal and mite-fungal associations, mutualistic and antagonistic effects of these fungal relations, and ecological and evolutionary consequences of beetle-fungal-mite relationships within the host complex.

Keywords: *Scolytinae*, phoresy, symbiosis, *Acari*, fungi, mycangia

1. Introduction

A wide diversity of symbionts has contributed to the success of bark and ambrosia beetles (Curculionidae: Scolytinae) [1]. All ambrosia beetles and most bark beetles have mutualistic fungi that serve as a nutrition source for young and adult beetles. However, some fungal (and other) symbiotic associations may be detrimental to beetles. Other beetle symbionts include mites that are frequently found traveling on adult beetles and live under the bark of beetle-infested trees [2]. Similar to fungi, the impacts of mites on beetles vary widely. Mites too can be beneficial or detrimental to beetles. Some mite species can contribute to the fungal diversity in beetle galleries by transporting spores within and between trees [3, 4] impacting the behavior, development, and population dynamics of beetles [5, 6]. Such complex systems are evolutionary unique and can be considered principal factors in the success of Scolytid beetles.

Ambrosia beetles possess symbiotic mutualistic relationships with fungi (*Ambrosiella* and *Raffaelea* spp.) [7–9]. Ambrosia beetles excavate tunnels within the wood of host trees, termed "galleries," within which female beetles oviposit eggs. The symbiotic fungi are carried by the female ambrosia beetle in a specialized pocket, termed "mycangium." The fungal inoculate is deposited along the walls of the gallery where it grows and serves as food for larval broods and adult beetles [10, 11]. Fungi allow ambrosia beetles to exploit the wood that would have otherwise been a nutritionally poor resource, as well as exploit living trees. In California and Israel, an invasive ambrosia beetle poses a serious threat to avocado plantations via the introduction of a *Fusarium* fungus species causing dieback, wilting, and ultimately host tree mortality [12]. Similarly, adult leaf-cutter ants (Attini: *Atta* or *Acromyremex*) use leaves to foster the cultivation of fungi inside their colonies, which they in turn use as a food source [13], resulting in large-scale defoliation of nearby trees [14]. Contradictory, fungi can serve an antagonistic function in insect development or survival. *Cordyceps* is an obligate, directly transmitted fungus that requires specific insects for reproduction [15], parasitizes tropical ants (among many other insects), and results in colony collapse and mortality by taking control of the host ants behavior [16]. Other fungal pathogens can result in the direct mortality of insects. For instance, blue-staining fungi, such as *Ophiostoma minus*, can cause the mortality of southern pine beetle larval broods by out-competing the beetle mycangial fungi [5]. *Beauveria bassiana* is a common insect pathogenic fungi that kills insects in a variety of habitats and has been demonstrated as an effective mortality agent against bark and ambrosia beetles such as the almond bark beetle, *Scolytus amygdali* [17], responsible for the destruction of almond plantations.

Mites (Subphylum: Chelicerata, Class: Arachnida, Subclass: Acari) share a common, but important relationship termed "phoresy" with insects and other organisms [18]. Phoresy refers to an interspecific relationship between two species where one acts as a host and the other acts as a "phoront," attaching itself to the host for the purpose of dispersal or migration. Consequently, Farish [19] explicitly defines "phoresy" as follows: a phenomenon in which one animal actively seeks out and attaches to the outer surface of another animal for a limited time during which the attached animal (the "phoretic" animal or "phoront") ceases both feeding and ontogenesis, presumably resulting in dispersal from areas unsuited for further development, either of the individual or its progeny. This sort of commensalism is common in species that live in rapidly changing environments, and/or in species that have limited mobility [20]. Mites are small-minute organisms ranging from 50 μm to 3 mm in body length [21, 22] and thus have limited dispersal ability, other than being blown by the wind or carried by other animals or objects. They generally feed on other small arthropods or larvae and eggs of other arthropods, nematodes, fungi, bacteria, or dead organic matter [21]. Due to their small size and limited mobility, mites are likely to be phoretic on insects that are able to transport them quickly and efficiently, i.e., flying insects such as bees (Hymenoptera), beetles (Coleoptera), and flies (Diptera). For instance, mites associated with ants tend to prefer alate queens (i.e., flying queens) over workers or any other type of ant [23], presumably because they are able to travel longer distances over a short span of time and can be introduced into future ant nests. Mites are usually found attached to setae (i.e., external hairs), to grooves in the tarsi, under the wings, under the elytra (hardened exoskeleton of beetles), or attached to any part of an insect's body that provides a firm holdfast (**Figure 1**).

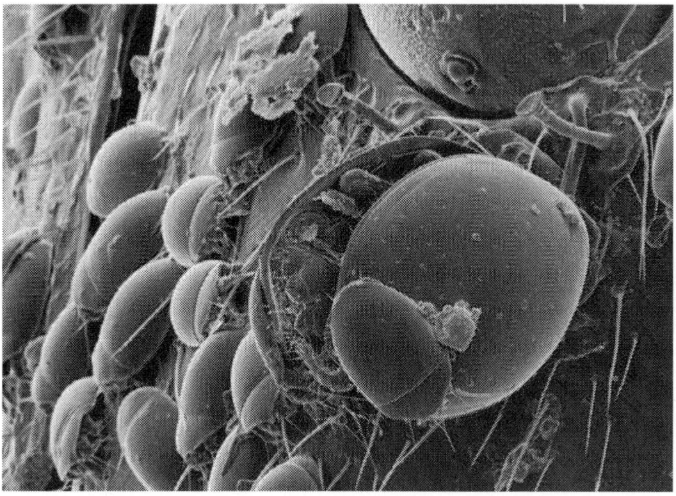

Figure 1. Scanning electron image of phoretic mites on *Dryocoetes confusus* (Coleoptera: *Scolytidae*). Picture taken at Northern Arizona University by R. Hofstetter; Nov. 2006. Image is 500× magnification.

Phoretic mites may have either advantageous, disadvantageous, or no effect on their insect host. During the act of phoresy itself, even though mites cease to feed, their mere presence may be antagonistic/parasitic as insects can be weighed down by the presence of numerous mites (e.g., southern pine beetles being weighed down by masses of Uropodine mites [24]) slowing down flight speed and increasing energy expenditure (**Figure 2**). The risk of antagonistic effects increase when mites are deposited at the insect's final destination. For example, varroa mites (*Varroa destructor*) associated with honeybees (Hymenoptera: Apidae) are known to carry the deformed wing virus (Iflaviridae); this causes new bees to emerge with crumpled dysfunctional wings rendering them useless to the colony, resulting in eventual colony collapse disorder [6]. Adult bees with varroa mites have also been shown to have significantly less protein, lipid, carbohydrate, and water content [25]. Other destructive mites are known to feed on insect larvae and eggs. For example, mites of the genus *Iponemus* are predatory on beetle eggs and larvae [26]. Other mites, such as *Tarsonemus ips* vector fungi useful for the development of southern pine beetle (Coleoptera: Curculionidae, Scolytinae) larvae (i.e., associated mycangial fungi) [22, 27]. Much like beetles, mites in turn act as vectors of microbes (fungi, bacteria, and viruses), which may present either a beneficial or a detrimental consequence for the insect and/or host plant(s). For example, the mite *Siteropsis reniformis* is associated with carrying the lint rot fungus, *Nigrospora,* which results in the rotting and death of cotton bolls in California [28]. Some *Tarsonemus* mite species associated with bark beetles (such as the southern pine beetle) carry the blue-stain fungus *O. minus,* which causes infection of phloem, host tree death, and ultimately beetle mortality (**Figures 3** and **4**) [2, 5, 29].

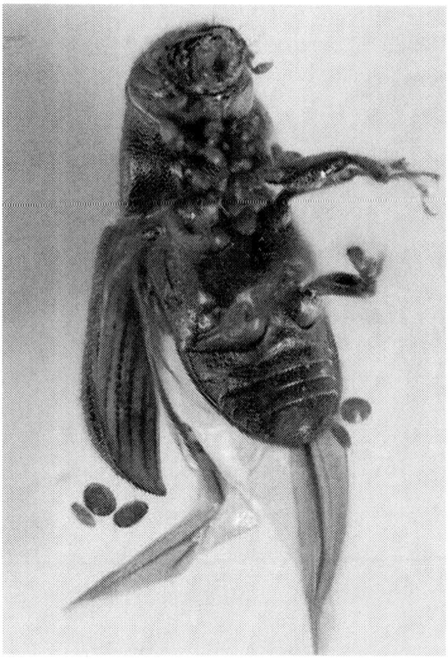

Figure 2. Phoretic *Trichuoropoda* sp. mites on the western pine beetle, *Dendroctonus brevicomis*. Picture taken by R. Hofstetter, Dec. 2016. Flagstaff, AZ, USA.

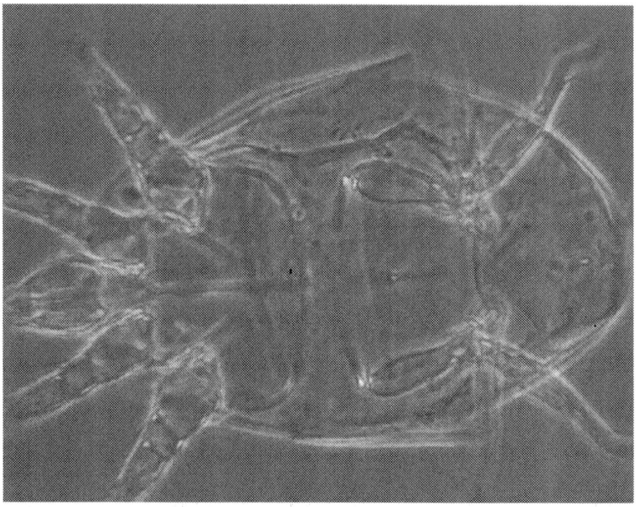

Figure 3. *Tarsonemus ips* mite with *O. minus* ascospores (dark banana shape spores). Mite was collected from the southern pine beetle, *Dendroctonus frontalis* in Alabama 2001 by R. Hofstetter.

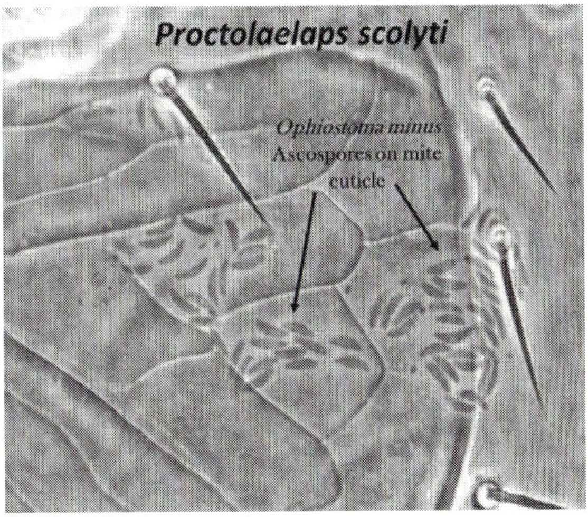

Figure 4. *Proctolaelaps scoltyi* mite with *O. minus* ascospores (dark banana shape spores). Mite was collected from the southern pine beetle, *D. frontalis* by John Moser.

2. Bark and ambrosia beetle-fungal interactions

Ambrosia beetles and bark beetles (Coleoptera: Curculionidae, Scolytinae) are known to share intimate relationships with fungi. The subfamily Scolytinae consists of bark beetles (~7000 species), ambrosia beetles (~3400 species), and other various seed and pith-feeding beetles [8, 30]. A prominent characteristic of Scolytinae is a widespread association with fungal species, particularly the bark beetles and ambrosia beetles. These fungal associations are mutualistic or commensal (i.e., benefiting the fungi via dispersal by beetles) with beetles and are primarily Ascomycetes in the genera *Ophiostoma, Fusaria, Grosmannia, Ceratocystiopsis,* and *Ceratocystis* [7, 31, 32]. Scolytinae are thought to have evolved in the late Jurassic—early Cretaceous period with these fungi, historically switching back and forth from angiosperm hosts to coniferous hosts [8]. The development of mutualisms with fungi helps explain diversification of Scolytinae to inhabit coniferous habitats and exploit a limiting resource such as living tree tissues by growing supplementary fungi within it.

2.1. Bark beetle-fungal associations

Bark beetles can be distinguished from ambrosia beetles and other Scolytinae by the fact that they exclusively invade the bark and phloem of trees, and not the woody tissue. Fungi associated with bark beetles are carried/introduced to the host in one of two ways—mycangially or nonmycangially—which indicate the specificity of the relationship between the fungus and the beetle. Some bark beetles such as the southern pine beetle (*Dendroctonus frontalis*) and the mountain pine beetle (*Dendroctonus ponderosae*) possess specialized structures called "mycangia" or "mycetangia," meaning "fungal vessel." Mycangia are strictly defined as specialized invagination of the integument lined with secretory glands used for the transport of fungi [33]. However, the term "mycangia" is often loosely used to describe invaginations that carry fungal spores [7]. These mycangial fungi (grown as a yeast-like structure or spores within the mycangia) are introduced by the beetle into the tree phloem where it eventually provides food for the developing offspring. The growth of fungi within the mycangia is supported by the glandular secretions of the mycangia [34]. Nine bark beetle species are known to have well-defined mycangia [32] (**Table 1**). There are several different kinds of mycangial fungal species: bionectrotrophic blue-staining ascomycetes, saprophytic ascomycetes, or saprophytic basidiomycetes [35]. For example, the southern pine beetle is associated with two types of mycangial fungi: *Ceratocystiosis ranaculosus* (ascomycete) and *Entomocorticium* sp. A (basidiomycete) [4, 36], and the mountain pine beetle is associated with three Ophiostomatoid mycangial fungi—*Ophiostoma montium* [37], *Grosmania clavigera,* and *Leptographium longiclavatum* [38, 39] but are also known to harbor *Ceratocystiopsis* and *Entomocorticium* [40]. However, only one fungal species exists within an individual beetle's mycangium and is inherited from the original parental pair [2, 41]. Other bark beetles such as (most) *Ips* spp. transport fungal spores without using a mycangia; i.e., spores are present in grooves or pits found on the beetle's exoskeleton. For example, *Ips typographus*, the spruce beetle, which is responsible for extensive

Bark Beetle	Tribe	Principal plant hosts	Mycangial type	Ascomycete associates	Basidiomycete associates
Dendroctonus frontalis	Tomicini	*Pinus* spp.	Prothoracic, glandular	*Ceratocysitis ranacuosus*	*Entomocorticium* sp. A
D. brevicomis	Tomicini	*P. ponderosae, P. coulteri*	Prothoracic, glandular	*C. brevicomis*	*Entomocorticium sp. B*
D. approximatus	Tomicini	*Pinus* spp.	Prothoracic, glandular	Unknown	*Phlebiopsis gigantea*
D. adjunctus	Tomicini	*Pinus* spp.	Prothoracic, glandular	*Leptographium pyrinum*	Unknown
D. ponderosae	Tomicini	*Pinus* spp.	Maxillary	*Ophiostoma clavigerum, O. Montium*	*Entomocorticium dendroctoni, E.* sp. D,E,F,G,H,*P. gigantea*
D. jeffreyi	Tomicini	*P. jeffreyi*	Maxillary	*O. clavigerum*	*Entomocorticium* sp. E
Tomicus minor	Tomicini	*Pinus* spp.	Unknown	*O. canum, Ambrosiella tingens*	Unknown
Ips acuminatus	Ipini	*Pinus* spp.	Mandibular	*O. clavatum, A. macrospora*	Unknown
I. avulsus	Ipini	*Pinus* spp.	Unknown	*O. ips*	*Entomocorticium* sp. 1
Pityoborus comatus	Corthylini	*Pinus* spp.	Prothoracic, pubescent	*Unknown*	*Entomocorticium* sp. C.

Table 1. Information adapted from Harrington, 2005 (Insect-fungal associations: ecology and evolution, p. 264)—Nine species of bark beetles associated with well-defined mycangial, and their associated fungi.

tree damage in Europe and in Asia [42], is associated with fungal species of *Ceratocystiopsis*, *Ceratocystis*, and *Grosmania* as well as *Pyxidiophora* sp. carried externally in grooves on the surface of the exoskeleton [43]. *Ips perterbatus* (the northern spruce engraver) is associated with 13 different Ophiostomatoid fungi [44] and thus also acts as a potential vector of tree-killing Ophiostomatoid fungal pathogens. *Ips sexdentatus* (a highly prevalent species in Europe) is associated with carrying a plethora of ascospores, some among the *Ophiostoma* genus [45]. Consequently, beetles that are mycangially associated with fungi seem to possess a stronger fidelity to their allied fungi, whereas those that are not mycangially associated are less specific as they are merely vectors [8]. For example, western pine beetle (*Dendroctonus brevicomis*) is known to exhibit strong fidelity to one of two inherited mycangial fungal symbionts, *Ceratocystiopsis brevicomi* and/or *Entomocorticum* sp. B, which is suggestive of coevolution of the beetle with their associated fungi [35, 41]. Other fungi associated with the western pine beetle, such as *O. minus* or *O. ips*, are not carried in the mycangia and are antagonistic to the beetle.

2.2. Ambrosia beetle-fungal associations

Ambrosia beetles (Family: Curculionidae; subfamily: Scolytinae) occur worldwide and share close novel symbioses with fungi, such as *Ambrosiella* spp. and *Raffaela* spp. [7–9, 46], which rely on arthropods for efficient dispersal. Beetle-fungal symbiosis in ambrosia beetles is one of the earliest known associations of insect-fungal relationships, and this specific association is a primary reason for their success [8, 30]. Ambrosia beetles are "xylomacetophagous" (feed on fungi grown in the xylem), and their activity is easily recognized by the distinct staining of wood by the inoculation of ectosymbiotic fungi, used as the sole nutrient source by both adults and larvae [10, 11, 47, 48]. Like bark beetles, ambrosia beetles transport these fungi via specialized mycangia [49], the location of which differs between beetle species [50]. Typically (as seen in European ambrosia beetles), females tend to have more well-developed mycangia [49], i.e., large and lined with glandular dermal cells containing a waxy substance [51] than males, and are responsible for depositing the fungi along the tunnel walls [10, 50]. The waxy secretions of the female mycangium contribute to the growth of ambrosia cells [52]. Ambrosia beetles may or may not exhibit specificity to ambrosia fungi. Batra [10] suggests that ambrosia beetles rely on a primary (mycangial) and auxiliary ambrosia fungi depending on the life stage and do not exhibit mycangial specificity to ambrosia fungi. However, further study shows ambrosia beetles in the genus *Xylosandrus* (which are exotic in the United States [46]) exhibit high specificity to a given ambrosia fungus in the genus *Ambrosiella* [53]. This type of fungal specificity highlights a unique obligate symbiosis between ambrosia beetles and ambrosia fungi and suggests coevolution of the beetle with its fungal associate [8].

Ambrosia beetles are not typically associated with significantly damaging tree diseases in the way bark beetles are, as they often attack trees that are already dead or severely stressed [41, 51, 54, 55]. Coevolution theory suggests that the ambrosia beetle-fungal association is adaptable to a given host environment. However, problems arise when atypical beetle-host relationships form, i.e., exotic ambrosia beetles attack foreign hosts [50, 51, 54]. A recent concern in North America is an increase in exotic species of ambrosia beetles attacking living trees with. In most cases, these beetles are inoculating trees with non-native ambrosia fungi, for which native North American trees have little resistance [46]. For example, the redbay ambrosia beetle (*Xyleborinus glabratus*) is an exotic beetle from southeast Asia, responsible for extensive mortality of redbay trees (*Persea borbonia*) [46, 55, 56]. These beetles are closely associated with a *Raffaela lauricola*, an asexually reproducing fungus, which grows within the xylem (wood) and causes affected branches to wilt, exhibiting a symptomatic black staining of the sapwood [55, 56]. In another example, an *Euwallacea sp.* ambrosia beetle is causing damage to more than twenty tree species in California, with particular concern to avocado [12]. The damage is caused by the spread of a novel *Fusarium* fungus causing tree wilting and dieback [12].

The full ecological potential of ambrosia beetle-fungal relationships for exotic invasives is still unknown [55]. There is little available detail on phoretic mite populations of ambrosia beetles as there is for bark beetles. Therefore, we are unaware of specific mite-fungal interactions for a given species of ambrosia beetle. There is a potential for similar complex mite-fungal interactions in ambrosia beetles as there are in bark beetles, particularly in situations with multiple beetle species attacking one host tree(s).

2.3. Factors affecting beetle-fungal associations

Both mutualistic and antagonistic fungal associations are context-dependent and highly affected on a combination of biotic and abiotic factors such as temperature, moisture, host tree defenses, presence of other microbes (i.e., bacteria, yeast, etc.), and other arthropods. Depending upon the beetle complex in question, such context-dependent conditions suggest that some fungi benefit from multipartite relationships in bark beetles more than others, and that the beetle's ecological niche is determined by the external factors that govern it [57, 58].

Host tree type and conditions (moisture, phytochemical composition/concentrations, etc.) have a significant influence on bark beetle-fungal associations. For instance, changes in the chemistry and nutritional content of the host tree have the ability to alter the distribution and relative prevalence of fungal associates within the tree [8, 59–62]. A conifer tree's primary defenses are resins and other induced secondary chemical compounds, which vary between tree species [57, 60, 63, 64]. In their study of the effect of tree defenses against fungi associated with the southern pine beetle, Hofstetter et al. [65] found that in general the phytochemistry of *Pinus* tree species negatively impacted the growth-rate of *O. minus* fungi vectored by beetles (and their accompanying mites) but (bar a few exceptions) did not hinder growth of the mycangial fungi. This indicates that the fungal microbiome of some bark beetles is dependent upon the geographically dominant host tree species and can accordingly benefit or pose a threat to beetle population dynamics [57]. Other host tree factors like moisture content can determine the success of a fungal inoculant as well (i.e., a moisture-rich tree typically serves as a better fungal host than a dry tree [33]). Ayres et al. also found that presence of basic elements such as high nitrogen concentration is dependent on the presence or absence of mycangial fungal associations [66].

Variations in temperature are one of the most vital factors affecting the relative abundance and diversity of fungal species, in turn affecting the ecological associations of the fungus' host bark beetle [3, 67–69]. The mountain pine beetle, *D. ponderosae,* for example, is associated with *Grosmannia clavigera* and *O. montium* mycangial fungi [37, 38]. The two fungi possess differential optimal temperatures with *G. clavigera* ceasing to grow past 25°C and *O. montium* dominating between 25 and 32°C [67]. Thus, the fungal association of *D. ponderosae* bark beetles is predictable over changes in season; i.e., the beetle will vector *G. clavigera* in cooler climates and *O. montium* in hotter climates. This in turn allows the beetle to occupy a wide geographical range (from southwestern North America into Canada) due to its ability to exploit not one, but two symbiotic resources that span a wide range of host environments. Hofstetter et al. [3] found that the growth rate of the symbiotic fungi (*O.minus, C. ranaculosus,* and *Entomocorticium* sp. A) of the southern pine beetle, *D. frontalis,* differs within a range of temperatures [3]. While the growth rate of *E.* sp. A. increased steeply between 8 and 22°C before reaching a stable constant, the growth rates of *O. minus* and *C. ranaculosus* were much less up to 22°C and increased in growth rate past 22°C. The growth of the two mycangial fungi associated with the western pine beetle, *D. brevicomis* (*C. brevicomi* and *Entomocorticum* sp. B) is also similarly affected by temperature and can vary between beetle populations [35]. *C. brevicomi* tends to thrive better than *E* sp. B. at cooler and hotter temperatures, but fair similarly in intermediate temperatures. This differential growth may thus potentially reduce competition between the two fungal symbionts and vary seasonally with the beetle's movement [69]. Such differential growth can also determine which associative symbionts are lost with climate change [35].

For bark beetles associated with multiple mycangial symbionts, the prevalence of a specific mycangial fungal associate depends upon the beetle's environmental temperature and consequently determines the beetle's ability to survive in a given geographic location. Often, the presence of multiple beetle-associated fungi results in competition between the different fungal species with one species dominating the other. Changes in external factors (such as host tree chemistry and temperature) can determine the competitive advantage of a resource between multiple fungal symbionts, and also determine the success of antagonistic tree-killing pathogens such as species of blue-stain (*O. minus, O. picae, O. penicillatum,* etc.). As blue-stain fungi thrive at higher temperature ranges (recall, *O. minus* growth rate increases past 22°C), the prevalence of blue-stain will depend on climate change and its consequential temperature conditions [70]). A thorough understanding of biotic and abiotic factors affecting pathogenic and mycangial fungi can thus allow us to predict which species of fungi will potentially establish itself within a major bark beetle's niche, and what effects it may have on host tree conditions and mortality. Thus, differential responses of mutualistic fungi to temperature could reduce competition between fungal symbionts by allowing each to dominate at different times as temperatures vary seasonally, annually, or geographically [69]. These differences are also likely to determine the future symbiotic complex of beetle populations as warming increases. However, mites have the potential to disrupt, alter, or magnify beetle-fungal symbiotic complexes.

3. Mite-fungal-beetle interactions

Mites are a prominent phoront of bark beetles and some ambrosia beetles [2, 58]. Mites (Arachnida: Acari) are subdivided into two main super orders: Parasitiformes and Acariformes [71]. The Parasitiformes are further divided into four orders—Ixodidae, Holothyrida, Opilioacaridae, and Mesostigmata, and the Acariformes are divided into two orders—Sarcoptiformes and Trombidiformes [22]. Mites phoretic on bark beetles are predominantly species of the order Mesostigmata and Trombidiformes [62, 72, 73] and range from 80 μm (e.g., *Iponemus* sp.) to 5 mm (*Mexecheles* sp.) in body length [22]. Mites are very efficient phoronts, often having specialized structures, such as suckers and tenet hairs to cling on to the host [20]. These organisms also have specialized adaptive stages/morphs for phoresy; e.g., many species (e.g., Acari: Astigmata) exhibit phoresy in the deutonymph (juvenile second-stage nymph) stage, and only one life stage is specialized to be phoretic for a given species of mite [21, 22]. Many mites phoretic on bark beetles predominantly tend to be female [26], i.e., mites exhibit sexual dimorphism where one (the female) is termed the "phoretomorph" (coined by Moser & Cross, 1975) and the other (the male) is "normal." This phoretomorph dimorphism is seen in Pyemotidae mite species (Acarina: Tarsonemoidea) [74] and (likely) provides the opportunity for females to develop in a suitable environment for egg laying. Mites can be found attached under the elytra, at the base of the elytra, within grooves in the tarsi, on external setae, under the wings, etc. However, each mite species is specific in the location on which they attach themselves to the beetle [75], depending upon what specialized clasping adaptations it may have (based on species). Like its host insect, mites also share complex interactions with fungi; however, the details of most mite-fungal interactions are relatively unknown and understudied [58]. Mycetophagous mites (mites which feed on fungi) such as *Tarsonemus krantzi* (specifically feed on the blue-staining *O. minus* Ophiostomatid fungus) and *Histiogaster* spp. (generalist fungal feeders) carry spores on their bodies to deposit in their designated host environment (usually a medium such as the inner tissues of vascular plants/trees from which the fungi may draw nutrition). *Tarsonemus* mites possess specialized structures known as sporothecae in which they carry the fungal spores (see **Figure 3**) [62, 76]. The sporothecae are similar to the beetle's mycangia except for the presence of gland cells. The fungi also do not multiply within the sporothecae but do so once it has been deposited in a suitable host environment [62].

The fungal symbionts of bark beetles often provide a food source for the mites, and the fungal associations of the mites provide a food source for the beetle. *Tarsonemus* mites associated with the southern pine beetle carry *C. ranaculosus* spores, which are mycangially associated with the beetle [77]. This enhances the growth of *C. ranaculosus* as a food source for both the beetle and the mite. However, between the southern pine beetle's two mycangial fungi, *C. ranaculosus* and *Entomocorticum* sp. A., the latter is a more beneficial food source for the beetle. And the presence of *C. ranaculosus* outcompeting *Entomocorticum* sp. as the dominant fungus [77], resulting in a minor reduction of relative fitness for the beetle, but serves as a compromise for both organisms. Inadvertently, not all mite-beetle interactions result in such harmonious coexistence between the two organisms. In many cases, the presence of mites results in decreased fitness, changes in reproductive success, and even beetle mortality.

3.1. Role of mites in altering population dynamics of bark beetles

We know little of the potential impact of mites on beetle population dynamics. A few studies indicate that mites may influence factors affecting beetle life history such as reproductive success, development, and mortality, for example, *Iponemus confuses* L. parasitizes bark beetle eggs resulting in mortality before the occurrence of the larval stage [26]. *Resinosa* mite spp. cause mortality in *Dendroctonus pseudotsugae* eggs and early instars of larvae [36]. *Cercoleipus coelonotus* and *Dendrolaelaps quadrisetus* have been known to feed on the eggs and larvae of *I. typographus* (pine engraver) bark beetles. As mentioned earlier, *Tarsonemus* mites can cause beetle mortality by introducing blue-stain fungi that outcompete the beetle's mutualistic mycangial fungi, causing host tree and beetle mortality [2, 5, 29, 77]. Similarly, *Cercoleipus* mites are known to carry strains of *Ophiostoma* (specifically *Ophiostoma bruneo-cilium* and *O. montium*) that cause damage to the pine hosts of carrier bark beetles [78]. Conversely, Pfammatter and Raffa assessed the effects of phoretic mites such as *Dendrolaelaps* spp., *Iponemus* spp., *Histiogaster* spp., *Tarsonemus* spp., etc., on the reproductive success of *Ips grandicollis* bark beetles (a nonmycangial beetle) and noticed a weak positive relationship between the abundance of mites and emergence of *I. grandicollis*. This suggests that there might also be a potential mutualistic relationship between the abundance of mites and the reproductive performance of beetles. This may be due to mutual suitability of the host substrate [36], but it is still uncertain whether mites truly encourage beetle performance. Most of our understanding of mite-beetle-fungal interactions come from the widely studied southern pine beetle. *Tarsonemus* mites carrying *O. minus* spores within sporothecae deposit the fungi in the host environment of the beetle; where at first, the beetles are able to coexist and feed on this fungus; however, over time the fungus outcompetes the beetles and invades the phloem of the tree killing the beetle brood and the tree (**Figure 5**). The presence or absence of a particular fungal species or strain can determine the success of development emergence of beetle progeny in mountain pine beetles. Six (1998) found that mountain pine beetles associated with *O. montium* mycangial fungi had a higher production of progeny adults with earlier emergence than those associated with *O. clavigerum* fungi [79]. Adult southern pine beetle individuals that developed with the presence of mycangial fungi were also found to be larger than individuals that developed in the absence of mycangial fungi [66, 80, 81].

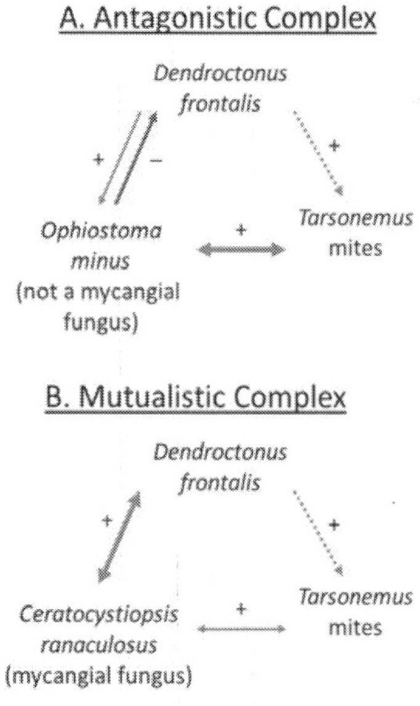

Figure 5. Models of interactions between the southern pine beetle, *D. frontalis*, phoretic mites of the genus *Tarsonemus*, and fungi. The fungal interaction includes either (a) the blue-stain fungus *O. minus*, which is not found associated with the bark beetle mycangium and is an antagonist of *D. frontalis*, or (b) the mycangial fungus *Ceratocystiopsis ranaculosus*, which is a mutualist of both *D. frontalis* beetles and *Tarsonemus* mites. Model A represents an indirect negative-feedback complex, and Model B represents an indirect positive-feedback complex.

The effect of mite-vectored fungi varies between mycangial and nonmycangial bark beetles. The presence of symbiont specific mycangia (e.g., *D. frontalis*, *D. ponderosae*, and *D. brevicomis*) indicates specificity for a fungal resource and a lack of adaptability to foreign fungal competition, whereas nonmycangial beetles (such as *Ips* spp.) are simply carriers and may possess greater adaptability to a variety of potential fungal introductions. While there is no literature supporting this hypothesis, it is a topic for future study to help us understand the role of mites in beetle life history.

3.2. Role of mites in spreading fungal pathogens

Bark and ambrosia beetles have been pinpointed as one of the primary vectors of tree disease-causing pathogens such as *Fusarium* dieback and Dutch elm disease. In some cases, tree pathogenic fungi have been isolated from the external surface of beetles, i.e., they are not mycangial fungi. Mite-beetle interactions suggest that it is more likely that the guilty party is the mites vectored by the beetle than the beetle itself (e.g., **Figure 6**) [82, 83].

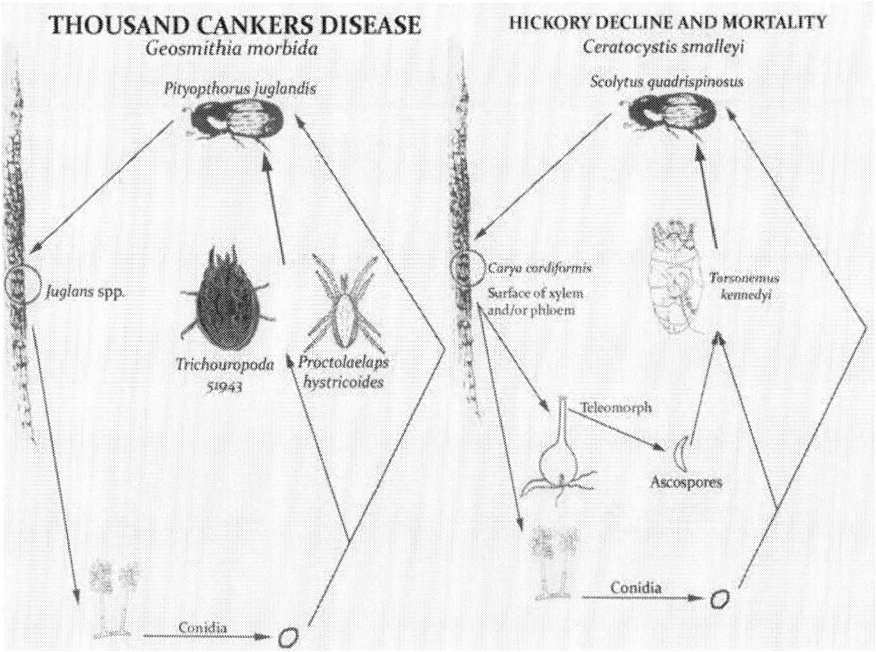

Figure 6. Hypothetical beetle-fungal-mite complex resulting in host tree disease or decline as seen in the Thousand Cankers Disease and Hickory Decline and Mortality. Created by J. Moser 2012.

Bark beetles are associated with an array of mites. The southern pine beetle alone is associated with 96 different species of mites [73]. The mountain pine beetle is associated with 57 different phoretic mite species [83], and the spruce beetle (*Dendroctonus rufipennis*) is associated with up to 10 different species of mites [84]. Other bark beetles such as *Ips* beetles (pine engravers) are also associated with mite species, though the diversity is less than that of *Dendroctonus* species. *Ips grandicollis*, a bark beetle prevalent in the Great Lakes area, was found to be associated with nine different phoretic mite species [36] and *Ips pini* is associated with up to five different species of phoretic mites [85]. *I. sexdentatus* have been associated with nine different mite species, which in turn were associated with ~16 morphologically distinct fungi [45]. Not all of these bark-beetle-associated mites are carriers of fungal pathogens. However, there are specific genera of mites that are associated with the spread of fungal pathogens and whose occurrence overlaps between the different species of bark beetles.

Dutch elm disease, a vascular wilt disease that affects elm trees, is caused by the introduction of the Ascomycete species (such as *Ophiostoma novo-ulmi* and *Ceratocystis ulmi* (Buis)) (Kinn, 1970). The spread of this deadly disease in Europe has been associated with elm bark beetles in the genus *Scolytus* [72]—*Scolytus multistriatus*, *Scolytus pygmaeus*, and *Scolytus scolytus*. However, like many other bark beetles, these elm bark beetles are associated with phoretic mite species such as *Elattoma fraxini*, *Proctolaelaps scolytii*, *Pseudotarsonemoides*, and *Tarsonemus crassus* [86]. Consequently, Moser et al. (2009) found that *P. scolytii* and *T. crassus* carried *O. novo-ulmi* spores, with *T. crassus* carrying the most number of spores, indicating that these beetles are only carriers of this disease because of its introduction by phoretic mites—the true carriers of the pathogen. Such interactions could occur in other beetle-fungal-mite complexes as well (**Figure 6**). Likewise, studies have shown the *O. minus*-associated southern pine beetle and the mountain pine beetle in North America [39, 87] are associated with a *Tarsonemus* mite sp. similar to the one found on elm bark beetles, *T. krantzi*, which carry the ascospores of *O. minus* [83, 88].

4. Conclusion

Complex interactions, particularly mutualisms like that of bark and ambrosia beetles and mycangial fungi have vast implications on the ability to exploit marginal resources and determining habitat range, which in turn affect the host tree environment, spread of fungal pathogens, changes in fungal community structure, etc. Mite-beetle-fungal interactions are likely to alter these ecological implications and effects. Phoretic mites (e.g., *Cercoleipus*, *Dendroctonus*, and *Iponemus* spp.) may reduce bark beetle success/fitness via antagonistic ecological interactions such as feeding on beetle eggs and beetle larvae [72], or feeding on other mutualist mites beneficial for the beetle. Many other mites serve as vectors of pathogenic fungi, such as *Histiostoma ovalis* (introduces *Ophiostoma bruneo-ciliatum* and *O. montium* (blue-stain fungi)) [78]. The introduction of antagonistic fungi by mites may ameliorate the effect of beetle-fungus mutualisms by outcompeting them. This has been documented in southern pine beetle populations in which associated mites carry antagonistic fungi in sporothecae and outcompete the beetle's mutualistic mycangial fungi [89]. Scolytinae have evolutionarily jumped back and forth from angiosperm to coniferous host types. The development of symbiotic associations with fungi has (potentially) assisted in the transition of using a tough resource such as pine bark [8, 32]. The evolutionary pressure exerted by a change in fungal environment introduced by mites (over time) may lead to yet another change in the fungal symbiont of emerging beetles. Mountain pine beetle (*D. ponderosae*), for example, are known to be associated with only one of three mycangial symbionts passed on from parent to offspring. Increased fungal competition introduced by mites in mountain pine beetle populations could cause an adaptive shift for the most successful fungal species or strain [39]. Therefore, utilizing microorganismal interactions is a promising area for future developments of biocontrol [62], particularly natural fungal inoculation via mite phoresy.

Scolytid beetles, mites, and fungi share a unique tripartite relationship that has the potential to affect entire ecosystems. Bark beetle-fungal relations are a primary cause for pine tree mortality, and mite-induced fungal complexity may potentially alter the effects/progress of such pathogenic fungi by either enhancing or diminishing them. For example, mites associated with the southern pine beetle cause blue-stain by vectoring *C. minor* (later renamed as *O.minus*). However, reference [81] indicates that the detrimental effect of *C. minor* naturally occurring within the tree is inhibited in the phloem where it coexists with the beetle and other microorganisms and is amplified when the fungus acts alone. It is very likely (though not experimentally clarified) that this is due to competitive diversity of the fungal microbiome being enhanced by spores that are introduced by the beetle and its associated mite population. Consequently, the beetle-associated timber loss leads to the need for control/management measures. Fungal associates and fungal biocontrol methods have been explored, such as the potential use of *B. bassiana* as a natural insecticide [79, 90]. However, the role of mites has generally been overlooked in these studies and thus the full ecological potential of biologically controlling beetle outbreaks remains unknown. If mites have the potential to alter beetle-fungal associations, factors such as optimal temperature range and geographic range of specific beetle species may also change. Some fungi are adaptive to higher temperatures and some to lower thus determining the geographic expansion of the beetle associated with it. In the past decade, the mountain pine beetle has expanded in range across North America, making its way into the northern forests of British Columbia [91], resulting in tree death. Climate change is the foremost theory for this expansion, as changes in thermal regime has a direct effect on factors such as host tree vigor, resource availability, etc. [91, 92]. However, it is important for managers to understand the consequences of climate and temperature change on bark beetle resource dependence. Mycangial (or other resource-dependent) association with fungi adaptable to colder temperatures will extend beetle populations up north into regions of colder climate and vice versa. The predominance of such a fungus may be influenced by its abundant dispersal by phoretic mites that favor it as a nutritional resource.

While bark and ambrosia beetle populations are typically monitored, their mite and fungal associations are not. We believe that geographic expansion, reproductive fitness, and other factors of beetle population dynamics rely on a thorough understanding of the mite and fungal diversity associated with beetles. The mite-beetle-fungal tripartite relationship is a relatively new realm of study in the field of multipartite symbioses, with a vast scope for new discoveries that expand or knowledge and understanding of ambrosia and bark beetle ecology.

Acknowledgements

We would like to acknowledge and thank Derek Uhey, PhD Student, Northern Arizona University, Flagstaff AZ 86011, USA, for his assistance with initial editing and formatting of the information presented.

Author details

Sneha Vissa and Richard William Hofstetter*

*Address all correspondence to: rich.hofstetter@nau.edu

School of Forestry, Northern Arizona University, Flagstaff, AZ, USA

References

[1] K.F. Raffa, J. Gregoire and B.S. Lindgren. 2015. Natural history and ecology of bark beetles. *In* Bark Beetles: Biology and Ecology of Invasive and Native Species. F.E. Vega and R.W. Hofstetter (eds). Academic Press, Elsevier, Oxford, U.K. pp. 1–28.

[2] R.W. Hofstetter, J. Dinkins-Bookwalter, K.D. Klepzig and T.S. Davis. 2015. Symbiotic associates of bark beetles. *In* Bark Beetles: Biology and Ecology of Native and Invasive Species. F.E. Vega and R.W. Hofstetter (eds). Academic Press, Elsevier Publishing. pp. 209–246.

[3] R.W. Hofstetter, T.D. Dempsey, K.D. Klepzig and M.P. Ayres. 2007. Temperature-dependent effects on mutualistic, antagonistic, and commensalistic interactions among insects, fungi and mites. *Community Ecology*, 8(1): 47–56.

[4] R.W. Hofstetter and J.C. Moser. 2014. Role of mites in insect-fungus associations. *Annual Review of Entomology*, 59(1): 537–557.

[5] R.W. Hofstetter, J.T. Cronin, K.D. Klepzig, J.C. Moser and M.P. Ayres. 2006. Antagonisms, mutualisms and commensalisms affect outbreak dynamics of the southern pine beetle. *Oecologia*, 147(4): 679–691.

[6] L. Wilfert, G. Long, H.C. Leggett, P. Schmid-Hempel, R. Butlin, S.J.M. Martin and M. Boots. 2016. Deformed wing virus is a recent global epidemic in honeybees driven by *Varroa* mites. *Science*, 351: 594–597.

[7] H.S. Whitney. 1982. Relationships between bark beetles and symbiotic organisms. *In* Bark Beetles in North American Conifers: A System for the Study of Evolutionary Biology. J.B. Mitton and K.B. Sturgeon (eds). University of Texas Press, Austin, TX. pp. 183–211.

[8] D.L. Six. 2012. Ecological and evolutionary determinants of bark beetle—Fungus symbioses. *Insects*, 3: 339–366.

[9] R.A. Beaver, N. Wilding, N. Collins, P. Hammond and J. Webber. 1989. Insect-fungus relationships in the bark and ambrosia beetles. *Insect-Fungus Interactions*, 121–143.

[10] L.R. Batra. 1966. Ambrosia fungi: extent of specificity to ambrosia beetles. *Science (New York, N.Y.)*, 153: 193–195.

[11] L. Batra. 1903. Ecology of ambrosia fungi and their dissemination. *Transactions of the Kansas Academy of Science*, 66: 213–236.

[12] S. Freeman, M. Sharon, M. Maymon, Z. Mendel, A. Protasov, T. Aoki, *et al.* 2013. *Fusarium euwallaceae* sp. nov—a symbiotic fungus of *Euwallacea* sp., an invasive ambrosia beetle in Israel and California. *Mycologia*, 105: 1595–1606.

[13] E.O. Wilson. 1986. The defining traits of fire ants and leaf-cutting ants. *In* Fire Ants and Leaf-Cutting Ants: Biology and Management. C.S. Lofgren and R.K. Van der Meer (eds). Westview Press, Boulder, CO. pp. 1–9.

[14] R. Wirth, W. Beyschlag, R. Ryel, H. Herz and B. Holldobler. 2003. The Herbivory of Leaf-Cutting Ants. A Case Study on *Atta Columbia* in the Tropical Rainforest of Panama. Spring Verlag, Berlin. 230 pp.

[15] D.P. Hughes, S.B. Andersen, N.L. Hywel-Jones, J.B. Winanda Himaman and J.J. Boomsma. 2011. Behavioral mechanisms and morphological symptoms of zombie ants dying from fungal infection. *BMC Ecology*, 11: 13.

[16] H.C. Evans and R.A. Samson. 1982. *Cordyceps* species and their anamorphs pathogenic on ants (Formicidae) in tropical forest ecosystems I. The Cephalotes (Myrmicinae) complex. *Transactions of the British Mycological Society*, 79: 431–453.

[17] Y.A. Batta. 2007. Biocontrol of almond bark beetle (*Scolytus amygdali* Geurin-Meneville, Coleoptera: Scolytidae) using *Beauveria bassiana* (Bals.) Vuill. (Deuteromycotina: *Hyphomycetes*). *Journal of Applied Microbiology*, 103: 1406–1414.

[18] R.A. Norton. 1980. Observations on phoresy by Oribatid mites (Acari: Oribatei). *International Journal of Acarology*, 6(2): 121–130.

[19] D.J. Farish. 1965. Some aspects of the sensory and phoretic behavior of macrocheles muscaedomesticae (Scopoli, 1772)(Acarina: Macrochelidae) with a consideration of phoresy as a phenomenon. *Biological Reviews*, 57(4).

[20] D.J. Farish and R.C. Axtell. 1971. Phoresy redefined and examined in *Macrocheles muscaedomesticae* (Acarina: Macrochelidae). *Acarologia*, 13(1): 16–29.

[21] D.E. Walters and H.C. Proctor. 1999. Mites: Ecology, Evolution and Behavior. CABI, Oxon, UK.

[22] R.W. Hofstetter. 2011. Mutualists and phoronts of the southern pine beetle. *In* Southern Pine Beetle II. K.D. Klepzig and R. Coulson (eds). United States Dept. of Agriculture Forest Service, Southern Research Station General Technical Report SRS-140, Asheville, NC, USA. pp. 161–181.

[23] K.U. Campbell, H. Klompen and T.O. Crist. 2013. The diversity and host specificity of mites associated with ants: the roles of ecological and life history traits of ant hosts. *Insectes Sociaux*, 60(1): 31–41.

[24] D.N. Kinn and J.J. Witcosky. 1978. Variation in southern pine beetle attack height associated with phoretic uropodid mites. *The Canadian Entomologist*, 110: 249–251.

[25] P.L. Bowen-Walker and A. Gunn. 2001. The effect of the ectoparasitic mite, *Varroa destructor* on adult worker honeybee (*Apis mellifera*) emergence weights, water, protein, carbohydrate, and lipid levels. *Entomologia Experimentalis et Applicata*, 101: 207–217.

[26] E. Lindquist. 1969. New species of *Tarsonemus* (Acarina:Tarsonemidae) associated with bark beetles. *Canadian Entomologist*, 101: 1291–1314.

[27] K.D. Klepzig, J.C. Moser, F.J. Lombardero, M.P. Ayres, R.W. Hofstetter and C.J. Walkinshaw. 2001. Mutualism and antagonism: ecological interactions among bark beetles, mites and fungi. *In* Biotic Interactions in Plant-Pathogen Associations. M.J. Jeger and N.J. Spence (eds). CABI Publishing, Wallingford, UK. pp. 237–267.

[28] F.F. Laemmlen and D.H. Hall. 1972. Interdependence of a mite, *Siteroptes reniformis*, and a fungus, *Nigrospora oryzae*, in the Nigrospora Lint Rot of Cotton. *Phytopathology*, 63: 308–315.

[29] M.J. Lombardero, M.P. Ayres, R.W. Hofstetter, J.C. Moser and K.D. Klepzig. 2003. Strong indirect interactions of *Tarsonemus* mites (Acarine: Tarsonemidae) and *Dendroctonus frontalis* (Coleoptera: *Scolytidae*). *Oikos*, 102: 243–252.

[30] B.D. Farrell, A.S. Sequeira, B.C. O'Meara, B.B. Normark, J.H. Chung and B.H. Jordal. 2001. The evolution of agriculture in beetles (Curculionidae: *Scolytinae* and *Platypodinae*). *Evolution*, 55: 2011–2027.

[31] D.L. Six. 2003. The bark beetle holobiont: why microbes matter. *Journal of Chemical Ecology*, 39: 989–1002.

[32] T.C. Harrington. 2005. Ecology and evolution of mycophagous bark beetles and their fungal partners. *In* Insect-Fungal Associations: Ecology and Evolution. F.E. Vega and M. Blackwell (eds). Oxford University Press, New York. p. 264.

[33] K.D. Klepzig and D.L. Six. 2004. Bark beetle-fungal symbiosis: context dependency in complex associations. *Symbiosis*, 37: 189–205.

[34] C. Yuceer, C. Hsu, N. Erbilgin and K.D. Klepzig. 2011. Ultrastructure of the mycangium of the southern pine beetle, *Dendroctonus frontalis* (Coleoptera: Curculionidae, Scolytinae): complex morphology for complex interactions. *Acta Zoologica*, 224: 216–224.

[35] J.C. Dysthe, R. Bracewell, D.L. Six and A.E. Hajek. 2015. Temperature effects on growth of fungal symbionts of the western pine beetle, *Dendroctonus brevicomis*. *Fungal Ecology*, 17: 62–68.

[36] J.A. Pfammatter and K.F. Raffa. 2015. Do phoretic mites influence the reproductive success of *Ips grandicollis* (Coleoptera: Curculionidae)? *Environmental Entomology*, 44: 1498–1511.

[37] C.T. Rumbold. 1941. A blue-stain fungus, *Cerato.stomella monfium* n. sp. and some yeasts associated with two species of *Dendroctonus*. *Journal of Agricultural Research*, 62: 589–601.

[38] S. Lee, J. Kim and C. Breuil. 2005. *Leptographium longiclavatum* sp. nov., a new species associated with the mountain pine beetle, *Dendroctonus ponderosae*. *Mycological Research*, 109(10): 1162–1170.

[39] J.E. Mercado, R.W. Hofstetter, D.M. Reboletti and J.F. Negrón. 2014. Phoretic symbionts of the mountain pine beetle (*Dendroctonus ponderosae* Hopkins). *Forest Science*, 60(3): 512–526.

[40] D.L. Six and R. Bracewell. 2015. Dendroctonus. *In* Bark Beetles: Biology and Ecology of Invasive and Native Species. F.E. Vega and R.W. Hofstetter (eds). Academic Press, Elsevier, Oxford, U.K. pp. 305–350.

[41] R. Bracewell and D.L. Six. 2015. Experimental evidence of bark beetle adaptation to a fungal symbiont. *Ecology and Evolution*, 5(21): 5109–5119.

[42] H. Solheim. 1992. Fungal succession in sapwood of Norway spruce infested by the bark beetle *Ips typographus*. *European Journal of Plant Pathology*, 22: 136–148.

[43] H. Solheim. 1988. Pathogenicity of Some *Ips typographus*-Associated Blue-Stain Fungi to Norway Spruce. Norsk Institutt for Skogforskning, Communications of the Norwegian Forest Research Institute.

[44] S. Massoumi, A.E. Kim, L.M. Humble, E.A. Uzunovic and C. Breuil. 2007. Ophiostomatoid fungi associated with the northern spruce engraver, *Ips perturbatus*, in western Canada. *Antonie van Leeuwenhoek*, 91(1): 19–34.

[45] B.J. Levieux and T.J. Perry. 1989. Transportation of phytopathogenic fungi by the bark beetle *Ips sexdentatus* Boerner and associated mites. *Journal of Applied Entomology*, 108: 1–11.

[46] C. Bateman, P.E. Kendra, R. Rabaglia and J. Hulcr. 2015. Fungal symbionts in three exotic ambrosia beetles, *Xylosandrus amputatus, Xyleborinus andrewesi*, and *Dryoxylon onoharaense* (Coleoptera: Curculionidae: Scolytinae: Xyleborini) in Florida. *Symbiosis*, 66: 141–148.

[47] J.M. Baker. 1963. Ambrosia beetles and their fungi, with particular reference to *Platypus cylindrus* Fab. *Symposium of the Society of General Microbiology*, 13: 232–265.

[48] R.D. Beaver. 1989. Insect-fungus relation-ships in the bark and ambrosia beetles. *In* Insect-Fungus Interactions. G. Meurant (ed). Academic, London. pp. 121–143.

[49] P. Cassier, J. Levieux, M. Morelet and D. Rougon. 1996. The mycangia of *Platypus cylindrus* Fab. and *P. oxyurus* Dufour (Coleoptera: Platypodidae) structure and associated fungi. *Journal of Insect Physiology*, 42: 171–179.

[50] H. Francke-Grosmann. 1963. Some new aspects in forest entomology. *Annual Review of Entomology*, 8: 415–438.

[51] L.R. Batra. 1963. Ecology of ambrosia fungi and their dissemination by beetles. *Transactions of the Kansas Academy of Science*, 66: 213–236.

[52] L.P. Abrahamson, H.M. Chu and D.M. Norris. 1967. Symbiotic interrelationships between microbes and ambrosia beetles. The organs of microbial transport and perpetuation in *Trypodendron betulae* and *T. retusum* (Coleoptera: Scolytidae). *Annuals of the Entomological Society of America*, 60: 1107–1110.

[53] T.C. Harrington, D. McNew, C. Mayers, S.W. Fraedrich and S.E. Reed. 2014. *Ambrosiella roeperi* sp. nov. is the mycangial symbiont of the granulate ambrosia beetle, *Xylosandrus crassiusculus*. *Mycologia*, 106(4): 835–845.

[54] R.C. Ploetz, J. Hulcr, M.J. Wingfield and Z.W. de Beer. 2013. Destructive tree diseases associated with ambrosia and bark beetles: black swan events in tree pathology? *Plant Disease*, 95: 856–872.

[55] D. Carrillo, C.A. Dunlap, P.B. Avery, J. Navarrete, R.E. Duncan, M.A. Jackson, *et al.* 2015. Entomopathogenic fungi as biological control agents for the vector of the laurel wilt disease, the redbay ambrosia beetle, *Xyleborus glabratus* (Coleoptera: *Curculionidae*). *Biological Control*, 81: 44–50.

[56] S.W. Fraedrich, T.C. Harrington, R.J. Rabaglia, M.D. Ulyshen, A.E. Mayfield, J.L. Hanula, *et al.* 2008. A fungal symbiont of the redbay ambrosia beetle causes a lethal wilt in Redbay and other Lauraceae in the southeastern United States. *Plant Disease*, 92: 215–224.

[57] T.S. Davis and R.W. Hofstetter. 2012. Plant secondary chemistry mediates the performance of a nutritional symbiont associated with a tree-killing herbivore. *Ecology*, 93: 421–429.

[58] R.W. Hofstetter and J.C. Moser. 2013. The role of mites in insect-fungus association. *Annual Review of Entomology*, 59: 537–557.

[59] J.A. Byers. 1995. Host tree chemistry affecting colonization in bark beetles. *In* Chemical Ecology of Insects. R.T. Carde and W.J. Bell (eds). Springer, U.S. pp. 154–213.

[60] T.D. Paine, K.F. Raffa and T.C. Harrington. 1997. Interactions among Scolytid bark beetles, their associated fungi, and live host conifers. *Annual Review of Entomology*, 42: 179–206.

[61] A.S. Adams, C.R. Currie, Y. Cardoza, K.D. Klepzig and K.F. Raffa. 2009. Effects of symbiotic bacteria and tree chemistry on the growth and reproduction of bark beetle fungal symbionts. *Canadian Journal of Forest Research*, 39: 1133–1147.

[62] J.C. Moser. 1985. Use of sporothecae by phoretic *Tarsonemus* mites to transport ascospores of coniferous bluestain fungi. *Transactions of the British Mycological Society*, 84: 750–753.

[63] K.F. Raffa and E.B. Smalley. 1995. Interaction of pre-attack and induced monoterpene concentrations in conifer defense against bark beetle-microbial complexes. *Oecologia*, 102: 285–295.

[64] C.I. Keeling and J. Bohlmann. 2006. Genes, enzymes and chemicals of terpenoid diversity in the constitutive and induced defence of conifers against insects and pathogens. *New Phytology*, 170: 657–675.

[65] R.W. Hofstetter, J.B. Mahfouz, K.D. Klepzig and M.P. Ayres. 2005. Effects of tree phytochemistry on the interactions among endophloedic fungi associated with the southern pine beetle. *Journal of chemical ecology,* 31(3): 539–560.

[66] M.P. Ayres, R.T. Wilkens, J.J. Ruel, M.J. Lombardero and E. Vallery. 2000. Nitrogen budgets of phloem-feeding bark beetles with and without symbiotic fungi. *Ecology,* 81: 2198–2210.

[67] D.L. Six and B.J. Bentz. 2007. Temperature determines symbiont abundance in a multipartite bark beetle-fungus ectosymbiosis. *Microbial Ecology,* 54: 112–118.

[68] L.M. Evans, R.W. Hofstetter, M.P. Ayres and K.D. Klepzig. 2011. Temperature alters the relative abundance and population growth rates of species within the *Dendroctonus frontalis* (Coleoptera: Curculionidae) community. *Environmental Entomology,* 40(4): 824–834.

[69] A.L. Addison, J.A. Powell, D.L. Six, M. Moore and B.J. Bentz. 2013. The role of temperature variability in stabilizing the mountain pine beetle–fungus mutualism. *Journal of Theoretical Biology,* 335: 40–50.

[70] B.S. Lindgren and K.F. Raffa. 2013. Evolution of tree killing in bark beetles (Coleoptera: Curculionidae): trade-offs between the maddening crowds and a sticky situation. *Canadian Entomologist,* 145: 471–495.

[71] G.W. Krantz, D. Evans and D.E. Walter. 2009. A Manual of Acarology (No. C/595.42 M35). Ed. 2.

[72] D.N. Kinn. 1970. The Life Cycle and Behavior of *Cercoleipus coelonotus* (Acarina: Mesostigmata) Including a Survey of Phoretic Mite Associates of California Scolytidae. University of California Publications in Entomology. p. 65.

[73] J.C. Moser and M. Rotom. 1971. Mites associated with southern pine bark beetles in Allen Parish, Louisiana. *The Canadian Entomologist,* 103: 1775–1971.

[74] J.C. Moser and E.A. Cross. 1975. Phoretomorph: a new phoretic phase unique to the Pyemotidae (Acarina: Tarsonemoidea). *Annals of the Entomological Society of America,* 68: 820–822.

[75] J.A. Pfammatter, K.M. Malas and K.F. Raffa. 2015. Behaviours of phoretic mites (Acari) associated with *Ips pini* and *Ips grandicollis* (Coleoptera: Curculionidae) during host-tree colonization. *Agricultural and Forest Entomology,* 106(02): 1–11.

[76] R.W. Hofstetter, J.C. Moser and S. Blomquist. 2013. Mites associated with bark beetles and their hypophoretic Ophiostomatoid fungi. *In* The Ophiostomatoid Fungi: Expanding Frontiers. K.A. Seifert, Z. Wilhelm de Beer and M.J. Wingfield (eds). CBS Biodiversity Series 12. KNAW Fungal Biodiversity Centre, Utrecht, The Netherlands. pp. 165–176.

[77] K.D. Klepzig and R.W. Hofstetter. 2011. From Attack to Emergence: Interactions between the Southern Pine Beetle, Mites, Microbes, and Trees. Southern Pine Beetle II.USDA Forest Service General Technical Reports, Asheville, NC, USA. SRS-140. pp. 141–152.

[78] M.L. Moraza, M. Fernández and M. Jurc. 2013. Phoretic mites of the six-spined engraver beetle, *Ips sexdentatus* (Böerner, 1776)(Coleoptera, Scolytinae), on *Pinus halepensis* in Slovenia. *International Journal of Acarology*, 39: 597–599.

[79] D.L. Six. 1998. Effects of mycangial fungi and host tree species on progeny survival and emergence of *Dendroctonus ponderosae* (Coleoptera: Scolytidae). *Environmental Entomology*, 27: 1393–1401.

[80] B.R. Coppedge, F.M. Stephen and G.W. Felton. 1995. Variation in female southern pine beetle size and lipid content in relation to fungal associates. *The Canadian Entomologist*, 127: 145–154.

[81] S.J. Barras. 1970. Antagonism between *Dendroctonus frontalis* and the fungus Ceratocystis minor. *Annals of the Entomological Society of America*, 63: 1187–1190.

[82] F. Roets, M.J. Wingfield, P.W. Crous and L.L. Dreyer. 2007. Discovery of fungus-mite mutualism in a unique niche. *Environmental Entomology*, 36: 1226–1237.

[83] M.J. Lombardero, K.D. Klepzig, J.C. Moser and M.P. Ayres. 2000. Biology, demography and community interactions of *Tarsonemus* (Acarina: Tarsonemidae) mites phoretic on *Dendroctonus frontalis* (Coleoptera: Scolytidae). *Agricultural and Forest Entomology*, 2: 193–202.

[84] Y.J. Cardoza, J.C. Moser, K.D. Klepzig and K.F. Raffa. 2008. Multipartite symbioses among fungi, mites, nematodes, and the spruce beetle, *Dendroctonus rufipennis*. *Environmental Entomology*, 37: 956–963.

[85] J.A. Pfammatter, J.C. Moser and K.F. Raffa. 2013. Mites Phoretic on *Ips pini* (Coleoptera: Curculionidae: Scolytinae) in Wisconsin Red Pine Stands. *Annals of the Entomological Society of America*, 106: 204–213.

[86] J.C. Moser, H. Konrad, S.R. Blomquist and T. Kirisits. 2010. Do mites phoretic on elm bark beetles contribute to the transmission of Dutch elm disease? *Naturwissenschaften*, 97: 219–227.

[87] A.V. Rice, M.N. Thormann and D.W. Langor. 2007. Mountain pine beetle associated blue-stain fungi cause lesions on jack pine, lodgepole pine, and lodgepole Â jack pine hybrids in Alberta. *Canadian Journal of Botany*, 85: 307–315.

[88] R.W. Hofstetter, J.C. Moser and S.R. Blomquist. 2005. Mites associated with bark beetles and their hypertrophic ophiostomatoid fungi. *Biodiversity Series*, 12: 165–176.

[89] K.D. Klepzig, J.C. Moser, F.J. Lombarder and R.W. Hofstetter. 2001. Review article symbiosis and competition: complex interactions among beetles, fungi and mites. *Symbiosis*, 30: 83–96.

[90] J. Kreutz, O. Vaupel and G. Zimmermann. 2004. Efficacy of *Beauveria bassiana* (Bals.) Vuill. against the spruce bark beetle, *Ips typographus* L., in the laboratory under various conditions. *Journal of Applied Entomology*, 128: 384–389.

[91] A. Carroll, S. Taylor, J. Regniere and L. Safranyik. 2003. Effect of climate change on range expansion by the mountain pine beetle in British Columbia. *In* Mountain Pine Beetle Symposium: Challenges and Solutions, Oct. 30–31, 2003. T.L. Shore, *et al.* (eds). Kelowna, BC. Natural Resources Canada, Infromation Report BC-X-399, Victoria. pp. 223–232.

[92] B.J. Bentz, J. Régnière, C.J. Fettig, E.M. Hansen, J.L. Hayes, J.A. Hicke, R.G. Kelsey, J.F. Negron and S.J. Seybold. 2010. Climate change and bark beetles of the western United States and Canada: direct and indirect effects. *BioScience*, 60(8): 602–613.

Section 3

Resistance and Defense Mechanisms

Chapter 7

Role of Carboxylesterases (ALiE) Regarding Resistance to Insecticides: Case Study of Colorado Potato Beetle (*Leptinotarsa decemlineata* Say)

Sladjan Stankovic and Miroslav Kostic

Additional information is available at the end of the chapter

http://dx.doi.org/10.5772/66254

Abstract

Colorado potato beetle is one of the most important pests because of rapid and strongly developed resistance to insecticides. Resistant insects' populations may detoxify or degrade the toxin faster than susceptible insects, or quickly rid their bodies of the toxic molecules. Resistant populations may possess higher levels or more efficient forms of these enzymes. Insecticide metabolic destruction inside the target organism is a common defensive mechanism, decreasing the duration and intensity of the exposure of the target site, lowering the probability of a lethal outcome. Three major mechanisms of metabolic transformation of insecticides underlie the vast majority of examples of biotransformation-based resistance: (i) oxidation; (ii) ester hydrolysis; and (iii) glutathione conjugation. Pyrethrins, pyrethroids, organophosphates, carbamates and other insecticides are degraded by hydrolysis. Insecticide detoxification primarily unfolds through molecule hydrolysis on different sites, thereby splitting ester, carboxyl-ester, amide and other chemical bonds. The most important hydrolytic enzymes are phosphoric triesters and carboxylesterases (ALiE esterases). Structural mutations in mutant carboxylesterases have now been widely described showing metabolic resistance to organophosphate and pyrethroid insecticides and relatively few cases of resistance to carbamates role. Carboxylesterases role in Colorado potato beetle resistance was confirmed by many authors.

Keywords: insecticide resistance, metabolism, hydrolysis, carboxylesterase, ALiE esterase

1. Introduction

Worldwide, *Leptinotarsa decemlineata* (Say.)—Colorado potato beetle (CPB) is one of the most important pests of potatoes, an insect extremely difficult to control due to rapid and strongly developed resistance against insecticides [1–4]. None of the control techniques, during the long-lasting history, developed against this pest has provided long-term protection for potato crops. CPB still remains a major threat to potato production because of its resistance to all major groups of insecticides [5–18]. Numerous alternative control strategies for Colorado potato beetle were investigated in the last few decades [19–31].

In insects, generally, the factors that lead to resistance are (1) morphological, (2) physiological and biochemical, and (3) behavioral [32]. Regarding CPB's resistance to insecticides, the most important, investigated, and best described are the metabolic changes. Populations of resistant insects may detoxify or degrade the toxin faster than susceptible insects or quickly rid their bodies of the toxic molecules. This type of resistance is the common mechanism and often presents the greatest challenge. Resistant populations may possess higher levels or more efficient forms of these enzymes. In addition to being more efficient, these enzyme systems also may have a broad spectrum of activity. The metabolic destruction of an insecticide inside the target organism is a common defensive mechanism that leads to a decrease in the duration and intensity of the exposure of the target site and thereby lowers the probability of a lethal outcome. Three major mechanisms of metabolic transformation of insecticides underlie the vast majority of examples of biotransformation-based resistance: (i) oxidation; (ii) ester hydrolysis; and (iii) glutathione conjugation.

In the first stage, the metabolism of insecticides is manifested through many reactions, most important of which are oxidation, reduction, and hydrolysis. In the second stage, conjugates are formed, which are practically nontoxic [33]. Selective toxicity of insecticides mostly comes from the balance of the reactions included in activation and in detoxification. Pyrethrins, pyrethroids, organophosphates (OP), carbamates, and other insecticides are degraded by hydrolysis. This is the basis for the selective effect of insecticides and for insects' resistance mechanisms. Insecticide detoxification primarily unfolds through molecule hydrolysis on different sites, thereby breaking ester, carboxylester, amide, and other chemical bonds [32, 34].

The most important hydrolytic enzymes are phosphoric triesters and carboxylesterases (ALiE esterases, nonspecific, or B-esterases).

Esterase-related insect resistance is based on the following:

- Increase in the total amount of esterase—by altering regulatory genes or regulatory loci combined with structural genes, which results in change in enzyme synthesis in the organism or amplification of genes responsible for DNA methylation.
- Change in their activity—by altering structural genes that directly determine the nature of enzymes [1].

The majority of widely used insecticides are esters. This includes virtually all carbamate and OPs, most pyrethroids, and others. In almost all the cases, the hydrolysis of the ester group leads to a significant decrease in or elimination of toxicity. Consequently, esterase activity often plays a key role in determining the comparative responses and resistance to current insecticides [1].

In insects, the primary groups of esterases of interest hydrolyze esters of carboxylic acids, and they are, therefore, termed carboxylesterases. The topic of their nature and significance in insecticide toxicology and resistance has been reviewed by different researchers [35–38]. A useful functional method with special relevance for insecticides was developed based on the ability of the esterase to either hydrolyze OPs (type A) or to become inhibited by them (type B). There are relatively a few cases of high-level esterase-mediated metabolic resistance to carbamates. Structural mutations in mutant carboxylesterases have now been described from four insect species, showing metabolic resistance to OP insecticides [39]. The role of esterase in CPB resistance was confirmed by many authors [4, 40–44].

2. Hydrolytic metabolic pathways

Insecticide detoxification is primarily performed by hydrolysis of molecules at different sites. Hydrolysis means splitting of molecules by adding water. This chemical reaction splits different chemical bonds, such as ester bonds (with phosphoric, carbamine, chrysanthemum, and other acids), carboxyl ester, amide, and other bonds [32–34].

Hydrolysis of such bonds is done enzymatically and nonenzymatically. Besides hydrolases (esterases, phosphatases, carboxylesterases, and amidases), in splitting some of the mentioned chemical bonds there are also some other enzymes involved, such as mixed function oxidases (MFO) and glutathione S transferases (GST) [33]. Hydrolytically, pyrethrins and pyrethroids, organophosphates, carbamates, and some other insecticides can decompose faster or more slowly. This type of metabolism often makes a basis for a selective mode of action of insecticides and a mechanism of insect resistance to insecticides [34].

2.1. Hydrolytic enzymes

Hydrolases are widely spread in diverse plant and animal tissues and different parts of cells. In vertebrates, they can be located in blood plasma, and they are able to attack a large number of xenobiotic esters, but natural substrates of these esterases are unknown. In mammals, A-type esterases (with –SH group) can be found in the serum, and they are associated with the lipoprotein fraction. Some hydrolases, especially B-type hydrolases (with –OH group), are bound to membranes in the microsomal fraction. Up to 7% of all microsomal proteins is membrane-bound esterases. Such esterases can also be found in the serum and pancreatic fluid. Hydrolases can be present in the cellular fluid [32].

Most of these enzymes are not purified. Hence, it is often unknown if the investigated hydrolytic reactions are catalyzed by one enzyme of weak specificity (for example, one enzyme with both esterase and amidase function) or by a mixture of two or more enzymes with higher specificities. Unlike oxidases and transferases, hydrolases do not require any coenzymes; however, from time to time, they require cations for activation [32].

There are still some difficulties in naming hydrolases and their specificities for substrates. The same bonds, especially in phosphates, can be attacked not only by hydrolases but also by other enzymes (MFO and transferases). In the broadest sense, one can identify phosphatases, carboxyl esterases, carboxyl amidases, and epoxide hydrolases [41]. Cyclic phenyl saligenin phosphate, S,S,S-tributyl phosphorotrithioate (DEF), and profenofos are stated as hydrolase inhibitors [45].

2.2. Hydrolysis of phosphorus esters

Enzymes that catalyze a hydrolytic attack on phosphorous esters or anhydride bonds are marked as hydrolases of phosphoric triesters or phospho-triesterases. In insects, these enzymes are not purified, so they have not been adequately compared with purified mammalian hydrolases. These hydrolases do not attack phosphoro-trithioates in any scope. The hydrolase reaction is activated with 1 mM Mn^{2+} and Co^{2+} [46].

In several cases, the activity of these enzymes is higher in resistant insect species. Phosphate triester hydrolysis results in forming an anion metabolite that is a weak AChE inhibitor, finally resulting in detoxification of the starting compound. There are two types of phosphate-ester bonds: an anhydride bond (P—O; P—S; P—C; P—N; and others) and an alkyl-ester bond (R—O—P). There is also an alkyl–nitrogen bond (R—N—P). These bonds are split not only by hydrolases but also by MFO and G-S transferases. Enzymatic hydrolysis is found in mammals, insects, plants, and microorganisms [34].

Some of these bonds can be split nonenzymatically, for example, when it comes to chlorpyrifos oxon in trout, hydrolysis is not stimulated by Ca^{2+} and EDTA does not deactivate this reaction. None of the known inhibitors has affected the hydrolysis. It is also known that trichlorfon can be nonenzymatically transformed into dichlorvos in some plant and animal species [34]. Hydrolases attack oxo forms of phosphates, such as paraoxon, diazoxon, malaoxon (**Figure 1**), and dichlorvos [46].

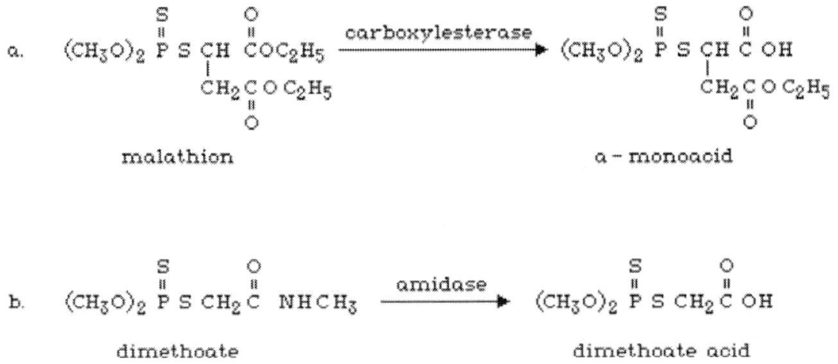

Figure 1. Different hydrolases and their specificities for substrates [47].

http://dx.doi.org/10.5772/66254

Hydrolases (A-esterases) split not only phosphate-ester bonds such as P—O and P—S but also anhydride bonds such as P—O—P, P—F, P—C. Metabolic splitting of the P—S—aryl bond is done at P–S, hydrolytically for thiolate esters and oxidatively for thiolothionates. Fonofos is hydrolyzed after oxidative desulphurization takes place. In some mammals, trichlorfon is hydrolytically metabolized by splitting the P—C bond, and in some others (rabbits), there is a different kind of metabolic reactions. Omethoate (thiolate) is also hydrolytically metabolized in rats by splitting of the P—S bond. However, splitting of the P—S bond in the P—S alkyl structure is not hydrolytic, while splitting of the S—C bond most probably is. The former can result in a P-OH product and the latter in a P—SH product. The ester bond in malathion is split into small scale by breaking the P—S bond (about 1%) and the S—C bond (about 0.5%) by a liver homogenate. Splitting of alkyl-ester bonds in organophosphates is done not only hydrolytically but also oxidatively and through a group transfer. Hydrolysis, for example, splits the ethyl-phosphorous bond in paraoxon. This reaction is conducted in a soluble liver fraction in mammals and in vivo in insects. In rats, there is also hydrolytic O-demethylation of omethoate [34].

2.3. Hydrolysis of carboxylic esters

Carboxyl esterases are also called ALiE esterases or B-esterases. The enzyme that catalyzes the splitting of the carboxyl ester bond in organophosphates is referred to as EC. 3.1.1.1. [34, 38].

It is widely spread in mammals (in the liver, kidneys, lungs, spleen, small intestine, and fluid) [34] and in insects, of both susceptible and resistant species. A purified carboxyl esterase in insects (ALiE) has the molecular weight of 16,000 Da [46].

After investigating the specificities of carboxylesterases in the liver of rats, it has been ascertained that nonphosphorous mono- and di-carboxyl esters serve as substrates. The α-carboxyl ester bond in malathion is hydrolyzed to form malathion α-monoacid. Similarly, the homogenate of the housefly gives α-monoacid more than β-isomer (α/β ratio is 3.5–5.0). However, esterases taken from the horse liver and rat liver microsomes primarily produce malathion β-monoacid (α/β ratio is 0.1). Malathion di-acid in rats does not emerge in vitro, but in vivo, under the influence of an unknown factor [35, 39].

Organophosphates of different structures inhibit carboxyl esterases or the metabolism of malathion and other insecticides from this group in vivo [34]. EPN oxygen analogue and n-propyl-paraoxon inhibit these enzymes and show a strong effect (I_{50} about 10^{-8} M) [46]. These compounds are also strong synergists for malathion and acethion in houseflies and mites. Tri-o-cresyl phosphate (TOCP), which is not an insecticide, is also tested as a carboxylesterase inhibitor. Compounds of such type are weak esterase inhibitors in vitro, but strong inhibitors of malathion metabolism in vivo. TOCP is in vivo transformed into saligenin cyclic-o-tolyl phosphate, marked as M-1, which selectively inhibits carboxyl esterases (pI_{50} for the mouse plasma esterase amounts to 7.2). TOCP increases malathion toxicity four times, and M-1 100 times. Other triaryl phosphates with a 0-alkyl group can be activated metabolically in a similar way. Moreover, increased toxicity of malathion for mice is also exhibited by chlorothion (insecticide) and S,S,S-tributyl-phosphoro-trithioate (DEF) (defoliant), inhibiting its metabolism in vivo. In resistant insect species, DEF also synergizes paraoxon, azinphos-methyl, carbamates, and DDT and with EPN it synergizes dicrotophos, dimethoate, and phorate in *Anthonomus grandis* [34].

Several cases have shown that the increased carboxylesterase activity is a mechanism of insect resistance to malathion. In general, the activity of these enzymes susceptibility to insects is low. Differences between the enzymes in susceptible and resistant insects are quantitative, and in *Tetranychus urticae,* they are also qualitative [46].

In the liver of mammals and products of several insect species, there are enzymes that hydrolyze pyrethroids. It is not excluded that carboxylesterase that hydrolyzes (+)-*trans*-resmethrin is the same one that hydrolyzes malathion. The enzymes in some insects, such as *Oncopeltus fesciatus* and *Trichoplusia ni,* degrade (+)-cis compounds, whereas isomer specificity is less pronounced in the enzymatic products from *Musca domestica* and *Blattella germanica.* Relative hydrolysis speed is much higher in mammals than in insects, which is most likely the basis of the selective toxicity of pyrethroids [34].

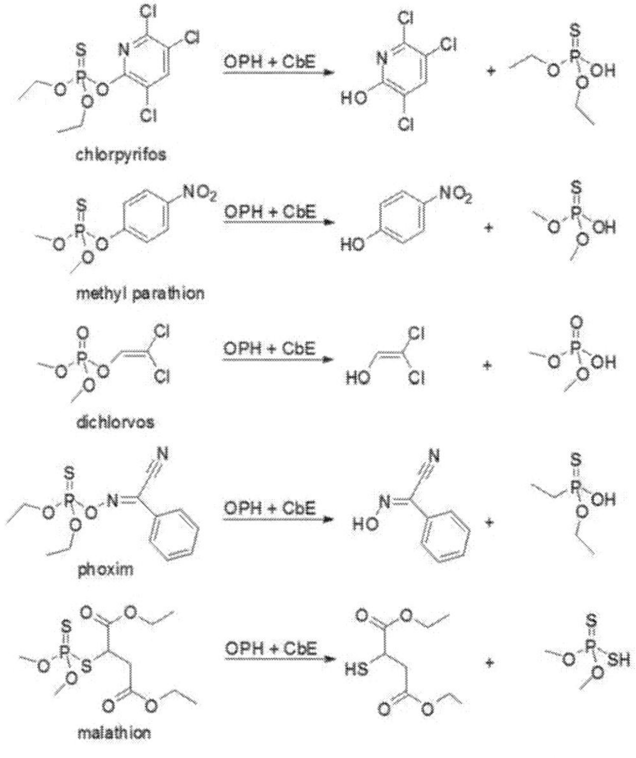

Figure 2. Hydrolysis of different OPs pesticides by OPH + CbE enzymes [48].

Little is known not only about the nature of esterases that hydrolase carbamates but also about biochemistry of these processes. Enzymes in the plasma of rabbits, sheep, and pigs are more efficient in carbaryl hydrolysis than enzymes in other mammals. In insects, perhaps, it is about a nonspecific esterase, aromatic esterase, primarily responsible for carbamate hydrolysis [34].

In some organophosphates (malathion, acethion, and phenthoate), the presence of the carboxyl ester group makes these compounds susceptible to hydrolysis at that site. In malathion, hydrolysis splits one carbethoxy group, forming nontoxic monoacid (**Figure 2**). The anionic charge of the carboxyl group shifts near phosphorous so that the electrophilicity is reduced by the effect of field. Hence, malathion-α-monoacid is inactive. The metabolism of malathion in insects and mammals is quantitatively similar, but qualitatively quite different. Due to its much faster metabolism in mammals, they are usually less susceptible to malathion than insects. Moreover, such type of malathion metabolism predominates in mammals. The presence of the carboxyl ester group in organophosphates, however, does not always lead to reduced toxicity. Mevinphos is much more toxic than it can be expected from its structure that includes the carbethoxy group. In detoxification of mevinphos, the carboxyl esterase plays no or a very little role, and this bond is hydrolyzed nonenzymatically [32, 34, 46].

Some arthropod species resistant to malathion have a much higher carboxylesterase activity than susceptible species. The carboxylesterase activity is, at least in part, responsible for resistance to organophosphorous insecticides in several insect and mite species (mosquitoes, houseflies, cicadas, and mites) [36, 40–43].

Carbamates can also be hydrolyzed by esterases (**Figure 3**). The isolated carbamic acid is immediately hydrolyzed into CO_2 and methyl or dimethyl-amine. The carbamate ester bond is quite stable in plants and insects, but easy to hydrolyze in the majority of animal species [32, 34].

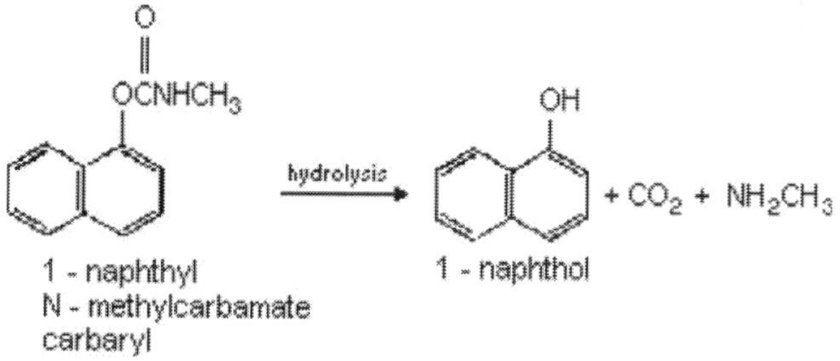

Figure 3. Hydrolytic pathway for carbaryl [49].

Besides, oxidative metabolites with a still intact carbamate ester bond are also subjected to hydrolysis by the same or perhaps different enzymes. In many cases, it is impossible to determine whether oxidation happens before or after hydrolysis [34]. N-substituted carbamates, the products of oxidative N-dealkylation are hydrolytically more unstable than N-methyl carbamates (about 100 times as much).

The rate of enzymatic hydrolysis depends on carbamate structure and the type of organism. Therefore, rats hydrolyze about 25% carbaryl, 33% propoxur, and 75% maxacarbate or isolan. Although most mammals hydrolyze carbamate ester bonds easily, these compounds are resistant in monkeys and pigs. On the other hand, many insects have difficulties in hydrolyzing carbamates, whereas hydrolysis is the main pathway to carbamate decomposing in *B. germanica*. The hydrolysis of carbaryl has been the most studied carbamate hydrolysis. It occurs in a large percentage in many mammals, in a small number of insect species (*B. germanica*, for example), and in a very few plants [34].

Pyrethrins and pyrethroids are quite differently hydrolyzed in living organisms. The differences are considerably due to the compound's structure and the type of organism. The hydrolysis is catalyzed by some kind of carboxylesterases. The basic metabolic pathway in pyrethroids (such as permethrin) is hydrolytic splitting of the ester bond, but oxidation is also important. Hydrolysis is, however, irrelevant and little important in the metabolism of pyrethrins (pyrethrin I, cynarin I, etc.) and related to older pyrethroids (allethrin, tralomethrin, tetramethrin, barthrin, etc.) [32, 34].

In pyrethroids with cyclopropane acid, the stereochemistry of 1,3-bond of this ring strongly affects the metabolism of these compounds. When the substituent is transposed to the carboxyl group at C-1, splitting of the ester bond is easier than when it is *cis*-positioned. The adding of a –CN group to α-carbon of 3-phenoxybenzyl alcohol decreases the susceptibility of molecules to hydrolytic (and oxidative) decomposition. Insects normally hydrolyze pyrethroids more slowly but split the ester bond of *trans*-isomers faster than mammals. However, there is an exception. The larvae of *Chrysopa* spp., which have unexceptionally high levels of esterase, hydrolyze *cis*-isomers of permethrin and cypermethrin faster than *trans*-isomers. This level of esterase activity is undoubtedly the main factor of resistance of these species to pyrethroids [50]. Like in mammals and in insects, the main primary metabolic process of pyrethroids (permethrin, deltamethrin, cypermethrin, etc.) in plants (beans, cotton, etc.) is the splitting of the ester bond. Fenvalerate acts similarly in different plants (tomatoes, tobacco, lettuce, cabbage, etc.). In all plants, the –CN group disappears [32, 34, 50].

Permethrin, a compound without the –CN group, is hydrolyzed in mammals primarily by splitting of the ester bond, during which the *trans*-isomer is about 100 times more susceptible (rats and mice). In goats and cows, about 30% metabolites of permethrin has the preserved ester bond (both *cis*- and *trans*-isomer). Splitting of the permethrin ester bond is more difficult in fish. Insects (cockroaches, houseflies, cabbage moth, and caterpillars) tear permethrin into acid and alcohol part, among others, which form conjugates with glucose and amino acids. In all three species, *trans*-permethrin is metabolized more easily. In vivo and in vitro results are similar. The metabolism of permethrin (**Figure 4**) is, in its basis, similar in other insect species and some mites species [34, 50]. The main pathway in the metabolism of deltamethrin, cypermethrin, and cyhalothrin in mammals (mice and rats) is the splitting of the ester bond, primarily by carboxylesterases, whereby *trans*-isomers are more susceptible. These compounds are basically similarly metabolized in insects [45, 50]. Generally speaking, esterases play a main role in the metabolism of pyrethroids in caterpillars and oxidases in houseflies. In *Spodoptera littoralis* and *T. ni*, for example, prophenophos inhibits the hydrolytic decomposition of *trans*-permethin for 65%, and *cis*-cypermethrin for more than 90%, thereby increasing the toxicity of the former four times and the latter 20 times (*T. ni*), i.e., three times for both compounds (*S. littoralis*).

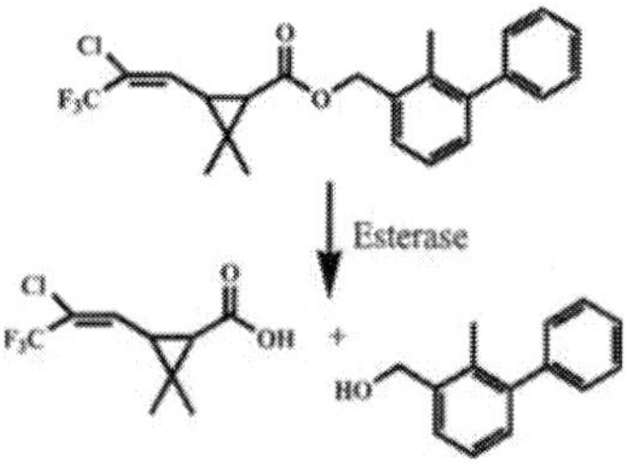

Figure 4. Esterase-mediated hydrolysis of the pyrethroid bifenthrin [51].

Phenyl-saligenin, a cyclic phosphate, increases the toxicity of *trans*-permethrin in *Chrysopa carnea* 68 times. Other compounds also undergo hydrolysis. For example, dinobuton (and similar compounds) is initially hydrolytically activated into dinoseb. Acaricide cycloprate is hydrolyzed into cyclopropane acid that afterward binds with carnitine, which eventually has a lethal effect [34, 50].

2.4. Hydrolysis of carboxamides

Amide bonds are relatively similar to ester bonds. Breaking of these bonds is catalyzed by carboxamidases. Carboxamidases are actually carboxylesterases capable of selecting amides as substrates [32]. In vertebrates, these enzymes are located only in the liver, and they are primarily related to the microsome fraction. Different divalent cations and nucleoids do not affect the enzymatic activity. The oxidative derivative of dimethoates is not hydrolyzed by carboxyamides but inhibits this enzyme. Dimethoate of amidases from different sources differs in susceptibility to EPN-induced inhibition in vivo. Hence, housefly and *Oncopeltus fasciatus* amidases are not susceptible to EPN, while mammal amidases are highly susceptible. On the whole, these enzymes are susceptible to organophosphates as inhibitors, including profenofos, TOCP, DEF, etc. [34].

Breaking the amide bond among organophosphates is primarily determined for dimethoate, dicrotophos, and vamidothion [32]. When compared to other pathways, splitting of the amide bond of dimethoates by hydrolysis amounts to 27.3% in houseflies, 2.5% in rat liver, and 0.0% of rice leaves in vitro. Splitting of the S—C bond in dimethoate is hydrolytic. Vamidothion is similarly metabolized. It has been determined that splitting of the amide bond of dimethoates and related insecticides is important for their selective toxicity. Besides, the amide bond of dimethoate is also hydrolyzed nonenzymatically on the leaf surface after oxidative desulphurization. However, the amide bond of phosphamidon is not split in plants and animals.

The metabolism of benzoyl phenyl urea (BPU) of derivatives is primarily conducted by the hydrolysis of amide bonds. In a large number of insect species (*Tribolium castaneum, S.

littoralis, Spodoptera exigua, etc.), it has been recorded that these compounds (diflubenzuron, chlorfluazuron, diflubenzuron, etc.) degrade rapidly. For example, diflubenzuron is rapidly eliminated from insects (t½ about 7 h), and chlorfluazuron slowly (t½ > 100 h). Adding hydrolase inhibitors (DEF and related compounds) in food prolongs the retention time (t½ from 7 to 18 h for diflubenzuron) and increases the toxicity of diflubenzuron for *T. castaneum* and *S. littoralis*. The main metabolites in BPU hydrolysis are chloroaniline, chlorpheniramine urea, and polar metabolites. The activity of these enzymes in hydrolysis of BPU is completely inhibited by DEF or prophenophos in the concentration of about 10^{-5} M [32]. Therefore, DEF and prophenophos express a synergetic activity with diflubenzuron in *S. exigua* that has developed resistance to them, ranging from 3.7 (to DEF) to 5.2 times to prophenophos [34, 45].

2.5. Hydrolysis of epoxides

Epoxide hydrolases (or epoxide hydrases, epoxide hydratases, EH), discovered in 1968, take part in the metabolism of epoxides. These enzymes (EH) are widely spread among mammals and insects. They are located in microsomes (MEH), solution (CEH), or other parts of mammal liver cells and different insect tissues [32, 34, 44, 46]. pH optimum for the EH activity is most often in the alkaline range, and in insects, it ranges from 7.9 to 9.0 [46]. They show a pronounced specificity for substrates, and there is a great variation among the different species. The mechanism of hydration is not clear enough. Perhaps it includes a nucleophilic attack of the OH group on oxirane carbon [46]. The EH activity from liver microsomes does not depend on NADPH or CO_2 and this enzyme is not affected by BDO-type synergists [34]. A certain number of EH inhibitors in insects include sesamex, piperonyl butoxide, a *Cecropia* hormone, and some organophosphates, which are partially in contrast to these enzymes from the mammal liver. Nevertheless, this enzyme is inhibited for about 80% with the same concentration of Cu^{2+} ions [46]. Specific EH inhibitors in mammals are phenyl-glycidoles. S,S-enantiomers are stronger inhibitors (I_{50} 1.6×10^{-6} M) than R,R-isomers. These compounds are also substrates for EH [44].

EHs catalyze the splitting of the epoxide ring of different insecticides and other compounds. Thereby, they form certain *trans*-diols that are less toxic, so these enzymes are included in epoxide detoxification processes. These processes are most studied in cyclodienes. Splitting of the epoxide ring of some cyclodienes has been shown in many insect and mammal species. In vitro, dieldrin and heptachlor epoxide are quite stable in this form of degradation. In vivo, however, especially in mammals, the main metabolite of dieldrin is *trans*-diol, which indicates the splitting of the epoxide ring [34]. Epoxides of various alkenes and arenas are enzymatically hydrolyzed, thus forming *trans*-di-hydro diols [46].

3. Determination of esterase activity in Colorado potato beetle

In our research, activity of Colorado potato beetle ALiE/Carboxylesterase was determined, using spectrophotometry at a wavelength of 585 μm, first described by Gomori [52]. Average enzyme was prepared out of 40 CBP fourth-instar larvae, using 40 ml of phosphate buffer

(0.02 M, pH7). Incubation of the average enzyme of Colorado potato beetle with different inhibitors (PBO, DEF, eserine sulfate (ES)) proved that decomposition of 1-naphtyl acetate (1-NA) to 1-naphtyl (1-N) is directly related to the activity of esterase. Incubation of an average enzyme with PBO gave, in all cases, much lower reduction in activity, than in the incubation with DEF or with eserine sulfate (ES). This indicates that the formation of 1-naphtyl does not occur due to the activity of oxidase or a glutathione transferase (GST), but due to the activity of esterases.

Activity of ALiE at varying concentrations of the substrate 1-NA showed that the increasing the concentration of the substrate affects the increase in 1-N amount, and this dependence is linear. Statistical analysis has obtained high-correlation coefficient (0.9279) and a very small statistical error (SE = 0.1601), indicating a high dependence of the examined parameters. The value of the Michaelis constant (Km) was low (6.664×10^{-3}). Km values for the activity of most enzymes typically range from 10^{-10} to 10^{-2} M dm^{-3}, which indicates quite high specificity of the tested enzyme for substrate.

3.1. Determination of the calibration line for 1-naphthol

In the experimental conditions, for the ALiE enzyme, 1-naphthylacetate (1-NA) is commonly used as a suitable substrate. Enzyme decomposes substrate 1-NA into 1 naphthol (1-N). In order to be able to determine the amount of generated 1-N, it is necessary to determine the calibration line. For these assays, it is also necessary to adjust the conditions of the experiment when examining the functioning of the enzyme. Tests were carried out in the visible region of wavelengths of light, for different concentrations of 1-N. Very low concentrations of 1-N does not give the expressed absorbance maximum. With increasing concentration, the maximum distinguishes more clearly, and most notably at higher concentrations. For the spectrophotometer device UV-VIS Perkin-Elmer 130, determined peak is at a wavelength of 585 nm and all subsequent determinations of 1-N were carried out at a wavelength of 585 nm.

Concentration (μM)	Absorbance per replication			Average absorbance
	1	2	3	
0.5999	0.205	0.201	0.233	0.213
1.1998	0.216	0.274	0.220	0.237
2.3995	0.219	0.234	0.233	0.229
4.7991	0.290	0.276	0.270	0.279
9.5981	0.323	0.349	0.358	0.343
19.1963	0.467	0.511	0.538	0.505
38.3925	0.642	0.643	0.610	0.632
76.7851	0.890	0.883	0.810	0.861
153.5702	0.822	0.822	0.815	0.820

Table 1. Data for the calibration line for 1-naphthol.

The initial concentration contained 22.14 mg/l of 1-N. Since 1 M solution of the compound is containing 144.17 gl^{-1}, this concentration expressed as a molar solution amounted to 153.57 μM of the compound. Absorbance values for the nine different concentration of 1-N are shown in **Table 1** and **Chart 1**. In order to avoid the effect of these concentrations of the field in which the readings are unreliable, for the calculation of the regression line only concentrations from of 76.8 to 4.8 μM were used.

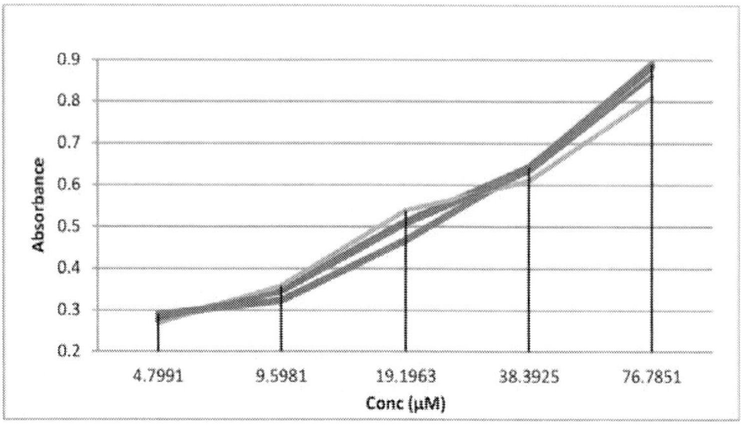

Chart 1. Data for the calibration line for 1 naphtol.

Statistical regression analysis calculated the following regression line for 1-N: $Y = 0.2918 + 0.0078 \times X$, as a basis for the calculation of the 1-N amounts for the appropriate absorbance. Statistical data analysis found out that the regression is linear and directly dependent. The correlation coefficient is 0.98, which indicates a high dependence of investigated parameters.

3.2. Determination of presence and activities of ALiE

Average enzyme was prepared out of 40 Colorado potato beetle fourth-instar larvae (L_4), from the locality of Dobanovci (Belgrade, Serbia), using 40 ml of phosphate buffer (0.02 M, pH7). Experimental conditions in terms of the average amount of enzyme in the required amount of the reaction mixture and the temperature were constant, but the substrate concentration was varied. It is found that the activity of Colorado potato beetle ALiE exists, since in all the examined substrate concentrations in the experimental conditions there is a degradation of the NA-1 1-N due to the enzyme (ALiE) activity (**Chart 2**).

Increasing the concentration of the substrate affects the increase in creation of 1-N, and this dependence is linear. Statistical analysis shows high correlation coefficient and a very small statistical error (as shown in **Table 2**), which indicates the high dependency of investigated parameters.

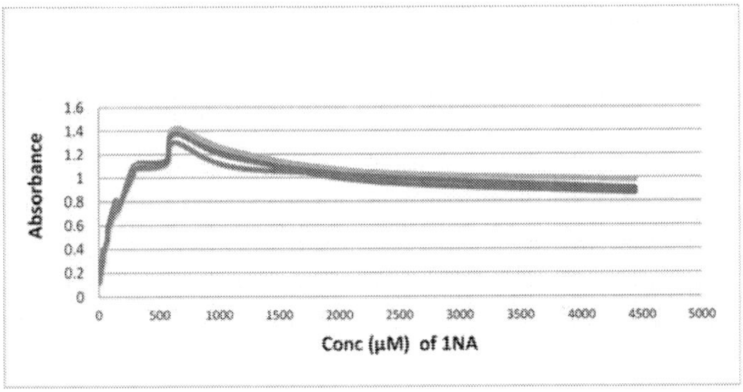

Chart 2. Activity of ALiE at varying concentrations of the substrate 1-NA.

The value of the Michaelis constant (Km) is small. These values of Km for most of the enzymes activity typically range from 10^{-2} to 10^{-10} M dm^{-3}, indicating quite high specificity of the investigated enzyme regarding substrate.

Correlation coefficient	Statistical error	Regression line	Km (Mdm^{-3})
0.9279	0.1601	$Y = 0.2849 + 0.00189 \times X$	6.664×10^{-3}

Table 2. ALiE activity at varying substrate (1-NA) concentrations—statistical parameters.

3.3. Determining the type of enzyme activity using inhibitors

For experiments with inhibitors, we used the average enzyme from Colorado potato beetle of the fourth-grade larvae (L4), population Dobanovci, which contains not only a complex of enzymes but also a variety of other compounds that can react with inhibitors. PBO is a typical oxidase inhibitor and therefore is often used as an insecticide synergist, which are subject to oxidative detoxification. DEF is a specific inhibitor of ALiE esterases, while the ES is specific cholinesterase inhibitor.

Pre-incubation of the enzyme with PBO showed minimal reduction in activity in the degradation of 1-NA. Statistical analysis showed that there are still significant differences in the results for the variant, which applies only to the enzyme and variant combinations of enzymes and piperonyl butoxide.

Pre-incubation of the enzyme with DEF resulted in significantly decreased activity of the enzyme. Differences in the enzyme activity without inhibitor and pre-incubation with DEF are very significant. There are very significant differences between variants, such as the enzyme pre-incubated with the PBO and variants when pre-incubated with DEF.

The greatest reduction in the enzyme activity was obtained when the enzyme is pre-incubated with eserine sulphate (ES). Statistical analysis showed that there are very significant differences between the basic enzyme activity and its activity after pre-incubation with ES. Duncan's test proved that all the variants belong to different groups.

Different concentrations of the CPB average enzyme (ALiE) from population Dobanovci, depending on the increase in the concentration of the enzyme, produce increasing amounts of 1 N, at a constant amount of substrate (1 NA). Similar instances happened with variable concentrations of substrates in the presence of a constant amount of average enzyme. Since the reactions are specific to this group of enzymes, these results indicate their distinguished activity in potato beetle. Incubating the average enzyme with different inhibitors has been proven in the case of Colorado potato beetle that degradation of 1-NA into 1-N comes under the influence of esterase. Incubation of average enzyme with PBO gave in all cases much lower lower reduction in activity, than the impairment of enzyme activities incubated with DEF or eserine sulphate (ES), which indicates that the creation of 1-N does not come due to the effect of oxidase or glutathione transferase (GST), but due to the effect of esterase (**Table 3**).

Treatment	Absorbance per replication					Average	Median	Variance	Standard deviation	Duncan test
	1	2	3	4	5					
Enzyme	0.82	0.88	0.82	0.82	0.90	0.848	0.82	1.52^{-3}	0.039	a
Enzyme + PBO	0.80	0.80	0.78	0.80	0.84	0.804	0.80	4.80^{-4}	0.022	b
Enzyme + DEF	0.72	0.70	0.70	0.75	0.72	0.718	0.72	4.20^{-4}	0.020	c
Enzyme + ES	0.66	0.65	0.69	0.66	0.67	0.665	0.66	1.75^{-4}	0.013	d

$LSD_{0.05} = 0.06$.
$LSD_{0.01} = 0.08$.

Table 3. Statistical parameters of the inhibitor effect on the activity of CPB larvae ALiE.

4. Conclusion

Resistant insect populations may detoxify or degrade the toxin faster than susceptible insects, or quickly rid their bodies of the toxic molecules. Resistant populations may possess higher levels or more efficient forms of these enzymes. The metabolic destruction of an insecticide inside the target organism is a common defensive mechanism. Three major mechanisms of metabolic transformation of insecticides underlie the vast majority of examples of biotransformation-based resistance: (i) oxidation; (ii) ester hydrolysis; and (iii) glutathione conjugation. Pyrethrins, pyrethroids, organophosphates, carbamates, and other insecticides are degraded by hydrolysis. The most important hydrolytic enzymes are phosphoric triesters and carboxylesterases (ALiE esterases, nonspecific, or B-esterases). Structural mutations in mutant carboxylesterases have now been widely described showing metabolic resistance to organophosphate and pyrethroid insecticides. There are relatively few cases of high-level esterase-mediated metabolic resistance to carbamates. The role of carboxylesterases in Colorado potato beetle's resistance to insecticides was confirmed by many authors. CPB is resistant to all major groups of insecticides, including organophosphates and carbamates. Insecticide resistance presence and level are measurable [53]. ALiE's role in the emergence of resistance to organophosphorus and other insecticides in insects, especially in Colorado potato beetle, is investigated. In most of the insect species in which the testing was performed, the dependence of the increase in the activity of the enzyme matched with the increase of the insecticide resistance. Probably, the primary role of this enzyme is its importance for the absorption of organophosphorus insecticides, which becomes nontoxic, and then to gradually decompose to nontoxic components. Increased concentrations of the average CPB enzymes produced increasing amounts of 1-naphthol (1-N) at a constant amount of substrate (1-NA). Similar results happened with variable concentrations of substrate (1-NA) in the presence of a constant amount of average enzyme. Since the reactions are specific to this group of enzymes, these results indicate their distinguished activity in Colorado potato beetle.

Acknowledgements

Work was supported by Ministry of Education and Science, Republic of Serbia, Grant No. III 46008. We would like to acknowledge Dr. Anton Zabel for his visionary work and also thank Ivana Stanimirovic and Dr. Slavica Colic.

Author details

Sladjan Stankovic[1*] and Miroslav Kostic[2]

*Address all correspondence to: sstankovic@ipn.bg.ac.rs

1 Institute for Science Application in Agriculture, Belgrade, Serbia

2 Institute for Medicinal Plant Research "Dr Josif Pancic", Belgrade, Serbia

References

[1] Kostic M., Stankovic S., Kuzevski J. (2015): Role of AChE in Colorado potato beetle (*Leptinotarsa decemlineata* Say) resistance to carbamates and organophosphates. In: Insecticides Resistance (Trdan S., ed.), InTech, Rijeka, Croatia, 19–40. http://dx.doi/10.5772/61460.

[2] Alyokhin A., Chen Y.H., Udalov M., Benkovskaya G., Lindström L. (2013): Evolutionary considerations in potato pest management. In: Insect Pests of Potato, 543–571.

[3] Stanković S., Kostić M., Sivčev I., Janković S., Kljajić P., Todorović G., Jevđović R. (2012): Resistance of Colorado potato beetle (Coleoptera: Chrysomelidae) to neonicotinoids,

pyrethroids and nereistoxins in Serbia. Romanian Biotechnological Letters, 17(5), 7599–7609.

[4] Zhao J.Z., Bishop B.A., Grafius E.J. (2000): Inheritance and synergism of resistance to imidacloprid in the Colorado potato beetle (Coleoptera: Chrysomelidae). Journal of Economic Entomology, 93, 1508–1514.

[5] Stanković S., Zabel A., Kostić M., Manojlović B., Rajković S. (2004): Colorado potato beetle (*Leptinotarsa decemlineata* Say) resistance to organophosphates and carbamates in Serbia. Journal of Pest Science, 77, 11–15.

[6] Zabel A., Stankovic S., Kostic M., Sivcev I., Kuzevski J., Kostic I., Krnjajic S. (2016): Acetylcholinesterase [AChE] activity of Colorado potato beetle populations in Serbia resistant to carbamates and organophosphates. Romanian Biotechnological Letters (in press).

[7] Alyokhin A., Baker M., Mota-Sanchez D., Dively G., Grafius E. (2008): Colorado potato beetle resistance to insecticides. American Journal of Potato Research, 85(6), 395–413.

[8] Stanković S., Zabel A., Kostić M., Šestović M. (2004): Comparative analysis of Colorado potato beetle (*Leptinotarsa decemlineata* Say) resistance monitoring methods. Pesticides, 18(3), 159–175.

[9] Zabel A., Rajkovic S., Manojlovic B., Stankovic S., Kostic M. (2000): Nenicotinoids in the control of *Leptinotarsa decemlineata* Say. 2nd Balkan Symposium on Vegetables and Potatoes. Thessaloniki, Greece. Book of Abstracts.

[10] Zabel A., Rajkovic S., Manojlovic B., Stankovic S., Veljkovic I. (2000): New pesticides in potato protection against the Colorado potato beetle (*Leptinotarsa decemlineata* Say.) and late blight (*Phytophtora infestans* Mont. de Bary) on potato. 2nd Balkan Symposium on Vegetables and Potatoes. Thessaloniki, Greece. Book of Abstracts.

[11] Stanković S., Kostić M., Sivčev I. (2003): Colorado potato beetle (*Leptinotarsa decemlineata* Say) resistance levels to endosulfan in Serbia. Plant Protection, 54(1–4), 105–113.

[12] Alyokhin A., Ferro D. (1999): Relative fitness of Colorado potato beetle (Coleoptera: Chrysomelidae) resistant and susceptible to the *Bacillus thuringiensis* Cry3A toxin. Journal of Economic Entomology, 92, 510–515.

[13] Jiang W.H., Wang Z.T., Xiong M.-H. (2010): Insecticide resistance status of Colorado potato beetle (Coleoptera: Chrysomelidae) adults in northern Xinjiang Uygur autonomous region. Journal of Economic Entomology, 103(4), 1365–1371.

[14] Szendrei S., Grafius E., Byrne A., Ziegler A. (2012): Resistance to neonicotinoid insecticides in field populations of the Colorado potato beetle (Coleoptera: Chrysomelidae). Pest Management Science, 68, 941–946.

[15] Węgorek P., Zamojska J., Mrówczyński M. (2011): Susceptibility level of the Colorado potato beetle (*Leptinotarsa decemlineata* Say) to chlorpyrifos and acetamiprid in poland

and resistance mechanisms of the pest to chlorpyrifos. Journal of Plant Protection Research, 51(3), 279–284.

[16] Alyokhin A., Dively G., Patterson M., Mahoney M., Rogers D., Wollam J. (2006): Susceptibility of imidacloprid-resistant Colorado potato beetles to non-neonicotinoid insecticides in the laboratory and field trials. American Journal of Potato Research, 83,485–494.

[17] Alyokhin A., Dively G., Patterson M., Castaldo C., Rogers D., Mahoney M., Wollam J. (2007): Resistance and cross-resistance to imidacloprid and thiamethoxam in the Colorado potato beetle. Pest Management Science, 63, 32–41.

[18] Zhou Z., Zhaoxu Z., Jinhuan P., Wenchao G, Zhong N., Tian Y., Xiaa G., Wua J. (2012): Evaluation of the resistance of transgenic potato plants expressing various levels of Cry3A against the Colorado potato beetle (*Leptinotarsa decemlineata* Say) in the laboratory and field. Pest Management Science, 68, 1595–1604.

[19] Zabel A., Manojlović B., Rajković S., Stanković S., Kostić M. (2002): Effect of Neem extract on *Lymantria dispar* L. (Lepidoptera: Lymantridae) and *Leptinotarsa decemlineata* Say. (Coleoptera: Chrysomelidae). Journal of Pest Science, 75, 19–26.

[20] Ascher K.R.S., Meisner J., Klein M., Rejesus R.S., Obra J.B., Zabel A., Kostic M., Manojlovic B., Stankovic S., Tomasi O., Sekulovic D., Beimain S.R., Golob P., Andan H.F., Atarigiya J., Chare F.A., Can P., Gupta P., Siddiqui M.R., Joseph S., Shivpuri A., Gupta R.B.L. (2000): Abstracts of presentations on selected topics at The XlVth International plant protection congress (IPPC) 2. NEEM (*Azadirachta indica*). Phytoparasitica, 28(1), 87–90.

[21] Gökçe A, Whalon M.E., Çam M.E., Yanar Y., Demirtaş İ., Gören N. (2006): Plant extract contact toxicities to various developmental stages of Colorado potato beetles (Coleoptera: Chrysomelidae), Annals of Applied Biology, 149: 197–202.

[22] Safaei Khorram M., Taher Nasabi N., Jafarnia S., Khosroshahi S. (2011): The toxicity of selected monoterpene hydrocarbons as single compounds and mixtures against different developmental stages of Colorado potato beetle, *Leptinotarsa decemlineata* say (Coleoptera: Chrysomelidae). Journal of Entomology, 8(5), 404–416.

[23] Trdan S., Vidrih M., Andjus L., Laznik Z. (2009): Activity of four entomopathogenic nematode species against different developmental stages of Colorado potato beetle, *Leptinotarsa decemlineata,* (Coleoptera, Chrysomelidae). Helminthologia, 46(1), 14–20.

[24] Kostić M., Zabel A., Popov V., Ristić M., Stanković S., Manojlović B., Rajković S. (2001): Effects of essential oils of some species of the genus Tanacetum on potato leaf attractiveness to larvae of the Colorado potato beetle. Food in the 21st Century. Book of Abstracts, 136–137.

[25] Kostić M., Kostić I., Marković T., Jevđović R., Stanković S., Todorović G., Nedić N. (2012): Disruption of attractant properties of potato foliage on *Leptinotarsa decemlineata*

Say by the use of *Salvia officinalis* L. essential oil. Proceedings of the 7th Conference on Medicinal and Aromatic Plants of Southeast European Countries, Serbia, 351–356.

[26] Zabel A., Manojlovic B., Stanković S., Thomasi O., Kostić M.(2000): The efficacy of essential oil of *Myristica fragrans* Hout to *Leptinotarsa decemlineata* Say. and *Lymantria dispar* L. Plant Protection, 233–234, 211–220.

[27] Stević T., Kostić M., Stanković S., Ristić M., Soković M. (2003): Effect of thujone to bioagents. Macedonian Pharmaceutical Bulletin. Book of Abstracts, 49(1,2), 165.

[28] Kostic M., Dražić S., Zabel A., Manojlovic B., Rajkovic S., Stankovic S. (2002): Disturbance of potato attraction for Colorado potato beetle larvae by essential oils of some species of the genus Tanacetum. Archive of Agricultural Sciences, 221–222, 87–99.

[29] Kostic M., Stanković S., Ristic M., Jevdjovic R., Rajkovic S. (2003): Effect of essential oils of genus Tanacetum plants to the attractiveness of potato foliage for adults of Colorado potato beetle. Medicinal Raw Materials, 23, 69–82.

[30] Kostic M., Stevic T., Stankovic S., Soković M., Ristic M. (2003): Effect of camphor on biological Agenses. VI Conference of Plant Protection. Abstracts, 2003, Number 112.

[31] Kostić M., Dražić S., Popović Z., Stanković S., Sivčev I., Živanović T. (2007): Developmental and feeding alternations in *Leptinotarsa decemlineata* Say. (Coleoptera: Chrysomelidae) caused by *Salvia officinalis* L. (Lamiaceae) essential oil. Biotechnology and Biotechnological Equipment, 21, 4.

[32] Hassall K.A. (1990): Biochemistry and Uses of Pesticides. Second Edition, VCH, New York.

[33] Šestović M., Peric I. (1995): Fundamentals of the metabolism of insecticides: 1. Enzymes that catalyze the oxidation processes of the first phase. Pesticides, 10, 7–23.

[34] Šestović M., Perić I., Vukša P. (1995): Fundamentals of the metabolism of insecticides: 3. Reductive and hydrolytic processes of the first phase. Pesticides, 10, 181–196.

[35] Oakeshott J.G., Claudianos C., Russell R.J., Robin G.C. (1999): Carboxyl/cholinesterases: a case study of the evolution of a successful multigene family. BioEssays, 21, 1031–1042.

[36] Wheelock C.E., Shan G., Ottea J.A. (2005): Overview of carboxylesterases and their role in metabolism of insecticides. Journal of Pesticide Science, 30, 75–83.

[37] Russell R., Claudianos C., Campbell P.M., Horne I., Sutherland T.D., Oakeshott J.G. (2004): Two major classes of target site insensitivity mutations confer resistance to organophosphate and carbamate insecticides. Pesticide Biochemistry and Physiology, 79, 84–93.

[38] Hollingwort R.M., Dong, K. (2008): The Biochemical and Molecular Genetic Basis of Resistance to Pesticides in Arthropods. In: Global Pesticide Resistance in Anthropods (eds M.E. Whalon, D. Mota-Sanchez and R.M. Hollingworth), pp. 40–50. CABI Publishing, U.K.

[39] Oakeshott J.G., Claudianos C., Campbell P.M., Newcomb R.D., Russell R.J. (2005): Biochemical genetics and genomics of insect esterases. In: Comprehensive Molecular Insect Science-Pharmacology (Gilbert L.I., Latrou K., Gill S.S. eds.), vol. 5, Elsevier, Oxford, 309–381.

[40] Anspaugh D.D., Kennedy G.G., Roe R.M. (1995): Purification and characterization of a resistance-associated esterase from the Colorado potato beetle, *Leptinotarsa decemlineata* (Say.). Pesticide Biochemistry and Physiology, 53, 84–96.

[41] Argentine A., Zhu K.Y., Lee S.H., Clark J.M. (1994): Biochemical mechanisms of azinphosmethyl resistance in isogenic strains of Colorado potato beetle. Pesticide Biochemistry and Physiology, 48 63–78.

[42] Zabel A. (1991): Esterase activity and resistance to organophosphate insecticides in potato beetle (*Leptinotarsa decemlineata* Say.). Faculty of Agriculture, Novi Sad.

[43] Feng Gong L., Fu K.-Y., Li Q., Guo W.-C., Tursun A., Li G.-Q. (2014): Identification of carboxylesterase genes and their expression profiles in the Colorado potato beetle *Leptinotarsa decemlineata* treated with fipronil and cyhalothrin. Pesticide Biochemistry and Physiology, doi: 10.1016/j.pestbp.2014.12.015.

[44] Argentine J.A., Clark J.M. (1990): Selection for abamectin resistance in Colorado potato beetle (Coleoptera: Chrysomelidae). Pestic. Sci., 28: 17–24. doi: 10.1002/ps.2780280104.

[45] Ishaaya I. (1990): Benzoylphenyl ureas and other selective control agents—mechanism and application. In: Pesticides and Alternatives (Casida J. E. ed.), Elsevier, Amsterdam, 365–376.

[46] Dauterman W.C., Hodgson E. (1978): Detoxication mechanisms in insects. In: Biochemistry of Insects (Rockstein M. ed.), Acad. Press, NY, 541–577.

[47] Dauterman W.C. (1971): Biological and non-biological modifications of organophosphorous compounds. Bulletin of the World Health Organization, 44, 133.

[48] da Silva N.A., Birolli W.C., Seleghim M.R., Porto A.L. (2013): Biodegradation of the organophosphate pesticide profenofos by marine fungi. In: Applied Bioremediation—Active and Passive Approaches (Patil Y. ed.), InTech, doi: 10.5772/56372.

[49] Dorough H.W., Ballard S.K. (1982): Degradation of pesticides by animals. In: Biodegradation of Pesticides (Matsumura F., Murti C.R.K. eds.), Plenum Press, N.Y., 3–20.

[50] Ishaaya I., Casida J.E. (1983): Pyrethroid detoxification and synergism in insects. In: Pesticide Chemistry: Human Welfare and the Environment (Miyamoto J., Kearney P.C. eds.), vol. 3, Pergamon Press, N.Y., 307–310.

[51] Wheelock C.E., Miller J.L., Miller M.J., Phillips B.M., Huntley S.A., Gee S.J., Tjeerdema R.S., Hammock B.D. (2006): Use of carboxylesterase activity to remove pyrethroid-associated toxicity to *Ceriodaphnia dubia* and *Hyalella azteca* in toxicity identification evaluations. Environmental Toxicology and Chemistry, 25(4), 973–984.

[52] Gomori G. (1953): Human esterases. Journal of Laboratory and Clinical Medicine, 42, 445–453.

[53] Miyata, T. (1983): Detection and monitoring methods for resistance in arthropod based on biochemical characteristics. In: Pest Resistance to Pesticides (Georghiou G.P., Saito T. eds.), Plenum Press, New York, 99–117.

Chapter 8

Cellular and Molecular Mechanisms of Insect Immunity

Carlos Rosales

Additional information is available at the end of the chapter

http://dx.doi.org/10.5772/67107

Abstract

Multicellular organisms constantly encounter potentially harmful microorganisms. Although insects lack an adaptive immune system, they do have powerful means of fighting infections. Cellular responses involve phagocytosis of bacteria and encapsulation of parasites. In addition, insects can mount a humoral response against pathogens. This is characterized by the secretion of antimicrobial peptides into the hemolymph. Recognition of foreign pathogens involves specific receptors for sensing infection. These include peptidoglycan recognition proteins (PGRPs) and β-glucan recognition proteins (βGRPs). Engagement of these receptors starts signaling pathways that activate the genes that encode antimicrobial peptides. These pathways include the Toll, the Imd, and the JAK-STAT. This chapter describes the innate immunity of insects including both the cellular and humoral responses to bacteria, fungi, and parasites. In addition, recent advances in insect antivirus immune responses are discussed.

Keywords: insect, phagocytosis, hemocyte, innate immunity, signal transduction, toll, Imd, JAK/STAT, TLR, siRNA, autophagy

1. Introduction

Multicellular organisms are constantly exposed to different microorganisms, many of which can be potentially harmful. To protect themselves from these microorganisms, multicellular organisms have evolved cellular and molecular defense mechanisms against infection. These defense mechanisms are known as immunity. At the beginning of an infection from viruses, bacteria, fungi, and protozoa, early mechanisms such as expression of antimicrobial products, recognition of microorganisms by pattern-recognition receptors (PRRs), and activation of phagocytic cells get engaged for eliminating pathogens. These early mechanisms are collectively known as innate immune systems. In vertebrates, such as mammals, cells facilitate the recognition of microorganisms at later times during the course of an infection with specific receptors for microbial antigens. The T- and B-lymphocytes are the cells responsible for the specific recognition of pathogenic antigens and together provide a more selective defense system, known as adaptive immunity, which provides a much better and faster response to the same pathogen during a second challenge.

Insect species live practically in every known habitat and ecological niche, except marine environments. This diversity exposes insects to all sorts of infectious agents. Yet, insects are clearly very successful organisms against infections. Although insects lack an adaptive immune system, they do have a powerful innate immune system for fighting infections. The innate immune system of insects consists of physical barriers, humoral responses, and cellular responses [1, 2].

Physical barriers include the integument and the peritrophic membrane. Integument, the outer surface of an insect, is formed by a single layer of cells covered by a multilayered cuticle [3]. The peritrophic membrane is a layer made of chitin and glycoprotein that covers the insect midgut. It functions as a physical barrier against abrasive food particles and digestive pathogens [4]. However, this membrane is semipermeable and therefore it is not an efficient barrier for viruses. These structures constitute the initial protection for the hemocele (the insect body cavity) and the midgut epithelium against microorganisms. When microorganisms enter these barriers, the humoral and cellular immune responses are activated. Humoral immune responses include production of antimicrobial peptides, activation of prophenoloxidase (proPO), and production of reactive oxygen species [5, 6]. Cellular immune responses include nodulation, encapsulation, and phagocytosis [7, 8].

Hemolymph, the liquid that fills the hemocele, transports nutrients throughout the insect body and also contains several types of free-moving cells or hemocytes. There are several types of hemocytes including granulocytes, plasmatocytes, spherulocytes, and oenocytoids [7, 9]. However, it is important to emphasize that not all these hemocyte types exist in all insect species [10, 11]. Hemocytes are essential for insect immunity, as shown in *Drosophila melanogaster* larvae where plasmatocytes, making up approximately 95% of circulating hemocytes, decrease in numbers during an infection [12]. Also, the genetic [13] or mechanical elimination [14, 15] of phagocytic hemocytes in adult *Drosophila* leads to an increase in infection susceptibility from various bacteria.

Upon infection of the hemocele, cellular immune responses are engaged almost immediately; while humoral responses take place several hours later. It is believed that invading microorganisms are first eliminated by hemocytes and later the humoral responses finish up the few microorganisms not eliminated by cells [16]. These defense mechanisms do not work independently from each other. For example, hemocytes produce molecules that promote hemocyte-microorganism interactions [17, 18]. These molecules function similarly to the opsonins (complement and antibodies) that increase phagocytosis of microorganisms by leukocytes [19]. Also, *Drosophila* plasmatocytes induce fat-body (insect equivalent of the liver) cells to produce antimicrobial peptides after a bacterial infection [14]. In addition, in adult flies, plasmatocytes contribute to reduce the infection susceptibility to various bacteria including *Escherichia coli, Bacillus subtilis,* and importantly *Staphylococcus aureus* [13, 15]. These findings clearly indicate that there is an effective cross-talk between humoral and cellular immunity in insects.

Here, I will describe insect cellular immune functions with emphasis on the innate immunity of insects including both the cellular and humoral responses to bacteria, fungi, and parasites. Specific receptors for sensing infection and the signaling pathways that activate genes for production of antimicrobial peptides will be described. In addition, recent advances in insect antivirus immune responses are discussed.

2. The inducible humoral response

One of the first identified defense mechanisms of insects is the production of antimicrobial peptides (AMPs). Upon microbial infection, a series of small peptides and proteins are produced and released into the hemolymph [20]. The production of AMPs is highly inducible following a microbial infection, the levels of AMPs change from mostly undetectable in uninfected animals to micromolar concentrations in hemolymph of infected individuals [21]. Expression of these AMPs comes mainly from fat-body although hemocytes also contribute to their production [5, 22]. The first identified antimicrobial protein of insects was the lysozyme from *Galleria mellonella*. This enzyme is structurally similar to the chicken C-type lysozyme [23] and is capable of degrading bacterial cell wall peptidoglycans of Gram-positive bacteria. It also has some activity against Gram-negative bacteria [24, 25] and against some fungi [26].

2.1. Antimicrobial peptides

Biochemical analysis of the hemolymph of the fruit-fly *D. melanogaster* and other Diptera has led to the discovery of seven groups of AMPs in insects. They present a wide variety of actions against microorganisms and can be grouped into three families based on their main biological targets [21]. Against Gram-positive bacteria, there are defensins. Against Gram-negative bacteria, there are cecropins, drosocin, attacins, and diptericin. Against fungi, there are drosomycin and metchnikowin.

2.1.1. Defensins

Insect defensins are characterized by having three or four stabilizing intramolecular disulfide bonds. The name comes from their molecular similarity to mammalian α and β defensins [27]. Insect defensins form two groups: one with peptides presenting α-helix/β-sheet mixed structure and the other with peptides presenting triple-stranded antiparallel β-sheets. Defensins with antibacterial and antifungal activity have been reported in many Lepidopteran species [28–30].

2.1.2. Cecropins

Cecropins are small basic peptides of about 31–37 amino acid residues with an amphipathic α-helix conformation [27]. The first amphipathic antimicrobial peptide from insects was identified in hemolymph of the silkworm *Hyalophora cecropia* and was named cecropin [31]. Amphipathic peptides present antimicrobial activity due to their capacity to damage pathogen cell membranes; they also inhibit proline uptake and cause leaky membranes. Now, several cecropin family genes from many lepidoptera species are known. In *Bomby mori*, 13 cecropin genes were found [32]. Moricins are another group of amphipathic α-helical antimicrobial peptides [33] found first in the silkworm *B. mori*. In the *B. mori* genome nine moricin genes were found [32], and in *G. mellonella* eight moricin homologs are reported to have activity against bacterial as well as against yeast and filamentous fungi [34]. Cecropins isolated from insects other than *H. cecropia* have been given various names, for example, bactericidin, lepidopterin, and sarcotoxin [21]. However, all of these peptides are structurally related.

2.1.3. *Drosocin*

Drosocin is a 19-mer cationic antimicrobial peptide from *D. melanogaster*. An O-glycosylated threonine residue has been identified as important for the antimicrobial activity of these peptides, since elimination of the disaccharide at this position renders them with activity several times lower than the native compound [35].

2.1.4. *Attacins*

Attacins are glycine-rich 20 kDa AMPs originally isolated from the hemolymph of *H. cecropia*. Two attacin isoforms, one acid and one basic, have been cloned from *H. cecropia* [36] and they induce an increase of permeability of the outer-membrane of bacteria, binding mainly to lipopolysaccharide (LPS). This explains why the basic attacin is more effective against *E. coli* than the acid attacin. Attacins also inhibit outer-membrane protein synthesis of bacteria at the transcriptional level [36]. Attacins have also been cloned from other Lepidoptera such as the beet armyworm, *Spodoptera exigua* [37].

Gloverins and lebocins are also glycine-rich AMPs found in the lepidoptera [11, 38, 39]. These peptides also inhibit bacterial growth by blocking outer-membrane protein synthesis [40]. In addition to their antibacterial activity, gloverins also present antifungal activity [38, 39], and recently, it has also been reported that they may have antiviral activity [41].

2.1.5. *Diptericin*

Diptericin is an AMPs rich in glycine synthesized by insects in response to a bacterial injection or to injury. It is a basic heat-stable peptide with a molecular weight of 8.6 kDa, containing high levels of Asx, Pro, and Gly. It is active only against a limited range of Gram-negative bacteria and seems to function by disrupting the cytoplasmic membrane of growing bacteria [21]. Recently, diptericin has been reported to be involved not only in inhibiting bacterial growth but also in protection from oxidative stress. Authors suggested that diptericin may trap or "scavenge" free radical anions and also attenuate oxygen toxicity by increasing antioxidant enzyme activities in *D. melanogaster* [42].

2.1.6. *Drosomycin*

Drosomycin is an inducible antifungal peptide of 44 residues initially isolated from bacteria-challenged *D. melanogaster*. It is synthesized in the fat-body and secreted into the hemolymph

http://dx.doi.org/10.5772/67107

of the insect. It exhibits potent antifungal activity but is inactive against bacteria. Drosomycin belongs to the cysteine-stabilized α-helical and β-sheet (CSαβ) superfamily and is composed of an α-helix and a three-stranded β-sheet stabilized by four disulphide bridges [43]. It also has a significant homology with a family of 5 kDa cysteine-rich plant antifungal peptides isolated from seeds of Brassicaceae [44]. Drosomycin exhibits a narrow antimicrobial spectrum and is only active against some filamentous fungi [45]. However, recent work using recombinant drosomycin expressed in *E. coli* revealed that it also has antiparasitic and antiyeast activities [46].

2.1.7. Metchnikowin

Metchnikowin is a 26-residue proline-rich peptide whose expression in *Drosophila* is inducible by infection [47]. This peptide is expressed in the fat-body after immune challenge and can be induced either by the Toll or the Imd pathways [48] (described later). The metchnikowin peptide is unique among the *Drosophila* antimicrobial peptides in that it is active against both Gram-positive bacteria and fungi. Recently, Metchnikowin has been shown to be able to protect a transgenic plant from fungal pathogens. Transgenic barley expressing the metchnikowin gene displayed enhanced resistance to several fungal ascomycetes pathogens, including powdery mildew and Fusarium head blight [49].

2.2. Signaling pathways activating genes that encode antimicrobial peptides

Once a microorganism is detected by PRRs, a series of signaling molecules are activated inside cells to instruct them for different responses. These molecules follow particular signaling pathways that determine the final cellular response. In insects, the signaling pathways involved in humoral immune responses are best described in *D. melanogaster* [50]. The humoral immune responses mainly involve the release of AMPs by the fat-body, via the Toll [51, 52], the immune deficiency (Imd) [53, 54], and the JAK-STAT [55] pathways. Gram-positive bacteria and fungi predominantly induce the Toll signaling pathway, whereas Gram-negative bacteria activate the Imd pathway.

2.2.1. The Toll pathway

The Toll pathway was initially identified as a developmental pathway *in D. melanogaster*. It involves signaling to nuclear factor kappa B (NF-κB) and is essential for embryonic development and immunity [51, 56]. The study of this pathway leads to the subsequent characterization of Toll-like receptors (TLRs) and using this it has reshaped our understanding of the mammalian immune system [57, 58]. Activation of the transmembrane receptor Toll requires a proteolytically cleaved form of an extracellular cytokine-like polypeptide, Spätzle [59], suggesting that Toll requires cooperation of other PRRs. This idea is supported by the fact that a mutation in a peptidoglycan recognition protein (PGRP-SA) blocks Toll activation by Gram-positive bacteria and significantly decreases resistance to this type of infection [60]. Toll activation is not only mediated by PGRPs, but it requires Gram-negative binding protein (GNBP) 1 for Gram-positive bacterial infections [61], and GNBP3 for fungal infections [62]. In addition, the *Drosophila* Persephone protease activates the Toll pathway when proteolytically matured by secreted fungal virulence factors [63] (**Figure 1**).

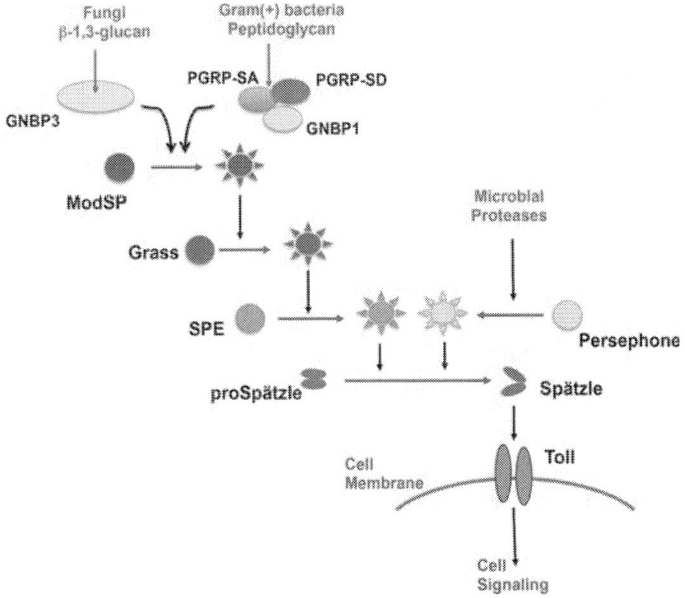

Figure 1. Protease cascades important for Toll activation. The Toll ligand Spätzle is formed when proSpätzle is cleaved by serine protease cascades. The fungi cell wall component β-1,3-glucan is recognized by the circulating pathogen recognition receptor Gram-negative binding protein 3 (GNBP3); while the receptors peptidoglycan recognition proteins PGRP-SA and PGRP-SD, together with GNBP1, recognize peptidoglycan of Gram-positive bacteria. These interactions initiate protease cascades that converge at the level of the serine protease ModSP, which then activates the protease Grass, which in turn activates the Spätzle processing enzyme (SPE). Some microbial proteases (virulence factors) released from pathogenic fungi or bacteria can also be detected by the protease Persephone. Cleavage of Persephone leads to activation of SPE and formation of active Spätzle. Horizontal blue arrows represent proteolytic conversion of the proenzymes (circles) into their active forms (stars). Vertical black arrows represent the site of action for the active proteases.

Toll signaling is activated when cleaved Spätzle binds the Toll receptor. This binding triggers dimerization of the intracytoplasmic TIR domains, inducing binding of the adaptor protein MyD88 through its own TIR domain. MyD88 binds the adaptor protein Tube, which in turn recruits the protein kinase Pelle. These interactions take place via contact of death domains in each protein. Recruitment of Pelle induces its autophosphorylation, triggering phosphorylation and degradation of cactus (an IκB inhibitor) and translocation to the nucleus of the NF-κB transcription factors Dorsal and Dif depending on the context [51, 52, 64, 65] (**Figure 2**).

2.2.2. The Imd pathway

The *D. melanogaster* Imd (immunodeficiency) pathway was discovered when adult flies carrying this mutation alone had impaired production of most AMPs after infection with *E. coli* and *Micrococcus luteus*. In these flies, however, the antifungal Drosomycin remained inducible [66]. It was later shown that Drosomycin induction, after fungal infection, was regulated by the Toll pathway, while the response to most Gram-negative bacteria was blocked by the Imd mutation [67]. The Imd pathway is activated when the receptors peptidoglycan recognition protein (PGRP)-LC and PGRP-LE bind meso-diaminopimelic acid (DAP)-type peptidoglycan [68, 69], which comprises the cell wall of most Gram-negative bacteria. These receptors initiate signaling to the NF-κB transcription factor Relish [70], via the Fas-associated protein with death domain (FADD)—death-related ced-3/Nedd2-like protein (DREDD), and the transforming growth factor beta (TGF-β)-activated kinase 1 (TAK1)—inhibitor of κB kinase (IKK) pathways [53, 68, 71] (**Figure 3**). Once bound to peptidoglycan, these receptors likely dimerize and connect to the adaptor protein Imd [72]. Imd recruits dFADD (Drosophila FADD) [73] and the DREDD caspase [74]. DREDD cleaves Imd, which is then further activated by K63-ubiquitination via the ubiquitination machinery component inhibitor of apoptosis 2 (IAP2) [75]. The K63-polyubiquitin chains are thought, recruit, and activate TAK1 via the ubiquitin-binding domain of its regulatory partner TAK1-associated binding protein 2 (TAB2). TAK1 is then responsible for activating the IKK complex to allow free Relish to translocate into the nucleus. DREDD is also required for mediating the cleavage of the precursor Relish [76] (**Figure 3**).

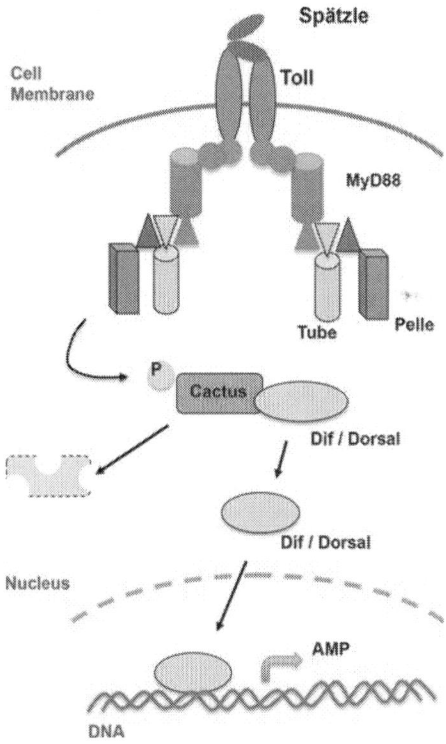

Figure 2. The Toll signaling pathway. Activation of the transmembrane receptor Toll requires a proteolytically cleaved form of Spätzle. Upon Spätzle recognition by a dimer of Toll molecules, a signaling complex is assembled. Toll binds Myd88 through TIR domains (circles), and in turn Myd88 binds Tube and Pelle through their death domains (triangles). The kinase Pelle gets activated by autophosphorylation and then phosphorylates cactus (an IκB inhibitor), marking it for degradation. The NF-κB transcription factors Dorsal or Dif get free and translocate to the nucleus, where they activate transcription of antimicrobial peptides (AMP). P represents a phosphate group.

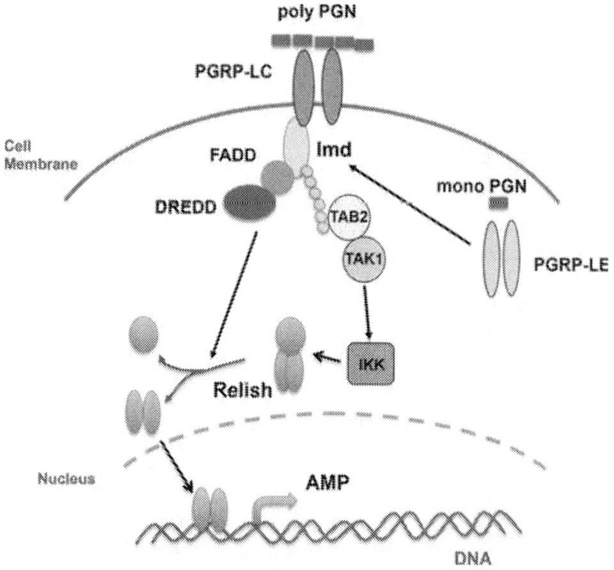

Figure 3. The Imd signaling pathway. In the case of Gram-negative bacteria and some Gram-positive species, polymeric DAP-type peptidoglycan (poly PGN) is recognized by a dimer of PGRP-LC to activate Imd signaling. Imd binds to FADD (Fas-associated protein with death domain), and then the caspase DREDD (FADD-death-related ced-3/Nedd2-like protein) is recruited. DREDD cleaves Imd, which is then activated by K63-ubiquitination. The K63-polyubiquitin chains (yellow circles) help connect to TAB2 (TAK1-associated binding protein 2) and to recruit and activate TAK1 (transforming growth factor beta (TGF-β)-activated kinase 1). TAK1 is then responsible for activating the IKK complex, which phosphorylates the NF-kB-like nuclear factor Relish. DREDD is also required for mediating the cleavage of the precursor Relish. Upon cleavage and phosphorylation, free Relish can translocate into the nucleus, where it activates transcription of specific antimicrobial peptides (AMP). Monomeric peptidoglycan can be recognized intracellularly by the receptor PGRP-LE, and also activate the Imd pathway.

2.2.3. The JAK-STAT pathway

As mentioned above, the Toll and Imd pathways were first described in *Drosophila* and then similar pathways were found in mammals, due to the fact that the central components of these pathways are conserved in evolution. In contrast, the Janus kinase-signal transducer and activator of transcription (JAK-STAT) signaling pathway was first recognized as important in regulating multiple processes of human immunity [77], including control of inflammation and activation of leukocytes, such as neutrophils and macrophages. Now, research is looking back to the fruit fly as a useful model system for elucidating the *in vivo* roles of the JAK-STAT pathway and its regulators, which are challenging to demonstrate in mammalian systems [55].

The canonical signaling model for the JAK-STAT pathway indicates that after binding of a cytokine to its receptor, the receptor dimerizes and JAKs that are constitutively associated with the cytoplasmic tail of the receptor get activated. Activated JAKs phosphorylate each other and specific tyrosine residues on the cytoplasmic part of the receptor. These phosphorylated tyrosines become docking sites for the Src homology 2 (SH2) domains of STAT molecules. The STATs are then tyrosine phosphorylated by JAKs, which allows them to form dimers and translocate into the nucleus, where they bind the promoters of their target genes [78]. In humans, this pathway is very complex due to the number of cytokines that can activate it, and the ability of the JAKs and STATs to form homo- and heterodimers and associate with multiple transcription factors and coactivators. There are four JAKs (JAK1, JAK2, JAK3, and TYK2) and seven STATs (STAT1, STAT2, STAT3, STAT4, STAT5A, STAT5B, and STAT6) [78]. In Drosophila, the known JAK-STAT pathway ligands consist of only three cytokine-like proteins called unpaired (upd), upd2, and upd3 [79]. All three upd molecule signal via a single receptor, Domeless (Dome) [80], which binds to a single JAK, hopscotch (hop) [81], and one STAT transcription factor, Stat92E [82] (**Figure 4**). In addition, in mammals, the JAK-STAT pathway is regulated at the receptor level by the membrane-spanning signal transducer protein gp130 [83], and by negative feedback loops involving the suppressor of cytokine signaling (SOCS) proteins [84]. In Drosophila, similar regulating mechanisms have been found. Eye transformer (ET), a no signaling protein that resembles gp130, is associated with the receptor complex, interacting with both Dome and hop. Thus, ET seems to inhibit intracellular signaling [85, 86] (**Figure 4**). Also, three members of the SOCS family are found in Drosophila, Socs16D, Socs36E, and Socs44A. Of these, Socs36E is the principal negative feedback loop regulator, and it is strongly induced by JAK-STAT signaling [87] (**Figure 4**).

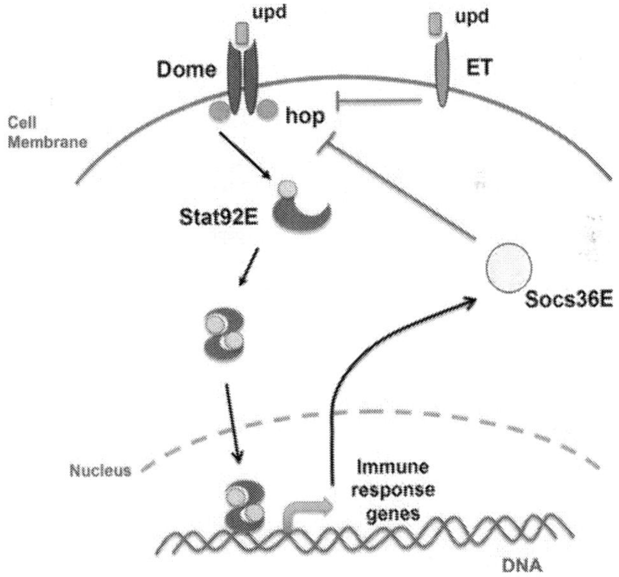

Figure 4. The JAK-STAT signaling pathway. Three cytokine-like proteins called unpaired (upd), upd2, and upd3 signal via the receptor Domeless (Dome), which binds to a single JAK, hopscotch (hop). Upon receptor activation, hopscotch phosphorylates itself and specific tyrosine residues on the cytoplasmic part of the receptor. These phosphorylated tyrosines become docking sites for the STAT transcription factor, Stat92E. Hopscotch also phosphorylates Stat92E at tyrosine residues, allowing it to form dimers and then translocate into the nucleus, where it binds the promoters of their target genes. This pathway is also regulated by a negative feedback loop involving the suppressor of cytokine signaling (SOCS) protein Socs36E, which is upregulated by STAT-JAK signaling. In addition, eye transformer (ET), a nonsignaling receptor for upd, is able to associate with the receptor complex, interacting with both Dome and hop. Thus, ET seems to inhibit intracellular signaling.

As described above, the humoral immune response in Drosophila is mainly controlled by the Toll and Imd pathways in cells of the fat-body and leads to the production of antimicrobial peptides [51, 54]. Also, the JAK-STAT pathway leads to production by the fat-body of other proteins, including cytokines and stress response proteins. This pathway is activated by the ligand upd3. Various stress conditions, such as injury, heat-shock, or dehydration, induce hemocytes to secrete upd3 [88] (**Figure 5**). Moreover, the JAK-STAT pathway has been shown to contribute to the Drosophila viral response. Established JAK-STAT pathway target genes, such as TotM, upd2, and upd3, are all induced by multiple viruses [89]. Finally, the JAK-STAT pathway also contributes to the antimicrobial defense in the gut by inducing the expression of a subset of antimicrobial peptides, such as drosomycin-like peptide (dro3). However, this response seems to be mediated by recognition of cell damage rather than the pathogen [90] (**Figure 5**).

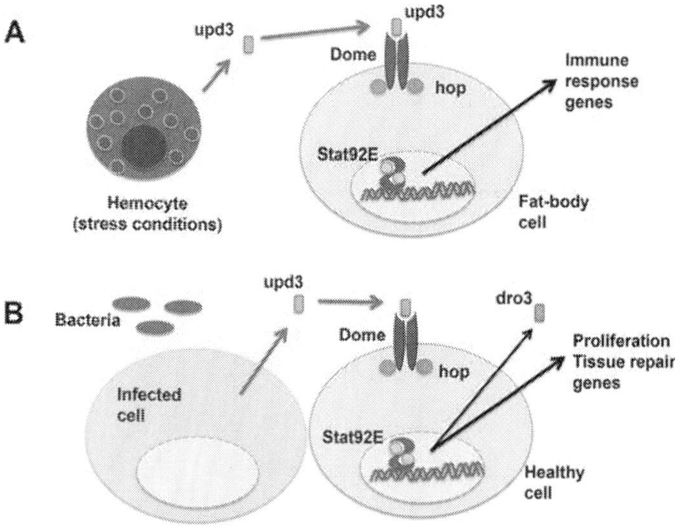

Figure 5. Activation signals for the JAK-STAT signaling pathway. (A) In *Drosophila*, hemocytes participate in recognizing stress conditions by secreting the cytokine-like protein upd3, which binds to the receptor Domeless (Dome). This activates the JAK, hopscotch (hop) and the STAT, Stat92E for induction of immune response genes. (B) In the fly gut epithelium, an infected (with pathogenic bacteria, for example) cell also produces upd3 for activating the JAK-STAT pathway in neighbor cells. These cells then produce antimicrobial peptides, such as drosomycin-like peptide (dro3). Also some proliferation and tissue repair responses are activated to protect the epithelium from infection.

3. Receptors sensing infections

Innate immune responses of insects can be triggered through the interaction of hemocyte receptors or plasma proteins with specific molecules, such as lipids or sugars, on the surface of many microorganisms [91]. Pattern-recognition proteins can be grouped into various types including peptidoglycan recognition protein (PGRP) [92], β-1,3-glucan recognition protein (βGRP), hemolin, and C-type lectins.

http://dx.doi.org/10.5772/67107

3.1. Peptidoglycan recognition proteins (PGRPs)

Peptidoglycan recognition proteins (PGRPs) are innate immunity proteins, conserved from insects to mammals, which recognize bacterial peptidoglycan, and function in antibacterial immunity and inflammation. Mammals have four PGRPs [93, 94]. They are secreted proteins expressed in polymorphonuclear leukocytes (PGRP1), in liver (PGRP2), or in secretions (PGRP3 and PGRP4). All PGRPs recognize bacterial peptidoglycan and three of them (PGRP1, PGRP3, and PGRP4) are directly bactericidal for both Gram-positive and Gram-negative bacteria [94]. Insects have up to 19 PGRPs, classified into short (S) and long (L) forms. The short forms are present in the hemolymph, cuticle, and fat-body cells, whereas the long forms are mainly expressed in hemocytes [95, 96]. The expression of insect PGRPs is often upregulated by exposure to bacteria. These receptors activate the Toll or the Imd signal transduction pathways (described above) or induce proteolytic cascades that generate antimicrobial products [94, 97].

Known functions of PGRPs in Drosophila are as follows: the PGRP-SA in hemolymph binds to Lys-type peptidoglycan and together with PGRP-SD and Gram-negative binding protein (GNBP) 1 leads to activation of the Toll pathway (**Figure 1**). GNBP3 also leads to activation of the Toll pathway in response to yeast. These pattern-recognition proteins initiate the serine protease cascades that lead to activation of the Spätzle-processing enzyme (SPE), which in turn cleaves proSpätzle to generate free Spätzle, the ligand for Toll (**Figure 1**). Similarly, the Imd pathway is activated when the PGRP-LCx homodimer complex binds DAP-type polymeric peptidoglycan, or the heterodimer PGRP-LCx/PGRP-LCa binds DAP-type monomeric peptidoglycan. PGRP-LE can bind both polymeric and monomeric DAP-type peptidoglycan. Extracellular PGRP-LE activates the Imd pathway through PGRP-LC transmembrane receptors and is also involved in activation of the prophenoloxidase (proPO) cascade upstream of the proPO-activating enzyme (PPAE) (**Figure 6**). Intracellular PGRP-LE can also activate the Imd pathway by recognizing intracellular bacteria with DAP-type peptidoglycan and binding to the Imd adaptor protein. In addition, intracellular PGRP-LE can activate autophagy in an Imd pathway-independent manner (**Figure 6**). PGRP-LF functions as an inhibitor of the Imd pathway, because it can bind to PGRP-LCx but not to peptidoglycan. In this manner, it prevents the formation of a PGRP-LC active dimer. PGRP-LB and -SC1a/1b/2 cleave DAP-type peptidoglycan to inactive fragments, thus preventing activation of the Imd pathway. In addition to its scavenger function, PGRP-SC1a is involved in the phagocytosis of bacteria as an opsonin. PGRP-SB1 is directly bactericidal due to its DAP-type peptidoglycan-specific amidase activity [98] (**Figure 6**).

3.2. Beta-1,3-glucan recognition proteins (βGRPs)

Insect β-1,3-glucan recognition proteins (βGRPs) and Gram-negative bacteria binding proteins (GNBPs) are a family of plasma proteins with an amino-terminal glucan-binding domain and a carboxyl-terminal region similar to β-1,3-glucanases [99]. All β βGRPs bind to β-1,3-glucans on bacteria and can activate the proPO cascade. *Manduca sexta* βGRP1 is constitutively expressed in fat-body, whereas βGRP2 gene expression is increased during the early wandering stage prior to pupation or after and immune challenge [5, 100]. Binding of these βGRPs to hemolymph proteinase-14 precursor (proHP14) induces autoactivation of HP14 to initiate a proteinase cascade leading to proPO activation [101]. A βGRP with glucanase activity was isolated from midgut extract of *Helicoverpa armigera* larvae. This enzyme hydrolyzes β-1,3-glucan but not β-1,4-glucan, and it probably functions more as a digestive enzyme than an immune activator [102].

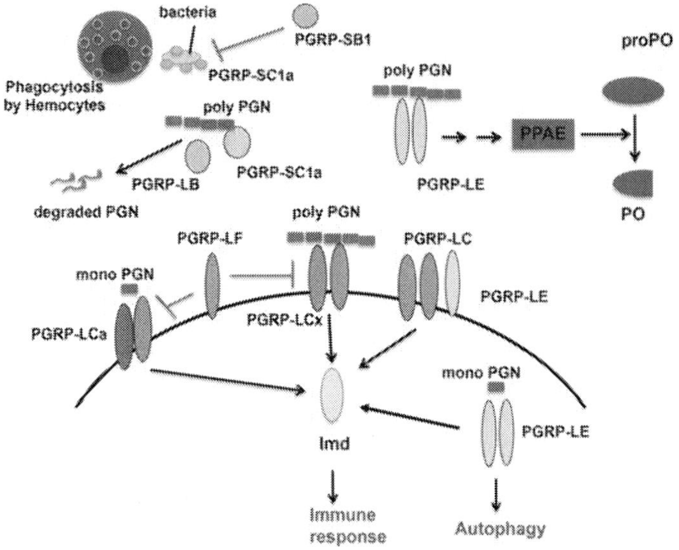

Figure 6. Known functions of PGRPs in Drosophila. Peptidoglycan recognition proteins (PGRPs) are innate immunity proteins, conserved from insects to mammals, which recognize bacterial peptidoglycan. The Imd pathway is activated when the PGRP-LCx homodimer complex binds polymeric peptidoglycan (poly PGN), or the heterodimer PGRP-LCx/PGRP-LCa binds monomeric peptidoglycan (mono PGN). PGRP-LE can bind both polymeric and monomeric peptidoglycan. Extracellular PGRP-LE activates the Imd pathway through PGRP-LC transmembrane receptors and is also involved in activation of the prophenoloxidase (proPO) cascade upstream of the proPO-activating enzyme (PPAE). Intracellular PGRP-LE can also activate the Imd pathway by recognizing intracellular monomeric peptidoglycan and inducing Imd signaling or autophagy independently of Imd. PGRP-LF functions as an inhibitor of the Imd pathway. PGRP-LB and -SC1a/1b/2 cleave DAP-type peptidoglycan to inactive fragments, thus preventing activation of the Imd pathway. In addition, PGRP-SC1a acts as an opsonin for phagocytosis of bacteria. PGRP-SB1 is directly bactericidal due to its DAP-type peptidoglycan-specific amidase activity.

3.3. Hemolin

Hemolin is a plasma protein with four immunoglobulin (Ig) domains commonly found in adhesion molecules of vertebrates [103]. Hemolin is a common protein in several Lepidopteran species, including *B. mori* [32], *Antheraea mylitta* [104], *Plutella xylostella* [105], and *Samia cynthia* [106], but it has not been identified in insects from other orders. Hemolin binds to bacterial LPS and lipoteichoic acid [23]. Hemolin also associates with hemocytes, thus serving as a bridge between microorganisms and hemocytes, and inducing phagocytosis or nodulation [107].

3.4. C-type lectins (CTLs)

C-type lectins (CTLs) from animals are a large group of carbohydrate-recognition molecules that bind ligands in a calcium-dependent manner. Several C-type lectins have been found in Lepidoptera including LPS-binding protein (LBP or CTL20), immulectins -1, -2, -3, and -4, [108, 109], CTL10 [110], CTL11, CTL19, and CTL21 [32]. All these lectins have two carbohydrate-recognition domains, and their genes suggest that these types of lectins are rather unique to Lepidoptera, since they have not been found in other insect species [5].

Most Lepidopteran CTLs bind to bacterial LPS and some also to lipoteichoic acid [108, 109, 111], inducing agglutination of bacteria and yeast [109, 110], probably because each of the two carbohydrate-binding domains bind to sugar residues on the surface of adjacent microbial cells [5]. This microbial aggregation may help hemocytes eliminate pathogens via phagocytosis and nodule formation.

4. The cellular response

Cellular immune responses are immediately after an invasion of the hemocele, while humoral responses appear several hours after an infection. Hemocytes are responsible for a variety of defense responses in insects. Many variations in hemocyte immune responses exist due to the presence of millions of insect species, and we are just beginning to understand these variations [7, 112]. However, a number of frequent cellular immune responses have been described in most insects studied. These responses include nodulation, encapsulation, melanization, and phagocytosis.

4.1. Hemocytes

There are various types of hemocytes described in insects, including granular cells, crystal cells, oenocytoids, and plasmatocytes [8]. These hemocytes are capable of adhesion and phagocytosis [2]. Other types of hemocytes like oenocytoids can produce proPO. This classification of hemocytes, based on morphology, does not always correlate well with cell function. Thus, other attempts have been made to classify hemocyte types. By flow cytometry, three major types of hemocytes can be separated: large granular cells, small semigranular cells, and small hyaline cells [113]. Also, there are some monoclonal antibodies that can distinguish hemocytes based on antigenicity rather than morphology [114, 115]. A number of those monoclonal antibodies could also inhibit some cellular responses [116, 117]. In *D. melanogaster*, three types of hemocytes have been described in greater detail: crystal cells, plasmatocytes, and lamellocytes [118].

Crystal cells are relatively large cells with crystalline inclusions, thus their name. They produce the zymogen proPO, which is activated during melanization. Melanin deposits are important for wound healing or encapsulation of parasites [119, 120]. Plasmatocytes comprise approximately 95% of the hemocyte pool. They are rather small cells (around 10 μm in diameter), but extend large lamellipodial protrusions and form dynamic filopodia [121, 122]. Plasmatocytes are long-lived cells that seem to persist through the entire life of a fly [122]. Mature plasmatocytes express Croquemort (Crq), a CD36 scavenger receptor ortholog, Peroxidasin, an extracellular matrix enzyme, and phagocytic receptors [123]. Lamellocytes are flat cells that appear during larval stages and only detectable when the larvae is infected by parasitic organisms. These hemocytes are mainly responsible for encapsulating the parasitoid wasp egg [119]. Lamellocytes seem to differentiate from a precursor pool of plasmatocytes [124], during a wasp egg infestation and also during sterile injury [125, 126] (**Figure 7**).

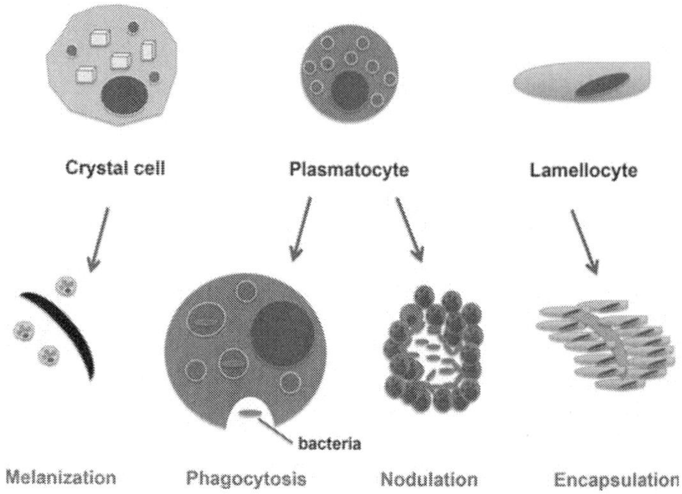

Figure 7. Types and functions of hemocytes in *Drosophila*. Crystal cells are relatively large cells with crystalline inclusions. They produce the zymogen prophenoloxidase, which is activated during melanization. Plasmatocytes are granular cells that comprise approximately 90% of all hemocytes. They express phagocytic receptors and eliminate most of the invading bacteria by phagocytosis or nodulation. Lamellocytes are flat cells that appear during larval stages and only detectable when the larvae are infected by parasitic organisms. These hemocytes are mainly responsible for encapsulating the parasitoid wasp egg. Images are not drawn to scale.

Independently of the type of hemocyte involved, insect immune responses initiate with adhesion of granular hemocytes and plasmatocytes to foreign surfaces or to other cells [127, 128]. Adhesion of hemocytes leads to phagocytosis and also to nodule formation and encapsulation. These cellular innate functions are described next.

4.2. Phagocytosis

Phagocytosis is the process by which cells recognize, bind, and ingest relatively large particles [19]. In insects, phagocytosis is performed by a subset of hemocytes in the hemolymph [7]. Professional phagocytes in Diptera and Lepidoptera have been described as plasmatocytes and granular hemocytes, respectively [129]. In agreement with this, plasmatocytes or granulocytes are the main phagocytic cells in most insects [7, 113, 130, 131]. Recognition of target particles for phagocytosis can be direct by specific cell-surface receptors, or indirect by opsonins that cover the particle so that it can be detected by phagocytic receptors. During development, phagocytic hemocytes eliminate many dying cells, which are detected by the scavenger receptors Croquemort [132], and Draper [133]. In the embryo, hemocytes phagocyte live bacteria but the receptors involved have not been yet identified [134]. In the larva and adult insects, recognition of microorganisms is mediated by the Nimrod family receptors Eater [135] and NimC1 [136], which bind to both Gram-positive and Gram-negative bacteria. Cytokines capable of activating hemocyte functions have also been reported in Lepidoptera insects. A hemocyte chemotactic peptide from *Pseudaletia separata* induces migration and aggregation of hemocytes [137]. This peptide belongs to a group of Lepidopteran cytokines called ENF peptides, which have various biological activities, including plasmatocyte adhesion and spreading, and release of proPO activation [138].

4.3. Nodulation

When the initial phagocytic immune response is not sufficient, hemocytes activate other mechanisms to control infections. To deal with large bacterial loads, hemocytes form nodules to control the infections. Nodulation involves the formation of multicellular hemocyte aggregates that entrap large numbers of bacteria. First, hemocytes surround bacteria and then join other hemocytes to form small aggregates. These cell aggregates continue growing by adding more hemocytes until large nodules are formed. At the end, the nodule is covered with layers of flattened hemocytes and it is melanized. Melanin-covered nodules efficiently isolate bacteria from the hemolymph. Although the process of nodule formation is not completely characterized, certain molecules such as eicosanoids, proPO, and dopa decarboxylase (Ddc) are important for nodule formation in many insect species [139–142]. In addition, screenings for novel immune genes from an Indian saturniid silkmoth (*A. mylitta*) larvae, and from *B. mori* larvae, identified two proteins, Noduler [143] and Reeler1 [144], respectively, as essential molecules in mediating nodulation against *E. coli* K12 and *B. subtilis* bacteria challenge.

4.4. Encapsulation

For larger pathogens such as parasites, protozoa, and nematodes, hemocytes respond by forming a capsule around the foreign organism. Lamellocytes are the effector cells of encapsulation. Lamellocytes bind to the target in multiple cell layers until they form a capsule around the invader. The capsule is normally melanized at the end by degranulation of crystal cells [145]. Inside the capsule the invading organism is killed by reactive cytotoxic products or by asphyxia [146]. Insect hemocytes aggregate in multiple layers during encapsulation and bind to microorganisms during phagocytosis. These functions can be mediated by integrins [147] and indeed various integrins have been found in insect hemocytes [129]. Integrins are also relevant for encapsulation. Various α and β integrins are required for microbial recognition by *M. sexta* hemocytes [148], and in *Drosophila*, the β2-integrin myospheroid is required for attachment to the wasp egg [149].

Interestingly, recent reports have shown that insect hemocytes can release chromatin in a controlled manner to form extracellular traps [150], similar to the NETs formed by mammalian neutrophils [151, 152]. Hemocytes release their nucleic acids in a process known as ETosis. The chromatin fibers participate in histone-mediated killing of microorganisms [150], and also in the process of encapsulation by creating a scaffold on which hemocytes can assemble [153].

4.5. Melanization

Melanization is the process of melanin formation. It is activated during wound healing and also in nodule and capsule formation against large pathogens or parasites in several insects [8, 154]. The enzyme phenoloxidase (PO) is a key in this process. Activation of proPO to PO [155] is mediated by a Serine proteinase cascade [156] and requires pattern-recognition proteins such as PGRP or βGRP. Then active PO binds to foreign surfaces including hemocyte membranes [157], where it initiates melanin formation. PO acts on tyrosine and converts it to dopa [22]. Dopa can then be decarboxylated by Ddc to dopamine or further oxidized by PO to dopaquinone. Both products are then further metabolized to eumelanin and finally melanin [22].

5. Antivirus insect response

Insects, like any other organism, are also infected by viruses. Some viruses are restricted to insect cells and are pathogenic to them; other viruses are transmitted to mammals by biting insects. Understanding the insect innate immune response against viruses thus has tremendous medical and economic importance.

The major mechanism of antiviral defense is the RNA interference (RNAi) pathway that recognizes virus-derived double-stranded RNA (dsRNA) to produce small, interfering RNAs (siRNAs). These siRNAs, in turn, target viral RNA for degradation and hence suppress virus replication. In addition, other innate antimicrobial pathways such as Imd, Toll, and JAK-STAT pathways have also been shown to play important roles in insect antiviral responses. In particular, the JAK-STAT pathway seems to function similarly to the mammalian interferon system. A virus-infected cell sends a signal that activates this pathway in uninfected bystander cells leading to antiviral activity. Finally, the autophagy pathway has also been suggested to be important in some viral infections.

5.1. The RNA interference (RNAi) pathway

When challenged with viruses, the most robust insect response is through the RNA interference (RNAi) pathway (**Figure 8**). Double-stranded viral RNA is detected by Dicer-2 (a member of the RNase III family of endoribonucleases) together with the protein R2D2 [158, 159]. Then, Dicer-2 cleaves the dsRNA into small (21-nucleotide) duplex DNA fragments [160, 161]. Unwinding of the duplex takes place and a guide strand is selected on the basis of complementarity. The siRNA guide strand is then loaded into the RNA-induced silencing complex (RISC), which includes the RNase Argonaute [162]. A target viral RNA pairs with the guide strand, and it is degraded by Argonaut (**Figure 8**).

The importance of the RNAi pathway for controlling virus infections is highlighted by the fact that several viruses have been found to produce RNAi suppressor proteins (1A proteins in Nodaviridae, or B2 proteins in Dicistroviridae) that block the action of the RISC during infection [163, 164]. The B2 protein from the flock house virus (FHV) is a dimer that binds to dsRNA and prevents the cleavage of dsRNA by Dicer-2 [165]. The A1 protein from Drosophila C virus (DCV) functions similarly to FHV B2, by binding to dsRNA and preventing cleavage [164]. In contrast, the 1A protein of cricket paralysis virus (CrPV) interacts with Argonaute and inhibits its RNAse activity [164] (**Figure 8**). When viruses do not have these proteins they replicate poorly and the insect is able to clear the infection completely. The RNAi pathway is clearly very important also for protecting mammalian cells against viruses. Recently, the NS4B protein of dengue virus 2 (DENV-2), flavivirus was found to inhibit the siRNA pathways both in mammalian and insect (Sf21) cells [166].

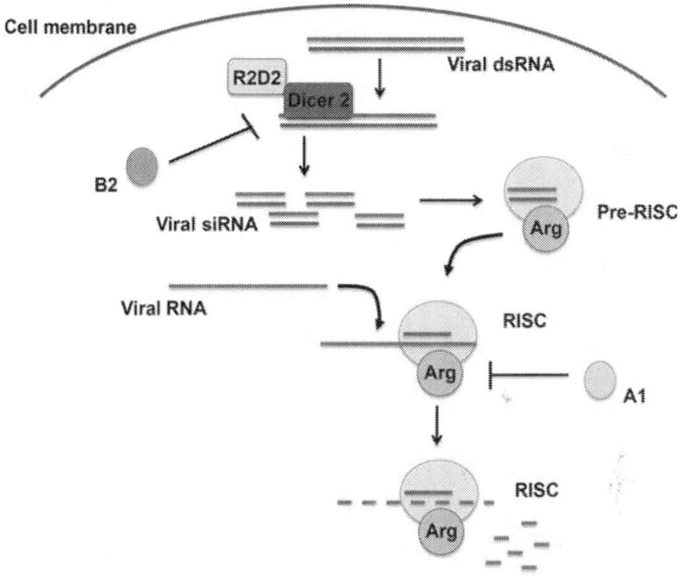

Figure 8. RNA interference (RNAi) pathway. Double-stranded viral RNA is detected by Dicer-2 together with the protein R2D2. Then, Dicer-2 cleaves the dsRNA into small duplex DNA fragments. These siRNA fragments are loaded into the preRNA-induced silencing complex (preRISC), which includes the RNase Argonaute (Arg). A target viral RNA pairs with the guide strand and Argonaut degrades it. RNAi suppressor proteins from some viruses can block the RNAi pathway. The B2 protein from the flock house virus (FHV) prevents the cleavage of dsRNA by Dicer-2; the 1A protein of cricket paralysis virus (CrPV) interacts with Argonaute and inhibits its RNAse activity.

5.2. The JAK-STAT pathway

In addition to the RNAi pathway, the Toll [167, 168] and Imd signaling [169, 170] pathways have been also reported to be involved in antivirus responses. In addition to AMPs, these pathways induce particular sets of genes that are distinct from the genes induced by bacteria or fungi, depending on the virus involved [171]. The actual mechanism for virus recognition and the particular response induced through these pathways is just beginning to be elucidated. In contrast, the JAK-STAT pathway response to viruses seems to be more relevant for preventing the spread of infection [172]. Recent reports also suggest that the JAK-STAT pathway may function similarly to the mammalian interferon system [173]. Infected cells produce factors that activate this pathway in uninfected bystander cells inducing an antiviral state in those cells [172, 173].

As mentioned earlier, the JAK-STAT pathway was initially characterized for its role in development and hemocyte proliferation [77]. The JAK-STAT pathway also gets activated in respond to bacterial infections leading to production of AMPs and other effector molecules [55, 174]. This pathway is activated in a paracrine fashion through the binding of secreted ligands. In the case of virus infections, a novel ligand for the JAK-STAT pathway has recently been identified. In fruit flies, DCV and Sindbis virus (SINV) infections result in increased expression of mRNA for Vago, an 18 kDa cysteine-rich protein with a single von Willebrand factor type C motif [175]. Vago was then shown to be secreted by West Nile virus (WNV)-infected *Culex quinquefasciatus* (southern house mosquito) cells [173]. In addition, Vago mRNA expression was dependent on Dicer-2 but no other RNAi pathway components [173]. Secreted Vago then goes and activates the JAK-STAT pathway in other cells, but interestingly it does not bind the Dome receptor. A different unknown receptor must be responsible for activation of this signaling pathway (**Figure 9**). This creates a new level of complexity to our understanding of the JAK-STAT pathway in insects [176]. The mechanism by which the JAK-STAT pathway creates an antiviral state in the cells is also not known. Future research will help understanding this complex immune response in insects.

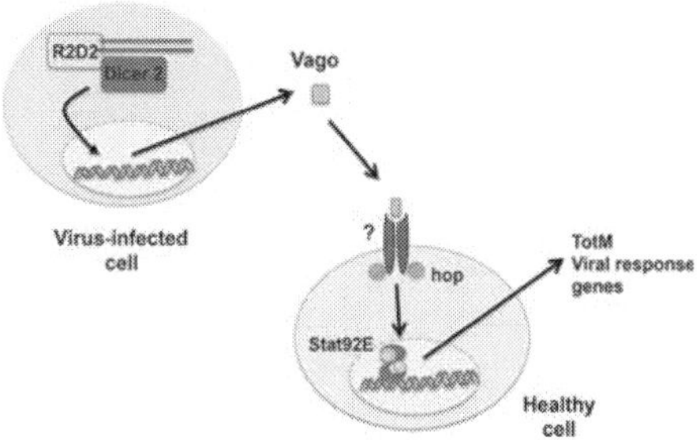

Figure 9. RNA interference (RNAi) pathway and JAK-STAT pathway in viral infections. In a virus-infected cell, an increased expression of mRNA for Vago is observed. The Vago mRNA expression is dependent on Dicer-2. Secreted Vago then goes and activates the JAK-STAT pathway in other cells, but interestingly it does not bind the Dome receptor. A different unknown receptor (?) must be responsible for activation of this signaling pathway to induce expression of viral response genes, including TotM, upd2, and upd3.

5.3. The autophagy pathway

Autophagy has also been proposed as another antiviral mechanism in insects that is independent of the Toll, Imd, or JAK-STAT pathways [177, 178]. Autophagy is the process by which double-membrane vesicles named autophagosomes are formed inside cells. These vesicles are formed with newly synthesized membranes that incorporate large cytoplasmic components including damaged organelles or protein aggregates. Then, the autophagosome fuses with lysosomes and degrades its content. Autophagy is induced by several stress signals including nutrient starvation, infection, and cellular repair mechanisms. In this manner, the degradative process of autophagy helps recycle nutrients and maintains cellular homeostasis [179]. The signaling pathway to autophagy involves the phosphoinositide 3-kinase (PI3K)-Akt pathway, which augments the level of TOR, a negative regulator of autophagy [180] (**Figure 10**). During growing conditions, TOR is active and phosphorylates Autophagy-related (Atg) 13 protein at multiple sites. This prevents Atg13 to bind with Atg1, a central regulator for autophagy [180], leading to decreased Atg1 kinase activity and blocking autophagy (**Figure 10**). During starvation conditions, TOR activity is reduced and Atg13 is rapidly dephosphorylated and forms a complex with Atg1, thus activating it. Atg1 in turn binds to other Atg proteins for assembly of the preautophagosomal structure (PAS) leading to autophagy (**Figure 10**). Different Atg proteins accumulate at the PAS under normal growing conditions to generate cytoplasm to vacuole targeting (Cvt) vesicles, or under starvation conditions to generate autophagosomes [181].

In an infection of *Drosophila* with vesicular stomatitis virus (VSV), the PI3K-Akt-TOR signaling pathway is inhibited. This activates autophagy and in turn decreases viral replication [178]. The viral surface glycoprotein, VSV-G, was proposed to be the pathogen-associated molecular pattern (PAMP) that initiated this cell response [178]. More recently, it was found that, the Drosophila TLR ortholog, Toll-7, was responsible for sensing VSV on the cell surface (**Figure 10**). Toll-7 signaling was activated upon VSV infection and knockdown of Toll-7 resulted in a higher viral protein level *in vitro* and greater pathogenesis *in vivo* [177].

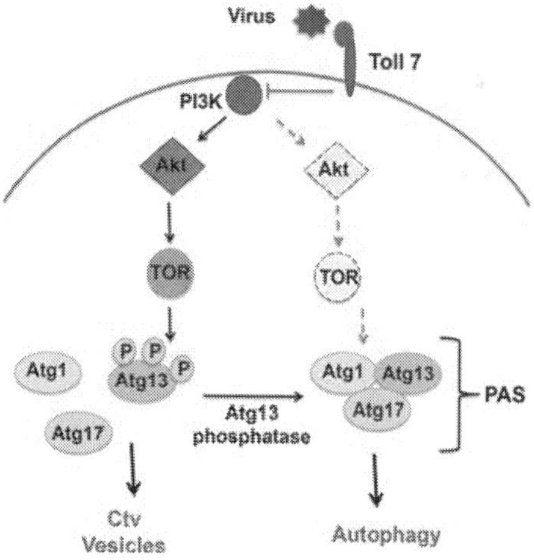

Figure 10. Autophagy response in viral infections. The signaling pathway to autophagy involves the phosphoinositide 3-kinase (PI3K)-Akt pathway, leading to activation of TOR. This kinase phosphorylates autophagy-related (Atg) 13 protein at multiple sites. This prevents Atg13 binding to Atg1 and other Atg proteins like Atg17, for assembly of the preautophagosomal structure (PAS), which leads to autophagy. Different Atg proteins accumulate with Atg1 under normal growing conditions to generate cytoplasm to vacuole targeting (Cvt) vesicles. During an infection with vesicular stomatitis virus (VSV), the receptor Toll-7 detects the virus, and the PI3K-Akt-TOR signaling pathway is inhibited. This activates autophagy and in turn decreases viral replication.

6. Conclusion

Insects clearly possess powerful defense mechanisms for fighting infections. Cellular responses involve phagocytosis of bacteria, and encapsulation of parasites, while humoral responses involve secretion of antimicrobial peptides into the hemolymph. Recognition of foreign pathogens involves specific receptors such as peptidoglycan recognition proteins (PGRPs), β-glucan recognition proteins (βGRPs), and Toll-related proteins. These receptors activate signaling pathways such as the Toll, the Imd, and the JAK-STAT pathways. The particular pathway activated by each pathogen and the final outcome in each case are still not completely known. This is particularly true for viral infections. Thus, future research in the area of insect immunity promises to be full of surprises.

Another fascinating aspect of insect defense mechanisms against infections is the current view that insects depend only on its innate immune response to fight invading microorganisms. By definition, innate immunity lacks adaptive characteristics. However, there are some reports showing that in *Drosophila*, an initial sublethal exposure to *Streptococcus pneumoniae* can protect flies from a second lethal exposure to the same bacteria [182]. Although not all microbial challenges generate this specific primed response, the fungus *Beauveria bassiana*, a natural fly pathogen, can also induce specific protection against a second exposure to the fungus [182]. These results indicate that insect immune responses can indeed adapt and suggest that insect hemocytes may also present an activation response similar to the one known in mammalian leukocytes.

Finally, most of what we know about insect innate immunity comes from studies of *Drosophila*, where genetics analysis has been instrumental in elucidating the antimicrobial peptide response, as well as to open the door for the study of Toll-like receptors, which are essential for the innate immune response of mammals. Similarly, future genetic screens will help identifying novel host antiviral genes and also the receptor molecules that sense viral infection. Yet, it is important to keep in mind that insect-pathogen interactions have coevolved. Thus, it is important to confirm findings from *Drosophila* studies in other insect species [8, 128, 176].

Acknowledgements

This work was supported by grant 254434 from Consejo Nacional de Ciencia y Tecnología, Mexico.

Author details

Carlos Rosales

Address all correspondence to: carosal@unam.mx

Immunology Department, Instituto de Investigaciones Biomédicas, Universidad Nacional Autónoma de México, Mexico City, Mexico

References

[1] Kanost MR, Jiang H, Yu XQ: Innate immune responses of a lepidopteran insect, *Manduca sexta*. Immunol. Rev. 2004; 198:97–105.

[2] Lavine MD, Strand MR: Insect hemocytes and their role in immunity. Insect Biochem. Mol. Biol. 2002; 32:1295–1309.

[3] Ashida M, Brey PT: Role of the integument in insect defense: pro-phenol oxidase cascade in the cuticular matrix. Proc. Natl. Acad. Sci. U S A 1995; 92:10698–10702.

[4] Hegedus D, Erlandson M, Gillott C, Toprak U: New insights into peritrophic matrix synthesis, architecture, and function. Annu. Rev. Entomol. 2009; 54:285–302. doi: 10.1146/annurev.ento.54.110807.090559

[5] Jiang H, Vilcinskas A, Kanost MR: Immunity in lepidopteran insects. Adv. Exp. Med. Biol. 2010; 708:181–204.

[6] Tsakas S, Marmaras VJ: Insect immunity and its signalling: an overview. Invertebrate Surv. J. 2010; 7:228–238.

[7] Browne N, Heelan M, Kavanagh K: An analysis of the structural and functional similarities of insect hemocytes and mammalian phagocytes. Virulence 2013; 4:597–603. doi: 10.4161/viru.25906

[8] Vlisidou I, Wood W: *Drosophila* blood cells and their role in immune responses. FEBS J. 2015; 282:1368–1382. doi: 10.1111/febs.13235

[9] Meister M: Blood cells of *Drosophila*: cell lineages and role in host defence. Curr. Opin. Immunol. 2004; 16:10–15.

[10] Manachini B, Arizza V, Parrinello D, Parrinello N: Hemocytes of *Rhynchophorus ferrugineus* (Olivier) (Coleoptera: Curculionidae) and their response to *Saccharomyces cerevisiae* and *Bacillus thuringiensis*. J. Invertebr. Pathol. 2011; 106:360–365. doi: 10.1016/j.jip.2010.12.006

[11] Wang Q, Liu Y, He HJ, Zhao XF, Wang JX: Immune responses of *Helicoverpa armigera* to different kinds of pathogens. BMC Immunol. 2010; 11:9. doi: 10.1186/1471-2172-11-9

[12] Williams MJ: *Drosophila* hemopoiesis and cellular immunity. J. Immunol. 2007; 178:4711–4716.

[13] Defaye A, Evans I, Crozatier M, Wood W, Lemaitre B, Leulier F: Genetic ablation of *Drosophila* phagocytes reveals their contribution to both development and resistance to bacterial infection. J. Innate Immun. 2009; 1:322–334. doi: 10.1159/000210264

[14] Charroux B, Royet J: Elimination of plasmatocytes by targeted apoptosis reveals their role in multiple aspects of the *Drosophila* immune response. Proc. Natl. Acad. Sci. U. S. A. 2009; 106:9797–9802. doi: 10.1073/pnas.0903971106

[15] Nehme NT, Quintin J, Cho JH, Lee J, Lafarge MC, Kocks C, Ferrandon D: Relative roles of the cellular and humoral responses in the *Drosophila* host defense against three Gram-positive bacterial infections. PLoS One 2011; 6:e14743. doi: 10.1371/journal.pone.0014743

[16] Haine ER, Moret Y, Silva-Jothy MT, Rolff J: Antimicrobial defense and persistent infection in insects. Science 2008; 322:1257–1259.

[17] Brivio MF, Mastore M, Nappi AJ: A pathogenic parasite interferes with phagocytosis of insect immunocompetent cells. Dev. Comp. Immunol. 2010; 34:991–998. doi: 10.1016/j.dci.2010.05.002

[18] Kim CH, Shin YP, Noh MY, Jo YH, Han YS, Seong YS, Lee IH: An insect multiligand recognition protein functions as an opsonin for the phagocytosis of microorganisms. J. Biol. Chem. 2010; 285:25243–25250. doi: 10.1074/jbc.M110.134940

[19] Rosales C (ed.): Molecular Mechanisms of Phagocytosis. Georgetown, Texas: Landes Bioscience/Springer Science; 2005.

[20] Cao X, He Y, Hu Y, Wang Y, Chen YR, Bryant B, Clem RJ, Schwartz LM, Blissard G, Jiang H: The immune signaling pathways of *Manduca sexta*. Insect Biochem. Mol. Biol. 2015; 62:64–74. doi: 10.1016/j.ibmb.2015.03.006

[21] Imler JL, Bulet P: Antimicrobial peptides in *Drosophila*: structures, activities and gene regulation. Chem. Immunol. Allergy. 2005; 86:1–21.

[22] Marmaras VJ, Lampropoulou M: Regulators and signalling in insect haemocyte immunity. Cell Signal. 2009; 21:186–195.

[23] Yu KH, Kim KN, Lee JH, Lee HS, Kim SH, Cho KY, Nam MH, Lee IH: Comparative study on characteristics of lysozymes from the hemolymph of three lepidopteran larvae, *Galleria mellonella, Bombyx mori, Agrius convolvuli*. Dev. Comp. Immunol. 2002; 26:707–713.

[24] Chapelle M, Girard PA, Cousserans F, Volkoff NA, Duvic B: Lysozymes and lysozyme-like proteins from the fall armyworm, *Spodoptera frugiperda*. Mol. Immunol. 2009; 47:261–269. doi: 10.1016/j.molimm.2009.09.028

[25] Wang WX, Wang YP, Deng XJ, Dang XL, Tian JH, Yi HY, Li YF, He XF, Cao Y, Xia QY et al.: Molecular and functional characterization of a c-type lysozyme from the Asian corn borer, *Ostrinia furnacalis*. J. Insect Sci. 2009; 9:17. doi: 10.1673/031.009.1701

[26] Vilcinskas A, Matha V: Effect of the entomopathogenic fungus *Beauveria bassiana* on the humoral immune response of *Galleria mellonella* larvae (Lepidoptera: Pyralidae). Eur. J. Entomol. 1997; 94:461–472.

[27] Bulet P, Stöcklin R: Insect antimicrobial peptides: structures, properties and gene regulation. Protein Pept. Lett. 2005; 12:3–11.

[28] Lee YS, Yun EK, Jang WS, Kim I, Lee JH, Park SY, Ryu KS, Seo SJ, Kim CH, Lee IH: Purification, cDNA cloning and expression of an insect defensin from the great wax moth, *Galleria mellonella*. Insect Mol. Biol. 2004; 13:65–72.

[29] Schuhmann B, Seitz V, Vilcinskas A, Podsiadlowski L: Cloning and expression of gallerimycin, an antifungal peptide expressed in immune response of greater wax moth larvae, *Galleria mellonella*. Arch. Insect Biochem. Physiol. 2003; 53:125–133.

[30] Volkoff AN, Rocher J, d'Alençon E, Bouton M, Landais I, Quesada-Moraga E, Vey A, Fournier P, Mita K, Devauchelle G: Characterization and transcriptional profiles of three *Spodoptera frugiperda* genes encoding cysteine-rich peptides. A new class of defensin-like genes from lepidopteran insects? Gene 2003; 319:43–53.

[31] Steiner H, Hultmark D, Engström A, Bennich H, Boman HG: Sequence and specificity of two antibacterial proteins involved in insect immunity. Nature 1981; 292:246–248.

[32] Tanaka H, Ishibashi J, Fujita K, Nakajima Y, Sagisaka A, Tomimoto K, Suzuki N, Yoshiyama M, Kaneko Y, Iwasaki T et al.: A genome-wide analysis of genes and gene families involved in innate immunity of *Bombyx mori*. Insect Biochem. Mol. Biol. 2008; 38:1087–1110. doi: 10.1016/j.ibmb.2008.09.001

[33] Dai H, Rayaprolu S, Gong Y, Huang R, Prakash O, Jiang H: Solution structure, antibacterial activity, and expression profile of *Manduca sexta* moricin. J. Pept. Sci. 2008; 14:855–863. doi: 10.1002/psc.1016

[34] Brown SE, Howard A, Kasprzak AB, Gordon KH, East PD: The discovery and analysis of a diverged family of novel antifungal moricin-like peptides in the wax moth *Galleria mellonella*. Insect Biochem. Mol. Biol. 2008; 38:201–212. doi: 10.1016/j.ibmb.2007.10.009

[35] Gobbo M, Biondi L, Filira F, Gennaro R, Benincasa M, Scolaro B, Rocchi R: Antimicrobial peptides: synthesis and antibacterial activity of linear and cyclic drosocin and apidaecin 1b analogues. J. Med. Chem. 2002; 45:4494–4504.

[36] Engström P, Carlsson A, Engström A, Tao ZJ, Bennich H: The antibacterial effect of attacins from the silk moth *Hyalophora cecropia* is directed against the outer membrane of *Escherichia coli*. EMBO J. 1984; 3:3347–3351.

[37] Bang K, Park S, Yoo JY, Cho S: Characterization and expression of attacin, an antibacterial protein-encoding gene, from the beet armyworm, *Spodoptera exigua* (Hübner) (Insecta: Lepidoptera: Noctuidae). Mol. Biol. Rep. 2012; 39:5151–5159. doi: 10.1007/s11033-011-1311-3

[38] Hwang J, Kim Y: RNA interference of an antimicrobial peptide, gloverin, of the beet armyworm, *Spodoptera exigua*, enhances susceptibility to *Bacillus thuringiensis*. J. Invertebr. Pathol. 2011; 108:194–200. doi: 10.1016/j.jip.2011.09.003

[39] Xu XX, Zhong X, Yi HY, Yu XQ: *Manduca sexta* Gloverin binds microbial components and is active against bacteria and fungi. Dev. Comp. Immunol. 2012; 38:275–284. doi: 10.1016/j.dci.2012.06.012

[40] Axén A, Carlsson A, Engström A, Bennich H: Gloverin, an antibacterial protein from the immune hemolymph of *Hyalophora* pupae. Eur. J. Biochem. 1997; 247:614–619.

[41] Moreno-Habel DA, Biglang-awa IM, Dulce A, Luu DD, Garcia P, Weers PM, Haas-Stapleton EJ: Inactivation of the budded virus of Autographa californica M nucleopolyhedrovirus by gloverin. J. Invertebr. Pathol. 2012; 110:92–101. doi: 10.1016/j.jip.2012.02.007

[42] Zhao HW, Zhou, Haddad GG: Antimicrobial peptides increase tolerance to oxidant stress in *Drosophila melanogaster*. J. Biol. Chem. 2011; 286:6211–6218. doi: 10.1074/jbc.M110.181206

[43] Gao B, Zhu S: The drosomycin multigene family: three-disulfide variants from *Drosophila takahashii* possess antibacterial activity. Sci. Rep. 2016; 6:32175. doi: 10.1038/srep32175

[44] Fehlbaum P, Bulet P, Michaut L, Lagueux M, Broekaert WF, Hetru C, Hoffmann JA: Insect immunity. Septic injury of Drosophila induces the synthesis of a potent antifungal peptide with sequence homology to plant antifungal peptides. J. Biol. Chem. 1994; 269:33159–33163.

[45] Zhang ZT, Zhu SY: Drosomycin, an essential component of antifungal defence in *Drosophila*. Insect Mol. Biol. 2009; 18:549–556. doi: 10.1111/j.1365-2583.2009.00907.x

[46] Tian C, Gao B, Rodriguez MdC, Lanz-Mendoza H, Ma B, Zhu S: Gene expression, antiparasitic activity, and functional evolution of the drosomycin family. Mol. Immunol. 2008; 45:3909–3916. doi: 10.1016/j.molimm.2008.06.025

[47] Levashina EA, Ohresser S, Bulet P, Reichhart JM, Hetru C, A HJ: Metchnikowin, a novel immune-inducible proline-rich peptide from *Drosophila* with antibacterial and antifungal properties. Eur. J. Biochem. 1995; 233:694–700.

[48] Levashina EA, Ohresser S, Lemaitre B, Imler JL: Two distinct pathways can control expression of the gene encoding the *Drosophila* antimicrobial peptide metchnikowin. J. Mol. Biol. 1998; 278:515–527. doi: 10.1006/jmbi.1998.1705

[49] Rahnamaeian M, Langen G, Imani J, Khalifa W, Altincicek B, von Wettstein D, Kogel KH, Vilcinskas A: Insect peptide metchnikowin confers on barley a selective capacity for resistance to fungal ascomycetes pathogens. J. Exp. Bot. 2009; 60:4105–4114. doi: 10.1093/jxb/erp240

[50] Lemaitre B, Hoffmann JA: The host defense of Drosophila melanogaster. Annu. Rev. Immunol. 2007; 25:697–743. doi:10.1146/annurev.immunol.25.022106.141615

[51] Valanne S, Wang JH, Rämet M: The *Drosophila* Toll signaling pathway. J. Immunol. 2011; 186:649–656. doi: 10.4049/jimmunol.1002302

[52] Lindsay SA, Wasserman SA: Conventional and non-conventional Drosophila Toll signaling. Dev. Comp. Immunol. 2014; 42:16–24. doi: 10.1016/j.dci.2013.04.011

[53] Kleino A, Silverman N: The *Drosophila* IMD pathway in the activation of the humoral immune response. Dev. Comp. Immunol. 2014; 42:25–35. doi: 10.1016/j.dci.2013.05.014

[54] Myllymäki H, Valanne S, Rämet M: The *Drosophila* imd signaling pathway. J. Immunol. 2014; 192:3455–3462. doi: 10.4049/jimmunol.1303309

[55] Myllymäki H, Rämet M: JAK/STAT pathway in *Drosophila* immunity. Scand. J. Immunol. 2014; 79:377–385. doi: 10.1111/sji.12170

[56] Ashok Y: *Drosophila* toll pathway: the new model. Sci. Signal. 2009; 2:jc1. doi: 10.1126/scisignal.252jc1

[57] Bryant CE, Gay NJ, Heymans S, Sacre S, Schaefer L, Midwood KS: Advances in Toll-like receptor biology: modes of activation by diverse stimuli. Crit. Rev. Biochem. Mol. Biol. 2015; 50:359–379. doi: 10.3109/10409238.2015.1033511

[58] Kawasaki T, Kawai T: Toll-like receptor signaling pathways. Front. Immunol. 2014; 5:461. doi: 10.3389/fimmu.2014.00461

[59] Shia AK, Glittenberg M, Thompson G, Weber AN, Reichhart JM, Ligoxygakis P: Toll-dependent antimicrobial responses in *Drosophila* larval fat body require Spätzle secreted by haemocytes. J. Cell Sci. 2009; 122:4505–4515. doi: 10.1242/jcs.049155

[60] Michel T, Reichhart JM, Hoffmann JA, Royet J: *Drosophila* Toll is activated by Gram-positive bacteria through a circulating peptidoglycan recognition protein. Nature 2001; 414:756–759. doi: 10.1038/414756a

[61] Wang Y, Jiang H: Interaction of β-1,3-Glucan with Its recognition protein activates hemolymph proteinase 14, an initiation enzyme of the prophenoloxidase activation system in Manduca sexta. J. Biol. Chem. 2006; 281:9271–9278. doi: 10.1074/jbc.M513797200

[62] Mishima Y, Quintin J, Aimanianda V, Kellenberger C, Coste F, Clavaud C, Hetru C, Hoffmann JA, Latgé JP, Ferrandon D et al.: The N-terminal domain of Drosophila Gram-negative binding protein 3 (GNBP3) defines a novel family of fungal pattern recognition receptors. J. Biol. Chem. 2009; 284:28687–28697. doi: 10.1074/jbc.M109.034587

[63] Gottar M, Gobert V, Matskevich AA, Reichhart JM, Wang C, Butt TM, Belvin M, Hoffmann JA, Ferrandon D: Dual detection of fungal infections in Drosophila via recognition of glucans and sensing of virulence factors. Cell 2006; 127:1425–1437. doi: 10.1016/j.cell.2006.10.046

[64] Imler JL, Hoffmann JA: Toll receptors in *Drosophila*: a family of molecules regulating development and immunity. Curr. Top. Microbiol. Immunol. 2002; 270:63–79.

[65] Tauszig-Delamasure S, Bilak H, Capovilla M, Hoffmann JA, Imler JL: *Drosophila* MyD88 is required for the response to fungal and Gram-positive bacterial infections. Nat. Immunol. 2002; 3:91–97. doi: 10.1038/ni747

[66] Lemaitre B, Kromer-Metzger E, Michaut L, Nicolas E, Meister M, Georgel P, Reichhart JM, Hoffmann JA: A recessive mutation, immune deficiency (imd), defines two distinct control pathways in the *Drosophila* host defense. Proc. Natl. Acad. Sci. U S A 1995; 92:9465–9469.

[67] Lemaitre B, Nicolas E, Michaut L, Reichhart JM, Hoffmann JA: The dorsoventral regulatory gene cassette spatzle/Toll/cactus controls the potent antifungal response in *Drosophila* adults. Cell 1996; 86:973–983.

[68] Kaneko T, N. S: Bacterial recognition and signalling by the *Drosophila* IMD pathway. Cell. Microbiol. 2005; 7:461–469. doi: 10.1111/j.1462-5822.2005.00504.x

[69] Kaneko T, Yano T, Aggarwal K, Lim JH, Ueda K, Oshima Y, Peach C, Erturk-Hasdemir D, Goldman WE, Oh BH et al.: PGRP-LC and PGRP-LE have essential yet distinct functions in the drosophila immune response to monomeric DAP-type peptidoglycan. Nat. Immunol. 2006; 7:715–723. doi: 10.1038/ni1356

[70] Choe KM, Werner T, Stöven S, Hultmark D, Anderson KV: Requirement for a peptidoglycan recognition protein (PGRP) in Relish activation and antibacterial immune responses in *Drosophila*. Science 2002; 296:359–362. doi:10.1126/science.1070216

[71] Aggarwal K, Silverman N: Positive and negative regulation of the *Drosophila* immune response. BMB Rep. 2008; 41:267–277.

[72] Choe KM, Lee H, Anderson KV: Drosophila peptidoglycan recognition protein LC (PGRP-LC) acts as a signal-transducing innate immune receptor. Proc. Natl. Acad. Sci. U S A 2005; 102:1122–1126. doi: 10.1073/pnas.0404952102

[73] Leulier F, Vidal S, Saigo K, Ueda R, Lemaitre B: Inducible expression of double-stranded RNA reveals a role for dFADD in the regulation of the antibacterial response in *Drosophila* adults. Curr. Biol. 2002; 12:996–1000.

[74] Leulier F, Rodriguez A, Khush RS, Abrams JM, Lemaitre B: The Drosophila caspase Dredd is required to resist gram-negative bacterial infection. EMBO Rep. 2000; 1:353–358. doi: 10.1093/embo-reports/kvd073

[75] Paquette N, Broemer M, Aggarwal K, Chen L, Husson M, Ertürk-Hasdemir D, Reichhart JM, Meier P, Silverman N: Caspase-mediated cleavage, IAP binding, and ubiquitination: linking three mechanisms crucial for *Drosophila* NF-κB signaling. Mol. Cell 2010; 37:172–182. doi: 10.1016/j.molcel.2009.12.036

[76] Ertürk-Hasdemir D, Broemer M, Leulier F, Lane WS, Paquette N, Hwang D, Kim CH, Stöven S, Meier P, Silverman N: Two roles for the *Drosophila* IKK complex in the activation of Relish and the induction of antimicrobial peptide genes. Proc. Natl. Acad. Sci. U S A 2009; 106:9779–9784. doi: 10.1073/pnas.0812022106

[77] O'Shea J, Plenge R: JAK and STAT signaling molecules in immunoregulation and immune-mediated disease. Immunity 2012; 36:542–550. doi: 10.1016/j.immuni.2012.03.014

[78] Kiu H, Nicholson SE: Biology and significance of the JAK/STAT signalling pathways. Growth Factors 2012; 30:88–106. doi: 10.3109/08977194.2012.660936

[79] Wright VM, Vogt KL, Smythe E, Zeidler MP: Differential activities of the Drosophila JAK/STAT pathway ligands Upd, Upd2 and Upd3. Cell. Signal. 2011; 23:920–927. doi: 10.1016/j.cellsig.2011.01.020

[80] Brown S, Hu N, Hombría JC: Identification of the first invertebrate interleukin JAK/STAT receptor, the *Drosophila* gene domeless. Curr. Biol. 2001; 11:1700–1705.

[81] Binari R, Perrimon N: Stripe-specific regulation of pair-rule genes by hopscotch, a putative Jak family tyrosine kinase in *Drosophila*. Genes Dev. 1994; 8:300–312.

[82] Yan R, Small S, Desplan C, Dearolf CR, Darnell JEJ: Identification of a Stat gene that functions in Drosophila development. Cell 1996; 84:421–430.

[83] Kishimoto T, Akira S, Narazaki M, Taga T: Interleukin-6 family of cytokines and gp130. Blood 1995; 86:1243–1254.

[84] Starr R, Willson TA, Viney EM, Murray LJ, Rayner JR, Jenkins BJ, Gonda TJ, Alexander WS, Metcalf D, Nicola NA et al.: A family of cytokine-inducible inhibitors of signalling. Nature 1997; 387:917–921. doi: 10.1038/43206

[85] Kallio J, Myllymäki H, Grönholm J, Armstrong M, Vanha-aho LM, Mäkinen L, Silvennoinen O, Valanne S, Rämet M: Eye transformer is a negative regulator of Drosophila JAK/STAT signaling. FASEB J. 2010; 24:4467–4479. doi: 10.1096/fj.10-162784

[86] Makki R, Meister M, Pennetier D, Ubeda JM, Braun A, Daburon V, Krzemień J, Bourbon HM, Zhou R, Vincent A et al.: A short receptor downregulates JAK/STAT signalling to control the Drosophila cellular immune response. PLoS Biol. 2010; 8:e1000441. doi: 10.1371/journal.pbio.1000441

[87] Callus B, Mathey-Prevot B: SOCS36E, a novel Drosophila SOCS protein, suppresses JAK/STAT and EGF-R signalling in the imaginal wing disc. Oncogene 2002; 21:4812–4821. doi: 10.1038/sj.onc.1205618

[88] Ekengren S, Tryselius Y, Dushay MS, Liu G, Steiner H, Hultmark D: A humoral stress response in Drosophila. Curr. Biol. 2001; 11:714–718.

[89] Kemp C, Mueller S, Goto A, Barbier V, Paro S, Bonnay F, Dostert C, Troxler L, Hetru C, Meignin C et al.: Broad RNA interference-mediated antiviral immunity and virus-specific inducible responses in Drosophila. J. Immunol. 2013; 190:650–658. doi: 10.4049/jimmunol.1102486

[90] Buchon N, Broderick NA, Poidevin M, Pradervand S, Lemaitre B: *Drosophila* intestinal response to bacterial infection: activation of host defense and stem cell proliferation. Cell Host Microbe. 2009; 5:200–211. doi: 10.1016/j.chom.2009.01.003

[91] Buchon N, Silverman, Cherry S: Immunity in Drosophila melanogaster—from microbial recognition to whole-organism physiology. Nat. Rev. Immunol. 2014; 14:796–810. doi: 10.1038/nri3763

[92] Kurata S: Peptidoglycan recognition proteins in *Drosophila* immunity. Dev. Comp. Immunol. 2014; 42:36–41. doi: 10.1016/j.dci.2013.06.006

[93] Dziarski R, Gupta D: Mammalian PGRPs: novel antibacterial proteins. Cell. Microbiol. 2006; 8:1059–1069. doi: 10.1111/j.1462-5822.2006.00726.x

[94] Dziarski R, Gupta D: Review: Mammalian peptidoglycan recognition proteins (PGRPs) in innate immunity. Innate Immun. 2010; 16:168–174. doi: 10.1177/1753425910366059

[95] Charroux B, Rival T, Narbonne-Reveau K, Royet J: Bacterial detection by *Drosophila* peptidoglycan recognition proteins. Microbes Infect. 2009; 11:631–636. doi: 10.1016/j.micinf.2009.03.004

[96] Royet J, Gupta D, Dziarski R: Peptidoglycan recognition proteins: modulators of the microbiome and inflammation. Nat. Rev. Immunol. 2011; 11:837–851. doi: 10.1038/nri3089

[97] Steiner H: Peptidoglycan recognition proteins: on and off switches for innate immunity. Immunol. Rev. 2004; 198:83–96.

[98] Mellroth P, Steiner H: PGRP-SB1: an N-acetylmuramoyl L-alanine amidase with antibacterial activity. Biochem. Biophys. Res. Commun. 2006; 350:994–999. doi: 10.1016/j.bbrc.2006.09.139

[99] Pal S, Wu LP: Lessons from the fly: pattern recognition in *Drosophila melanogaster*. Adv, Exp. Med. Biol. 2009; 653:162–174.

[100] Jiang H, Ma C, Lu ZQ, Kanost MR: Beta-1,3-glucan recognition protein-2 (betaGRP-2) from *Manduca sexta*; an acute-phase protein that binds beta-1,3-glucan and lipoteichoic acid to aggregate fungi and bacteria and stimulate prophenoloxidase activation. Insect Biochem. Mol. Biol. 2004; 34:89–100.

[101] Wang Y, Sumathipala N, Rayaprolu S, Jiang H: Recognition of microbial molecular patterns and stimulation of prophenoloxidase activation by a β-1,3-glucanase-related protein in *Manduca sexta* larval plasma. Insect Biochem. Mol. Biol. 2011; 41:322–331. doi: 10.1016/j.ibmb.2011.01.010

[102] Pauchet Y, Freitak D, Heidel-Fischer HM, Heckel DG, Vogel H: Immunity or digestion: glucanase activity in a glucan-binding protein family from Lepidoptera. J. Biol. Chem. 2009; 284:2214–2224. doi:10.1074/jbc.M806204200

[103] Rosales C, O'Brien V, Kornberg L, Juliano RL: Signal transduction by cell adhesion receptors. Biochim. Biophys. Acta 1995; 1242:77–98.

[104] Gandhe AS, Arunkumar KP, John SH, Nagaraju J: Analysis of bacteria-challenged wild silkmoth, *Antheraea mylitta* (lepidoptera) transcriptome reveals potential immune genes. BMC Genomics. 2006; 7:184. doi: 10.1186/1471-2164-7-184

[105] Eum JH, Seo YR, Yoe SM, Kang SW, Han SS: Analysis of the immune-inducible genes of *Plutella xylostella* using expressed sequence tags and cDNA microarray. Dev. Comp. Immunol. 2007; 31:1107–1120. doi: 10.1016/j.dci.2007.02.002

[106] Bao Y, Yamano Y, Morishima I: Induction of hemolin gene expression by bacterial cell wall components in eri-silkworm, *Samia cynthia ricini*. Comp. Biochem. Physiol. Part B. Biochem. Mol. Biol. 2007; 146:147–151. doi: 10.1016/j.cbpb.2006.10.092

[107] Eleftherianos I, Gökçen F, Felföldi G, Millichap PJ, Trenczek TE, ffrench-Constant RH, Reynolds SE: The immunoglobulin family protein Hemolin mediates cellular immune responses to bacteria in the insect *Manduca sexta*. Cell. Microbiol. 2007; 9:1137–1147. doi: 10.1111/j.1462-5822.2006.00855.x

[108] Yu X-Q, Kanost MR: Immulectin-2, a Lipopolysaccharide-specific lectin from an insect, *Manduca sexta*, is induced in response to Gram-negative bacteria. J. Biol. Chem. 2000; 275:37373–37381. doi:10.1074/jbc.M003021200

[109] Yu XQ, Ling E, Tracy ME, Zhu Y: Immulectin-4 from the tobacco hornworm *Manduca sexta* binds to lipopolysaccharide and lipoteichoic acid. Insect Mol. Biol. 2006; 15:119–128. doi: 10.1111/j.1365-2583.2006.00618.x

[110] Watanabe A, Miyazawa S, Kitami M, Tabunoki H, Ueda K, Sato R: Characterization of a novel C-type lectin, *Bombyx mori* multibinding protein, from the *B. mori* hemolymph: mechanism of wide-range microorganism recognition and role in immunity. J. Immunol. 2006; 177:4594–4604.

[111] Shin SW, Park DS, Kim SC, Park HY: Two carbohydrate recognition domains of *Hyphantria cunea* lectin bind to bacterial lipopolysaccharides through O-specific chain. FEBS Lett. 2000; 467:70–74.

[112] Schmid-Hempel P: Evolutionary ecology of insect immune defenses. Annu. Rev. Entomol. 2005; 50:529–551. doi: 10.1146/annurev.ento.50.071803.130420

[113] Garcia-Garcia E, Garcia-Garcia PL, Rosales C: An fMLP receptor is involved in activation of phagocytosis by hemocytes from specific insect species. Dev. Comp. Immunol. 2009; 33:728–739. doi: 10.1016/j.dci.2008.12.006

[114] Gardiner EM, Strand MR: Monoclonal antibodies bind distinct classes of hemocytes in the moth Pseudoplusia includens. J. Insect Physiol. 1999; 45:113–126.

[115] Nardi JB, Pilas B, Bee CM, Zhuang S, Garsha K, Kanost MR: Neuroglian-positive plasmatocytes of *Manduca sexta* and the initiation of hemocyte attachment to foreign surfaces. Dev. Comp. Immunol. 2006; 30:447–462. doi: 10.1016/j.dci.2005.06.026

[116] Wiegand C, Levin D, Gillespie J, Willott E, Kanost M, Trenczek T: Monoclonal antibody MS13 identifies a plasmatocyte membrane protein and inhibits encapsulation and spreading reactions of *Manduca sexta* hemocytes. Arch. Insect Biochem. Physiol. 2000; 45:95–108. doi: 10.1002/1520-6327(200011)45:3<95::AID-ARCH1>3.0.CO;2-0

[117] Zhang S, Clark KD, Strand MR: The protein P23 identifies capsule-forming plasmatocytes in the moth *Pseudoplusia includens*. Dev. Comp. Immunol. 2011; 35:501–510. doi: 10.1016/j.dci.2010.12.006

[118] Parsons B, Foley E: Cellular immune defenses of *Drosophila melanogaster*. Dev. Comp. Immunol. 2016; 58:95–101. doi: 10.1016/j.dci.2015.12.019

[119] Crozatier M, Meister M: Drosophila haematopoiesis. Cell. Microbiol. 2007; 9:1117–1126. doi: 10.1111/j.1462-5822.2007.00930.x

[120] Hillyer JF: Insect immunology and hematopoiesis. Dev. Comp. Immunol. 2016; 58:102–118. doi: 10.1016/j.dci.2015.12.006

[121] Honti V, Csordas G, Kurucz E, Markus R, Ando I: The cell-mediated immunity of *Drosophila melanogaster*: hemocyte lineages, immune compartments, microanatomy and regulation. Dev. Comp. Immunol. 2014; 42:47–56. doi: 10.1016/j.dci.2013.06.005

[122] Wood W, Faria C, Jacinto A: Distinct mechanisms regulate hemocyte chemotaxis during development and wound healing in *Drosophila melanogaster*. J. Cell Biol. 2006; 173:405–416. doi: 10.1083/jcb.200508161

[123] Nagaosa K, Okada R, Nonaka S, Takeuchi K, Fujita Y, Miyasaka T, Manaka J, Ando I, Nakanishi Y: Integrin βν-mediated phagocytosis of apoptotic cells in *Drosophila* embryos. J. Biol. Chem. 2011; 286:25770–25777. doi:10.1074/jbc.M110.204503

[124] Stofanko M, Kwon SY, Badenhorst P: Lineage tracing of lamellocytes demonstrates *Drosophila* macrophage plasticity. PLoS One 2010; 5:e14051. doi: 10.1371/journal.pone.0014051

[125] Shaukat Z, Liu D, Gregory S: Sterile inflammation in Drosophila. Mediators Inflamm. 2015; 2015:369286. doi: 10.1155/2015/369286

[126] Markus R, Kurucz E, Rus F, Ando I: Sterile wounding is a minimal and sufficient trigger for a cellular immune response in *Drosophila melanogaster*. Immunol. Lett. 2005; 101:108–111. doi: 10.1016/j.imlet.2005.03.021

[127] Schmidt O, Söderhäll K, Theopold U, Faye I: Role of adhesion in arthropod immune recognition. Annu. Rev. Entomol. 2010; 55:485–504. doi: 10.1146/annurev.ento.54.110807.090618

[128] Ratheesh A, Belyaeva V, Siekhaus DE: *Drosophila* immune cell migration and adhesion during embryonic development and larval immune responses. Curr. Opin. Cell Biol. 2015; 36:71–79. doi: 10.1016/j.ceb.2015.07.003

[129] Rosales C: Phagocytosis, a cellular immune response in insects. Inv. Surv. J. 2011; 8:109–131.

[130] Castillo JC, Robertson AE, Strand MR: Characterization of hemocytes from the mosquitoes Anopheles gambiae and Aedes aegypti. Insect Biochem. Mol. Biol. 2006; 36:891–903. doi: 10.1016/j.ibmb.2006.08.010

[131] Lamprou I, Mamali I, Dallas K, Fertakis V, Lampropoulou M, Marmaras VJ: Distinct signalling pathways promote phagocytosis of bacteria, latex beads and lipopolysaccharide in medfly haemocytes. Immunology 2007; 121:314–327. doi: 10.1111/j.1365-2567.2007.02576.x

[132] Franc NC, Dimarcq JL, Lagueux M, Hoffmann J, Ezekowitz RA: Croquemort, a novel Drosophila hemocyte/macrophage receptor that recognizes apoptotic cells. Immunity 1996; 4:431–443.

[133] Manaka J, Kuraishi T, Shiratsuchi A, Nakai Y, Higashida H, Henson P, Nakanishi Y: Draper-mediated and phosphatidylserine-independent phagocytosis of apoptotic cells by *Drosophila* hemocytes/macrophages. J. Biol. Chem. 2004; 279:48466–48476. doi: 10.1074/jbc.M408597200

[134] Vlisidou I, Dowling AJ, Evans IR, Waterfield N, ffrench-Constant RH, Wood W: Drosophila embryos as model systems for monitoring bacterial infection in real time. PLoS Pathog. 2009; 5:e1000518. doi: 10.1371/journal.ppat.1000518

[135] Kocks C, Cho JH, Nehme N, Ulvila J, Pearson AM, Meister M, Strom C, Conto SL, Hetru C, Stuart LM et al.: Eater, a transmembrane protein mediating phagocytosis of bacterial pathogens in Drosophila. Cell 2005; 123:335–346. doi: 10.1016/j.cell.2005.08.034

[136] Kurucz E, Márkus R, Zsámboki J, Folkl-Medzihradszky K, Darula Z, Vilmos P, Udvardy A, Krausz I, Lukacsovich T, Gateff E et al.: Nimrod, a putative phagocytosis

receptor with EGF repeats in Drosophila plasmatocytes. Curr. Biol. 2007; 17:649–654. doi: 10.1016/j.cub.2007.02.041

[137] Nakatogawa S, Oda Y, Kamiya M, Kamijima T, Aizawa T, Clark KD, Demura M, Kawano K, Strand MR, Hayakawa Y: A novel peptide mediates aggregation and migration of hemocytes from an insect. Curr. Biol. 2009; 19:779–785. doi: 10.1016/j.cub.2009.03.050

[138] Strand MR: The insect cellular immune response. Insect Science 2008; 15:1–14.

[139] Shrestha S, Kim YJ: Various eicosanoids modulate the cellular and humoral immune responses of the beet armyworm, *Spodoptera exigua*. Biosci. Biotechnol. Biochem. 2009; 73:2077–2084. doi: 10.1271/bbb.90272

[140] Shrestha S, Park Y, Stanley D, Kim Y: Genes encoding phospholipases A2 mediate insect nodulation reactions to bacterial challenge. J. Insect Physiol. 2010; 56:324–332. doi: 10.1016/j.jinsphys.2009.11.008

[141] Stanley D, Miller J, Tunaz H: Eicosanoid actions in insect immunity. J. Innate Immun. 2009; 1:282–290. doi: 10.1159/000210371

[142] Zhao F, Stanley D, Wang Y, Zhu F, Lei CL: Eicosanoids mediate nodulation reactions to a mollicute bacterium in larvae of the blowfly, *Chrysomya megacephala*. J. Insect Physiol. 2009; 55:192–196. doi: 10.1016/j.jinsphys.2008.10.018

[143] Gandhe AS, John SH, Nagaraju J: Noduler, a novel immune up-regulated protein mediates nodulation response in insects. J. Immunol. 2007; 179:6943–6951.

[144] Bao YY, Xue J, Wu WJ, Wang Y, Lv ZY, Zhang CX: An immune-induced reeler protein is involved in the *Bombyx mori* melanization cascade. Insect Biochem. Mol. Biol. 2011; 41:696–706. doi: 10.1016/j.ibmb.2011.05.001

[145] Ling E, Yu XQ: Cellular encapsulation and melanization are enhanced by immulectins, pattern recognition receptors from the tobacco hornworm *Manduca sexta*. Dev. Comp. Immunol. 2006; 30:289–299. doi: 10.1016/j.dci.2005.05.005

[146] Carton Y, Frey F, Nappi AJ: Parasite-induced changes in nitric oxide levels in *Drosophila paramelanica*. J. Parasitol. 2009; 95:1134–1141. doi: 10.1645/GE-2091.1

[147] Rosales C: Fc receptor and integrin signaling in phagocytes. Signal Transduction 2007; 7:386–401.

[148] Zhuang S, Kelo L, Nardi JB: Multiple α subunits of integrin are involved in cell-mediated responses of the *Manduca* immune system. Dev. Comp. Immunol. 2008; 32:365–379. doi: 10.1016/j.dci.2007.07.007

[149] Irving P, Ubeda JM, Doucet D, Troxler L, Lagueux M, Zachary D, Hoffmann JA, Hetru C, Meister M: New insights into *Drosophila* larval haemocyte functions through genome-wide analysis. Cell Microbiol. 2005; 7:335–350.

[150] Altincicek B, Stotzel S, Wygrecka M, Preissner KT, Vilcinskas A: Host-derived extracellular nucleic acids enhance innate immune responses, induce coagulation, and prolong survival upon infection in insects. J. Immunol. 2008; 181:2705–2712.

[151] Branzk N, Papayannopoulos V: Molecular mechanisms regulating NETosis in infection and disease. Semin. Immunopathol. 2013; 35:513–530. doi: 10.1007/s00281-013-0384-6

[152] Brinkmann V, Zychlinsky A: Neutrophil extracellular traps: is immunity the second function of chromatin? J. Cell Biol. 2012; 198:773–783. doi: 10.1083/jcb.201203170

[153] Robb CT, Dyrynda EA, Gray RD, Rossi AG, Smith VJ: Invertebrate extracellular phagocyte traps show that chromatin is an ancient defence weapon. Nat. Commun. 2014; 5:4627. doi: 10.1038/ncomms5627

[154] Nappi A, Poirié M, Carton Y: The role of melanization and cytotoxic by-products in the cellular immune responses of *Drosophila* against parasitic wasps. Adv. Parasitol. 2009; 70:99–121. doi: 10.1016/S0065-308X(09)70004-1

[155] Eleftherianos I, Revenis C: Role and importance of phenoloxidase in insect hemostasis. J. Innate Immun. 2011; 3:28–33. doi: 10.1159/000321931

[156] Krem MM, Di Cera E: Evolution of enzyme cascades from embryonic development to blood coagulation. Trends Biochem. Sci. 2002; 27:67–74.

[157] Ling E, Yu XQ: Prophenoloxidase binds to the surface of hemocytes and is involved in hemocyte melanization in Manduca sexta. Insect Biochem. Mol. Biol. 2005; 35:1356–1366. doi: 10.1016/j.ibmb.2005.08.007

[158] AliyariR,DingSW:RNA-basedviralimmunityinitiatedbytheDicerfamilyofhostimmune receptors. Immunol. Rev. 2009; 227:176–188. doi: 10.1111/j.1600-065X.2008.00722.x

[159] Bernstein E, Caudy AA, Hammond SM, Hannon GJ: Role of a bidentate ribonuclease in the initiation step of RNA interference. Nature 2001; 409:363–366. doi: 10.1038/35053110

[160] Liu Q, Rand TA, Kalidas S, Du F, Kim H-E, Smith DP, Wang X: R2D2, a bridge between the initiation and effector steps of the Drosophila RNAi pathway. Science 2003; 301:1921–1925. doi: 10.1126/science.1088710

[161] Okamura K, Chung WJ, Ruby JG, Guo H, Bartel DP, Lai EC: The Drosophila hairpin RNA pathway generates endogenous short interfering RNAs. Nature 2008; 453:803–806. doi: 10.1038/nature07015

[162] Rand TA, Ginalski K, Grishin NV, Wang X: Biochemical identification of Argonaute 2 as the sole protein required for RNA-induced silencing complex activity. Proc. Natl. Acad. Sci. U S A 2004; 101:14385–14389. doi: 10.1073/pnas.0405913101

[163] Aliyari R, Wu Q, Li HW, Wang XH, Li F, Green LD, Han CS, Li WX, Ding SW: Mechanism of induction and suppression of antiviral immunity directed by virus-derived small RNAs in Drosophila. Cell Host Microbe. 2008; 4:387–397. doi: 10.1016/j.chom.2008.09.001

[164] Nayak A, Berry B, Tassetto M, Kunitomi M, Acevedo A, Deng C, Krutchinsky A, Gross J, Antoniewski C, Andino R: Cricket paralysis virus antagonizes Argonaute 2 to modulate antiviral defense in Drosophila. Nat. Struct. Mol. Biol. 2010; 17:547–554. doi: 10.1038/nsmb.1810

[165] Lingel A, Simon B, Izaurralde E, Sattler M: The structure of the Flock House virus B2 protein, a viral suppressor of RNA interference, shows a novel mode of double-stranded RNA recognition. EMBO Rep. 2005; 6:1149–1155. doi: 10.1038/sj.embor.7400583

[166] Kakumani PK, Ponia SS, S RK, Sood V, Chinnappan M, Banerjea AC, Medigeshi GR, Malhotra P, Mukherjee SK, Bhatnagar RK: Role of RNA interference (RNAi) in dengue virus replication and identification of NS4B as an RNAi suppressor. J. Virol. 2013; 87:8870–8883. doi: 10.1128/JVI.02774-12

[167] Zambon RA, Nandakumar M, Vakharia VN, Wu LP: The Toll pathway is important for an antiviral response in Drosophila. Proc. Natl. Acad. Sci. U S A 2005; 102:7257–7262. doi: 10.1073/pnas.0409181102

[168] Xi Z, Ramirez JL, Dimopoulos G: The *Aedes aegypti* toll pathway controls dengue virus infection. PLoS Pathog. 2008; 4:e1000098. doi: 10.1371/journal.ppat.1000098

[169] Avadhanula V, Weasner BP, Hardy GG, Kumar JP, Hardy RW: A novel system for the launch of alphavirus RNA synthesis reveals a role for the Imd pathway in arthropod antiviral response. PLoS Pathog. 2009; 5:e1000582. doi: 10.1371/journal.ppat.1000582

[170] Costa A, Jan E, Sarnow P, Schneider D: The Imd pathway is involved in antiviral immune responses in Drosophila. PLoS One 2009; 4:e7436. doi: 10.1371/journal.pone.0007436

[171] Hedges LM, Johnson KN: Induction of host defence responses by Drosophila C virus. J. Gen. Virol. 2008; 89:1497–1501. doi: 10.1099/vir.0.83684-0

[172] Souza-Neto JA, Sim S, Dimopoulos G: An evolutionary conserved function of the JAK-STAT pathway in anti-dengue defense. Proc. Natl. Acad. Sci. U S A 2009; 106:17841–17846 doi: 10.1073/pnas.0905006106

[173] Paradkar PN, Trinidad L, Voysey R, Duchemin JB, Walker PJ: Secreted Vago restricts West Nile virus infection in Culex mosquito cells by activating the Jak-STAT pathway. Proc. Natl. Acad. Sci. U S A 2012; 109:18915–18920. doi: 10.1073/pnas.1205231109

[174] Lamiable O, Imler JL: Induced antiviral innate immunity in Drosophila. Curr. Opin. Microbiol. 2014; 20:62–68. doi: 10.1016/j.mib.2014.05.006

[175] Deddouche S, Matt N, Budd A, Mueller S, Kemp C, Galiana-Arnoux D, Dostert C, Antoniewski C, Hoffmann JA, Imler JL: The DExD/H-box helicase Dicer-2 mediates the induction of antiviral activity in Drosophila. Nat. Immunol. 2008; 9:1425–1432. doi: 10.1038/ni.1664

[176] Kingsolver MB, Huang Z, Hardy RW: Insect antiviral innate immunity: pathways, effectors, and connections. J. Mol. Biol. 2013; 425:4921–4936. doi: 10.1016/j.jmb.2013.10.006

[177] Nakamoto M, Moy R, Xu J, Bambina S, Yasunaga A, Shelly SS, Gold B, Cherry S: Virus recognition by Toll-7 activates antiviral autophagy in Drosophila. Immunity 2012; 36:658–667. doi: 10.1016/j.immuni.2012.03.003

[178] Shelly S, Lukinova N, Bambina S, Berman A, Cherry S: Autophagy is an essential component of Drosophila immunity against vesicular stomatitis virus. Immunity 2009; 30:588–598. doi: 10.1016/j.immuni.2009.02.009

[179] Kuma A, Mizushima N: Physiological role of autophagy as an intracellular recycling system: with an emphasis on nutrient metabolism. Semin. Cell Dev. Biol. 2010; 21:683–690. doi: 10.1016/j.semcdb.2010.03.002

[180] Stephan JS, Herman P: The regulation of autophagy in eukaryotic cells: do all roads pass through Atg1? Autophagy 2006; 2:146–148.

[181] Mizushima N: The role of the Atg1/ULK1 complex in autophagy regulation. Curr. Opin. Cell Biol. 2010; 22:132–139. doi: 10.1016/j.ceb.2009.12.004

[182] Pham LN, Dionne MS, Shirasu-Hiza M, Schneider DS: A specific primed immune response in *Drosophila* is dependent on phagocytes. PLoS Pathog. 2007; 3:1–8. doi: 10.1371/journal.ppat.0030026

Section 4

The Edible Insect: Used as Food Sources

Chapter 9

Potential of Insect-Derived Ingredients for Food Applications

Daylan Amelia Tzompa Sosa and Vincenzo Fogliano

Additional information is available at the end of the chapter

http://dx.doi.org/10.5772/67318

"The supreme irony is that billions are spent every year to save crops that contain no more than 14% of plant protein by killing another food source (insects) that may contain up to 75% of high quality animal protein". [1]

Abstract

Insects are a sustainable and efficient protein and lipid source, compared with conventional livestock. Moreover, insect proteins and lipids are highly nutritional. Therefore, insect proteins and lipids can find its place as food ingredients. The use of insect proteins and lipids as food ingredients requires a deep understanding on the chemical and physical characteristics of these ingredients, as well as its functionality. Information on the chemical and physical characteristics of insect proteins and oils will help to assess the possibilities of its use on different food applications. In this chapter, we briefly review the nutritional aspects of insect proteins and lipids, insect processing, protein and lipid extraction as well as the perspectives of food applications of insect protein and lipids. Future studies should delve into extraction methods and into intrinsic properties of insect ingredients. This knowledge will be useful to introduce insect ingredients into various food preparations. Also valuable will be the study of other insect species with perspectives for its commercial rearing.

Keywords: fractionation, insect proteins, insect fats and oils, food ingredient, food applications

1. Introduction

Sustainability is becoming increasingly important in the world. Alternative sources of proteins and oils have to be found to replace traditional less sustainable ingredients. One of the possible alternatives to replace traditional proteins and oils in food products is insects. About 1900 insect species are traditionally consumed as a whole or low processed by approximately two billion people worldwide [2]. In some cases, insects have been consumed as emergency food, in other circumstances as staple food, and in other cultures, they are even considered delicacies [3]. Insects are in general a healthy food source with a high content of protein, fat, vitamins, minerals and fibres. Furthermore, there is a wide range of edible insects which contributes to a high variation in terms of protein and fat profiles. Using insects as a source of protein and lipids can contribute to global food security through feed or as a direct food. Edible insects have been traditionally harvested from natural forests or in the fields, providing food in rural areas for self-consumption or local markets.

Despite the proven nutritional aspects of insects, in the western world, the acceptability of insects as food is low due to its association as pests and disease transmitters [4], which is often described as "dirty image." In other countries with entomophagy tradition, the consumption of insects has decreased due to the growing urban society. FAO has pointed out the need to examine modern food science practices to increase insect trade, consumption and acceptance. Food scientists and food technologists have been innovating and have been looking for alternative solutions for postharvest handling, to improve processing and increase shelf life of insect products to increase availability and consumer acceptance. One of the proposed solutions to increase consumer acceptance is isolation of insect proteins and fats to be used as food ingredient. This requires a deep understanding on the chemical and physical characteristics of insect proteins and lipids, its functionality as well as an assessment on the consumer's perception and motivations to accept this novel source. In this chapter we describe some nutritional aspects of insect proteins and lipids, insect processing, protein and lipid extraction as well as the perspectives of food applications of insect-derived ingredients. Since the use of insect-derived ingredients is still in its infancy, there are more questions than answers. In this chapter, we also point out the need of information on the chemical and physical characteristics of insect proteins and oils that will help to assess the possibilities of its use for different food applications.

2. Insect proteins

Food scientists need to tackle in the next future the challenge of food security: how to ensure enough protein production to the 2–3 billions of additional people that will populate the planet in the coming decades [5]. There are different possibilities that could be explored as alternative source to obtain a sustainable production of proteins in the future. In this vein, several studies have been done looking at the potentiality of protein from yeast [6], microscopic fungi [7] and microalgae [8].

Recently, insects have been proposed as one of the most promising alternative source of proteins to solve the global issue of protein shortage: the main advantage of insects over other protein sources is the low environmental costs of production, which becomes essential to satisfy the rise in the global protein demand [2, 9]. In **Table 1**, a short summary of the main

Organism	Pro	Cons	Reference
Insects	Thousands of species High conversion rate of feed into edible biomass Protein and lipid production	Low consumer acceptability Scale up of rearing facilities	[10]
Fungi (Quorn)	Easy to grow and harvest Good consumer acceptability	Lower growth rate and relatively low protein content	[7]
Microalgae	Metabolic versatility Environmental friendly Easy to grow and harvest High-quality proteins and lipids	Nondigestible cell wall Easily contaminated by heavy metals	[8]
Yeast	High consumer acceptability Production easy to scale up Low DNA amount	Slow growth rate Low protein content	[6]

Table 1. Intrinsic advantage and disadvantage related to the use of different organisms for feed and food production.

advantage and disadvantages related to the use of insect proteins in comparison with other alternative protein sources is provided.

Table 1 focuses on the intrinsic advantage and disadvantage of the different organisms; however, it is clear that new technological solutions as well as changes in consumer awareness can probably solve many of the present disadvantages. It is very likely that in the next few years, the scenario will be significantly different.

In particular, it is important to consider that the sustainability aspects are gaining more and more importance and for this reason, they are already shaping the production systems in many food chains. In this respect, insects and microalgae have now the competitive advantage that can be produced using nonsophisticated technological infrastructure. Moreover, the possibility to feed insect using by-products of food productions or organic municipality waste could represent the perfect solution to close the circle of food and feed production favouring the transition towards a zero waste system.

As for the sustainability and also food security standpoint, the best way to take advantage from the insect proteins is to use the whole insect eventually grinded to favour consumer acceptance. However, protein-rich ingredients, which can be used in various food preparations, are also very important, and their development would require a better knowledge of the intrinsic properties of these proteins and also to understand the behaviour of insect proteins during extraction and processing.

2.1. Insect processing and protein extraction

Protein extraction and fractionation are necessary steps to produce insect protein-rich ingredients. The protein concentration in the various insect species is usually very high (typically 50–70% of the dry matter), and this facilitates the isolation process.

Grinding of fresh insect is mechanically complicated, and it would result in the production of a slurry, which could be difficult to store and be further processed. Therefore, the first necessary processing step immediately after the insect harvesting is their drying. The conditions of the biomass preliminary treatment are critical: time/temperature can be adjusted according to the species and their initial water content. The damage of the tissue should be limited to avoid the contact of proteolytic and browning enzymes with their substrates. These enzymes are confined in specific organelles surrounded by membranes. The main objective of the drying step is to reduce the water content up to 5–10%. In this condition, the microbiological growth is ruled out, and the quality of the starting material can be better preserved. Fat oxidation is another main factor that could bring to quality decay during this step and during the following storage. In production, plants dedicated to insect protein production, drying and defatting conditions can be adjusted according to the need of the following steps.

In fact, for many purposes and especially to produce insect protein-rich ingredients, extraction of fats and grinding into small particles are essential prerequisite. Also when protein extraction is not the ultimate goal, it must be considered that the particle size of the final powder has a big impact on the technological properties of the insect-based ingredients. In particular grinding conditions will influence the:

- Dispersibility and solubility
- Water and lipid holding capacity
- Rheological performance when rehydrated

Therefore, a careful selection of the grinding conditions according to the species should be performed particularly when adult insect (such as crickets or grasshoppers) rich in chitins and with a structured exoskeleton must be processed.

Up to now mainly wet fractionation processes have been proposed in the literature, although in principle dry fractionation could potentially lead to a better quality, and it is also intrinsically more sustainable because you do not need to eliminate water after extractions.

Data on the amount of proteins that can be extracted by grinded insect powder in aqueous solutions showed an enormous variability [11]. Factors like solid/water ratio as well as the pH of the media and the temperature play an important role. Using alkaline pHs (i.e. far from the protein isoelectric point), higher extraction yield was obtained [12]. Also ionic strength, i.e. the presence of sodium chloride in the solution, can increase the solubility, and as above mentioned, the finer the granulometry, the higher the extraction yield. However, the most important factor affecting protein extractability is the thermal treatment insects undergo before extraction. When the biomass is heated before extraction, the yield of soluble protein extraction drops dramatically moving from about 50% of the total proteins below the 20%. **Figure 1** suggests that the denaturation of insect proteins decreases their water solubility in a similar way observed for meat proteins.

There are some good reasons to perform a thermal treatment before extraction. First of all, the higher the temperature, the faster is the drying step. Secondly, a treatment at high temperature can be used to prevent enzymatic browning. Enzymatic browning catalysed by polyphenol oxidases (PPOs) is a major phenomenon taking place during insect protein extraction on aqueous solutions [12]. The browning reaction has negative effects on proteins by affecting the solubility and other technological functionalities. Moreover, brown colour is generally not well accepted by consumers, and consequently brown ingredients are always more difficult to use in several food preparations. Browning can be prevented also with not thermal treatment adding chemical agent such as sodium bisulphite or ascorbic acid in the extraction buffer [12].

The design of an insect protein extraction and fractionation process could often result in a trade-off between yield and purity. A higher yield of the extraction always comes with a lower purity of the final product and *vice versa*. For instance, an aqueous extraction followed by an acidic precipitation step could bring to a fraction containing more than 85% of proteins, which is a remarkably high purity, but this comes at the expenses of a final yield below 20% [13].

In such a condition, it is useful to follow an approach driven by the ingredient final destination of use, i.e. when the final goal is to get the maximum nutrient value, it is logical to prioritise the protein yield. On the other hand, when specific techno-functionality of proteins is desired, such as gelling or foaming properties, it is better to elaborate strategies that allow to obtain more purified preparation also at the expenses of a low yield.

2.2. Insect protein techno-functional properties and their food applications

Unfortunately, due to lack of standardised fractionation procedure, studies about the technological functionality of insect proteins are still at their infancy. Looking at what happened first with the dairy proteins and later on with legumes proteins (soybean, pea, lupine), the capability to fractionate them in defined preparations opened many possibilities of use as ingredients. We can assume that also for insect proteins, their use in different foods will be mainly driven by the added value they can bring to specific food preparations.

The main techno-functional properties that should be considered for insect proteins are:

- Water and lipid holding capacity
- Thickening capacity
- Emulsification capacity
- Foaming capacity
- Gelation capacity
- Structuring capacity

Interestingly, all these properties depend on the nature of the starting material (for instance, the species and maturity stage of the insects); however, they can be also modulated by the extraction and fractionation process especially by heating and the consequent protein denaturation.

Looking at the properties that can be influenced by the hydration of the proteins (i.e. the water holding and the thickening capacities), they can be particularly useful to improve the quality of meat and bakery products or even to modify the sensory properties of thick beverages such as the electrolyte-rich solutions for sportsman or the satiating products. As far as the emulsification and foaming capacities, they are particularly interesting for drinkable dairy products, desserts and salad dressing, while the gelation and structuring capacities are relevant for products such as cheese and tofu.

These potential applications of various insect protein fractions in real food product should also take into account the potential negative effects associated to their use. Although the increase of protein concentration is often desired parameters, insect proteins are often brown or even dark. The colour change in the final products caused by protein addition could be perceived as a very negative sensory attribute especially in dairy products. On the other hand, in many bakery products or when insect proteins will be used as meat replacers, this is not expected to be a significant problem.

2.3. Nutritional and healthy properties

A large part of the literature focused on the nutritional quality and the benefit related to the whole insect consumption. The attention is focused on proteins and the most important parameter is their total concentration. When measured on fresh weight basis, the insect protein concentration is within the range of that of the other animals, i.e. between 10 and 30%. So the real nutritional advantage of using insect proteins can be only understood when looking at the bigger picture as it was done by the consortium of the SUSFAN project. What makes the insect proteins particularly interesting from the nutritional point of view is not only their high concentration and the favourable amino acid composition but also their potential to meet all the so-called sustainable, healthy, affordable, reliable, palatable (SHARP) principles [14]. To have a 360 degree evaluation of the potentiality, the protein sources of the future should confront with these five criteria. In this respect, insects will perform very well.

In the previous paragraph, the reasons highlighting the sustainability, affordability, reliability and palatability of insect proteins have been discussed. Regarding the healthiness which includes also the nutritional properties, it is essential to say that insect proteins are rich in all essential amino acids. A diet based on insect proteins can perfectly sustain the harmonious growth of laboratory and husbandry animals also indicated that there are no major limiting or anti-nutritional factors affecting the growth of mice, fish and poultry [15].

Interestingly, insect proteins can be a good vehicle to increase mineral bioavailability. Insect contain much more iron, zinc and calcium than beef, pork and chicken [16]. As observed for the casein-calcium system in milk, it can be hypothesised that thanks to the interaction with specific peptides formed during insect protein digestion, these minerals can be more bioavailable than in protein-free food matrix.

Recent papers also confirmed that insect proteins are well digested during gastroduodenal digestion: in vitro data showed that the extent of the digestion depends also on the fraction of the insect proteins considered. In particular, the water soluble proteins are digested more efficiently than those remaining in the insoluble moiety [13]. No consisting data have been still published about the effect of browning and thermal treatment on the insect protein digestibility. Both phenomena can have opposite effects: on one hand they determine protein aggregation and crosslinking which decrease protein digestibility; on the other they favour protein denaturation, thus improving their degradation.

In addition to their use as nutrients, proteins from different sources have emerged as precursors of specific peptides, which are now known to be bioactive. A wide range of biological properties have been demonstrated with protein-derived bioactive peptides including anti-hypertensive, immunomodulatory, antimicrobial and antioxidant. These effects can be relevant in human and animal health promotion and can also be applied in food preservation for extended shelf life. The production of bioactive peptides has relied heavily on the use of food proteins, which further contributes to the depletion of their primary food sources. Edible insects can also be sustainable sources of proteins for bioactive peptide production. The availability of the gene sequence of major proteins in several insects such as *Tenebrio molitor* or *Hermetia illucens* allows the theoretical calculations of the linear structure, conformation and biological significance of peptide motifs released by specific proteinases used in industrial food processing. Open-access web-based tools can be used to verify the presence of these potential bioactive peptides, and enzymatic and microbiological methodology can be used to generate these peptides from the different insect biomasses.

We could image that the same variety of products and applications now available for dairy or soybean proteins will be soon available also for insects.

3. Insect lipids

Lipids in insects are either obtained from the diet or de novo synthesised. These lipids are stored in the insects in the fat body, where lipids are stored, degraded, transformed and further transported to the site of utilisation [17, 18]. Lipids are used by insects for several physiological functions, among them for reproduction, development, flight, buoyancy, integumental waterproofing, communication via pheromones, structure of cell membranes, etc. [17, 19]. Insects are rich in lipids; its content ranges from 10 to up to 50% in dry basis [20]. The fat content and the fatty acid (FA) composition of insect are related to its species, sex, life stage, diet, environmental temperature, diapause and migratory flight [17, 21, 22]. Therefore, it is not strange to find a wide variability of FA profiles among insects.

Insect lipids are composed mainly of triacylglycerols. Other types of lipids present in minor amounts include cholesterol, partial glycerides, free fatty acids (FFA), phospholipids and wax esters [18, 23]. When insect lipids are extracted for its consumption as edible oils, the main types of lipids are triacylglycerols, which reflect the original composition of the lipids in the insect. The presence of other minor compounds in the lipid extraction greatly depends on the extraction process.

The extraction of insect lipids aiming its use as edible oils has been investigated using Soxhlet extraction, an aqueous method and supercritical CO_2 extraction [24, 25]. The lipid extraction process does not have a major impact in the FA composition of the extract, but it strongly influences the lipid extraction yield and the types of lipids extracted. For instance, when aqueous extraction is used, only triacylglycerols are extracted. In contrast, when organic solvents are used, phospholipids, partial glycerides and triacylglycerols are extracted [25]. Partial glycerides, FFA and phospholipids are undesired in edible oils and fats. To eliminate these undesired compounds, the fats and oils undergo refining processes, which increases the cost of the oils and fats. Insect lipid aqueous extraction provides a high oil quality similar to that of virgin oils. However the yield is lower than that obtained using organic solvents or supercritical CO_2 extraction in which more than 95% of the lipids are extracted [24, 25]. Therefore, the extraction process should be carefully selected according to desired application and the costs of each extraction process. Other method for insect lipid extraction with industrial potential is extrusion and expelling, which is a method commonly used in oil extraction of seeds. However, this method remains unexplored in the scientific literature.

Most insect lipids are liquid at room temperature (20°C); thus, they are called "insect oils." Insect oils are rich in unsaturated fatty acids (UFA > 60% of total FA), being the most abundant C18:1 cis 9 (>30% of total FA) and C18:2 cis 9,12 (>20% of total FA). The most abundant saturated fatty acid (SFA) found in these oils is C16:0 (20–30% of total FA) [10, 26]. In general, insect oils are rich in essential fatty acids such as C18:2 cis 9,12 (LA) and contain other FA associated to health benefits such as C18:3 cis9,12,15 (ALA) and ω-3 FA, which further points its nutritional value. The concentration of these FA varies among species [21], and it is strongly influenced by the insect diet. Studies showed that the insects caught in the wild are richer in ω-3 than those commercially reared.

There are a few known insect lipids that are solid at room temperature, such as the lipids extracted from black soldier fly larvae (*H. illucens*). In this case, they are called "insect fats." The solid state of this insect fat reflects its high content in saturated FA, which ranges from 57 to 75% of total FA. The most abundant SFA in this fat are C12:0 (≈45% of total FA), C14:0 (≈9% of total FA) and C16:0 (≈12% of total FA) [22, 27]. This insect fat is especially interesting for food and feed applications since it is rich in C12:0, with a concentration similar to that of coconut oil. C12:0 is more water soluble and is readily digestible than long-chain FAs. Moreover, after digestion C12:0 is transported to the liver and is immediately converted to energy; thus, it is not stored as fat [28]. Other properties related to lauric acid and monolaurin are antimicrobial activity against gram-positive bacteria and a number of fungi and viruses [28, 29]. Therefore, the fat from black soldier fly larvae could be used as a functional ingredient in food applications.

In **Figure 1**, we show the spatial arrangement of vegetable oils and fats, animal fats and insect oils. This plot was obtained using the FA composition of different fats and oils and was analysed using principal component analysis. When the FA of insect oils are compared with that of vegetable oils, vegetable fats and animal fats, it appears that insect oils have a FA profile similar to that of vegetable oils and animal fats. This is due to the increase content of SFA, namely, C16:0 and C18:0, and at the same time a high content of mono- and polyunsaturated FA, namely, unsaturated C18 FA. Saturated C18 and C16 FA are increased in most animal fats, whereas unsaturated C18 FA are typical of vegetable oils.

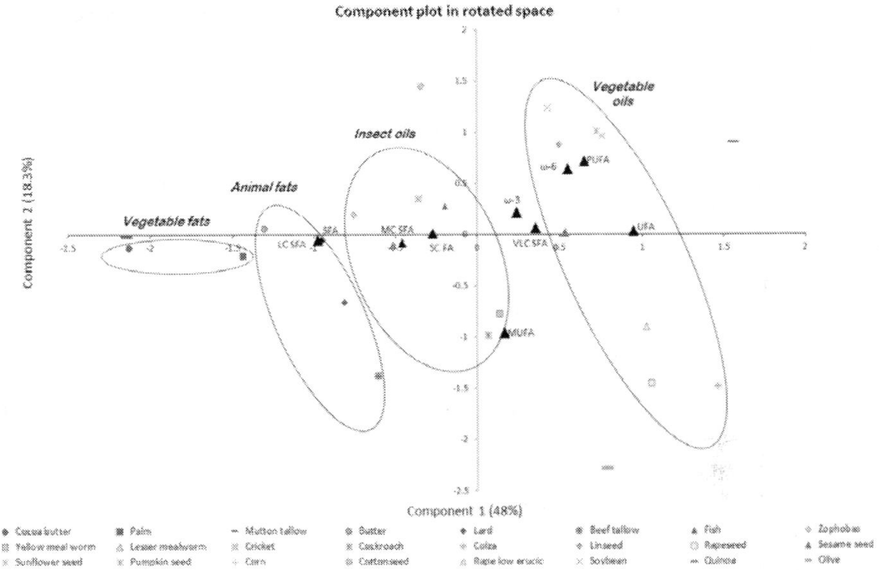

Figure 1. The first and second principal components of the principal component analysis. The amount of explained variance is provided in parenthesis in each axis. Fatty acids per group; LC SFA, long-chain saturated fatty acids; SFA, saturated fatty acids; MC SFA, middle-chain saturated fatty acids; SC FA, short-chain fatty acids; MUFA monounsaturated fatty acids; VLC SFA, very-long-chain saturated fatty acids; PUFA, polyunsaturated fatty acids; and UFA, unsaturated fatty acids. Source: Refs. [25, 30, 31] and the authors.

3.1. Potential food uses of insect oils

The liquid nature of insect oils makes them ideal for its use in mayonnaise, in vinaigrettes, as frying oils and as food grade lubricants, among others. Applications such as bakery, spreads, or confectionery are ideal for insect fats, such as the fat extracted from black soldier fly larvae, because these applications require a certain percentage of solid fat. In order to evaluate the specific applications of these oils, a detailed chemical and physical characterisation of the oils is necessary. Ekpo et al. [32] performed detailed chemical analysis in the extracted oil of four insect species consumed in Nigeria. They concluded that these insect oils had potential used in the pharmaceutical industry due to their low melting point, specific gravity (<0.89) and refractive index (<1.3) [32]. Other insect application oils remain unexplored in the scientific literature.

Traditionally, insect oils and fats have only being characterised in terms of its FA composition, mainly because this determines the nutritional value of fats and oils. However, other chemical and physical analyses are required in order to evaluate its potential applications as food ingredient. For instance, crystallisation, texture analysis and solid fat content at different temperatures are required to evaluate the use of a fat as a spread, in confectionary or its use as an ingredient in bakery. Rheological properties help to determine spreadability, ease of cutting and stand-up in margarines. Volatile compound analysis and sensory evaluations are required to assess its taste, mouth feel and aroma. Detailed triacylglycerol profiling and thermal behaviour analysis are required to evaluate the possibilities of fractionation of an insect fat or oil. Profiling of minor lipid compounds such as phospholipids and sterols help to fully assess its nutritional value. Currently, none of these information is available. However, information on the physical and chemical characteristics of insect fats and oils will become available as the attention of the consumers and the food industry for insect lipids increases.

3.2. Fractionation of an insect oil

Dry fractionation is a thermal procedure use to separate oils and fats into two or more components with different melting points. The aim of this separation is to extend the range of applications of the original oil or fat as well as to increase its commercial value. Fractionation is applied mainly to palm oil but also to coconut, palm kernel, butter oil and beef tallow, among others. Dry fractionation can be applied to insect oils showing different melting fractions. To show the possibilities of this process in insect oils, we applied dry fractionation to oil extracted from yellow mealworm (*T. molitor*). Then we performed physical and chemical analysis to the original oil and its fractions.

The first step of this study included a thermal analysis of the original oil. This analysis was performed to study the possibilities for fractionation of the yellow mealworm oil. In this analysis, the oil was heated to 70°C to eliminate all the crystals present in the oil; then the oil was cooled at a rate of 5°C/min from 70 to −60°C. In this step, the oil is crystallised and the crystallisation points of the fat are obtained. Then, the fat was melted again at 5°C/min from −60 to 70°C. In this step, all the formed crystals are melted and the melting points are obtained. The results of yellow mealworm oil showed four crystallisation points and three melting points (**Figure 2**). Each peak indicates the crystallisation or melting of crystals formed by triacylglycerols with similar structure. Therefore, when more than one crystallisation/melting point is obtained, it indicates that the triacylglycerols in the oil are crystallising/melting independently. This is because the structure and/or FA within the triacylglycerols are different and they cannot co-crystallise into one single crystal lattice. The results from this first thermal analysis showed that the yellow mealworm oil can be fractionated because several crystallisation and melting peaks were obtained and these peaks were clearly separated from each other.

In the second step of this study, we aimed to fractionate yellow mealworm oil into a high melting fraction (or stearin) and a low melting fraction (or olein). We applied two cooling temperatures, namely, 2 and 4°C, and then the solid and the liquid fractions were separated. These temperatures were selected because in the previous analysis, it was shown that the highest crystallisation point was 3.6°C. The first step of dry fractionation was to melt the fat at 60°C for 30 min to eliminate the crystals present. Then, the oil was placed in a water bath, at the crystallisation temperature (2, or 4°C) for 24 h. During this crystallisation process, only triacylglycerols capable of crystallising at this temperature nucleate and grow into crystals (solid fat); the rest of the triacylglycerols having lower crystallisation temperatures will remain liquid (liquid fat). Finally, the liquid fraction was separated from the solid fraction by centrifugation at 4800 g for 20 min at the cooling temperature. Both fractions were placed in a separated tube and weighted to obtain the fractionation yield. The physical (thermal analysis and colour) and chemical characteristics (FA composition) of the original oil and the obtained solid and liquid fractions were performed to assess the effect of fractionation in the oil.

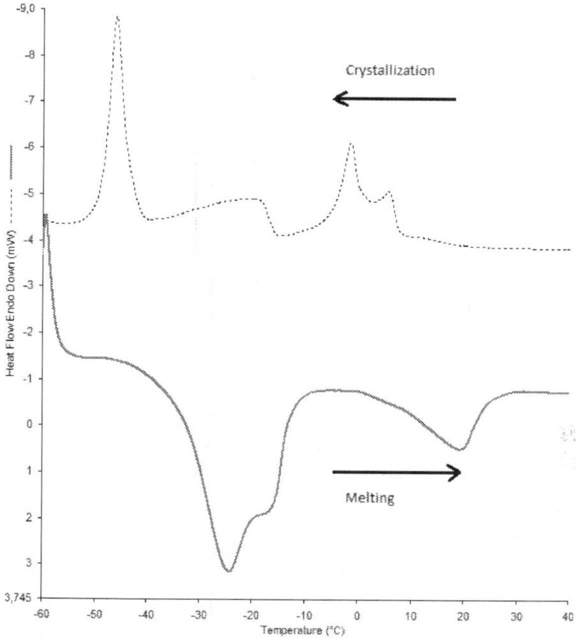

Figure 2. Non-isothermal crystallisation and melting of unfractionated yellow mealworm oil. Cooled from 70 to −60°C at 5°C/min and heated from −60 to 70°C at the same rate.

3.2.1. Physical characteristics of the yellow mealworm oil and its fractions

Separation into solid and liquid fat was possible at 2 and 4°C. Fractionation changed the colour of the liquid and solid fractions as compared with the unfractionated oil as analysed by Hunter Lab colorimeter in L*a*b* scale (**Table 2**). In general, the solid fractions had a more red-yellow colour when compared with the liquid fractions. The liquid fraction had a bright appearance, and the red and yellow tones were lower in the liquid oil fractionated at 2°C than oil separated at 4°C (**Figure 3**). As expected, the amount of liquid fat fraction increased as the crystallisation temperature increased (**Table 3**).

Sample	**L***	**a***	**b***
Unfractionated oil	15.72	1.35	12.86
Solid fraction 4°C	34.27	3.26	25.33
Solid fraction 2°C	21.01	3.14	18.97
Liquid fraction 4°C	5.12	0.15	2.81
Liquid fraction 2°C	4.40	−0.56	1.74

L*a*b* scale was used. The data shown are the average of two repetitions. Taken from [33]

Table 2. Colour of unfractionated and fractionated yellow mealworm oil measured at 20°C.

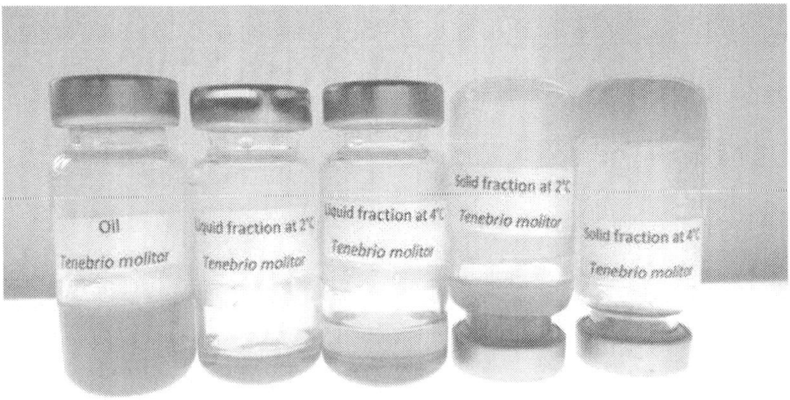

Figure 3. Unfractionated and fractionated yellow meal worm oil (*Tenebrio molitor*). This picture was taken after stabilising the oil for 2 h at 24°C.

The unfractionated oil of yellow mealworm showed four crystallisation points at −45.4, −19.9, -5 and 3.6°C and three melting points at −24.7, −17.4 and 16.7°C (**Table 3**). The liquid fractions lack the two highest crystallisation points at −5 and at 3.6°C, and the last melting point was reduced from 16.7 to 1.7 and 3.2°C (L4 and L2, respectively). The solid fractions lack the crystallisation point at −5°C and showed an increase in the highest crystallisation points from 3.6 to up to 10.7°C (S4 and S2, respectively). The final melting point of the solid fractions was increased from 16.7 to 21.5 and 27.1°C (S2 and S4, respectively).

Sample	Yield (%)	Crystallisation points (°C)				Melting points (°C)		
YMW		-45.4	-19.9	-5.0	3.6	-24.7	-17.4	16.7
S4	54	-46.3	-19.3	10.7	–	-24.4	-18.0	27.1
S2	81	-46.3	-19.8	2.7	–	-24.4	-16.9	21.5
L4	46	-49.4	-19.7	–	–	-25.5	-18.3	1.7
L2	19	-44.7	-19.5	–	–	-24.3	-17.4	3.2

Non-isothermal analysis cooled from 70 to −60°C and then heated from −60 to 70°C at 5°C/min.
YMW: original yellow meal worm oil extracted by Soxhlet using petroleum ether
S4, solid fraction at 4 °C; S2, solid fraction at 2 °C; L4, liquid fraction at 4 °C; and L2, Liquid fraction at 2 °C
Adapted from Ref. [33]

Table 3. Yield, crystallisation and melting points of unfractionated and fractionated yellow meal worm oil (N = 2).

http://dx.doi.org/10.5772/67318

After fractionation the solid and the liquid fraction did show differences in their physical properties, as shown by the change in colour and by the changes in crystallisation and melting points. The liquid fractions obtained after fractionation had bright and transparent appearance and were liquid even at refrigeration temperatures (4°C). Therefore, these liquid fractions can be used for sauces, dressings, vinaigrettes, mayonnaise, etc. Regarding the solid fractions obtained, the solid fraction obtained after fractionation at 4°C was solid when it was kept at room temperature (24°C). This solid fraction can be potentially used as margarine. Further studies on crystal form (α, β′ and β) are necessary to fully assess the use of this solid fraction as a margarine. To produce high-quality spreads, such as margarines, the crystals should be in the β′ form because these crystals are relatively small and can incorporate large volumes of oil; moreover, β′ crystals give the product a glossy surface and a smooth lustre [34, 35]. In contrast, β crystals are less desirable in spreads because they grow into large needle-like crystals producing a sandy mouthfeel and are less able to incorporate liquid [36].

3.2.2. *Chemical characteristics of the yellow mealworm oil and its fractions*

The FA composition of the unfractionated oil and its fraction was determined by GC-FID. Similar FA composition of the unfractionated oil was found in previous studies done in our laboratory [25]. It is interesting that the FA composition between the unfractionated oil and its fractions did not change (**Table 4**). The solid fat separated at 4°C was highly unsaturated (74% of FA). In theory, this fraction should have remained liquid. However the experimental results showed a different behaviour. Several possibilities exist to explain this behaviour. It is likely that the arrangement of FA within the triacylglycerol molecule is different in the liquid than in the solid fat, if the arrangement differs, so will the physical properties of the triacylglycerol. The second possibility is that the solid fraction contains a compound that is structuring the oil and turning it into a solid-like material. It is known that natural waxes and partial glycerides are used to structure highly unsaturated oils, forming oleogels [37]. Waxes and partial glycerides are present in insect tissues and in the oil fraction [18, 25] and could remain in the solid fraction after fractionation. Therefore, it can be suggested that these two components can be structuring the solid fraction of the yellow mealworm oil and turning it into a solid-like material or oleogel.

The differences seen in physical properties of the fractions cannot be explained by its FA composition. To understand the differences in physical properties seen in this study, it will be necessary to analyse the profile and structure of the triacylglycerols as well as to analyse the concentration of other minor compounds such as waxes.

The fractionation process presented in the present study succeeded in changing the physical properties of yellow mealworm oil. This fractionation can broaden the applications of this oil as a food ingredient. However, further physical, chemical and sensory analyses are necessary to fully assess the potential use of yellow mealworm oil as a food ingredient.

FA	Unfractionated oil	Fractionated			
		Solid 2 °C	Solid 4 °C	Liquid 2 °C	Liquid 4 °C
Total CLA	<0.10	<0.10	0.13	0.14	<0.10
Total ω-3	1.68	1.66	1.65	1.69	1.71
Total ω-6	30.96	30.63	30.75	31.19	31.18
ω-6/ω-3 Ratio	18.43	18.45	18.64	18.46	18.23
Total SFA	21.35	21.05	21.80	21.85	21.00
Total UFA	73.98	73.36	73.28	74.72	74.70
Total MUFA	41.23	40.87	40.70	41.73	41.70
Total PUFA	32.71	32.45	32.54	32.94	32.96

Adapted from Ref. [33]

Table 4. Fatty acid composition (g/100 g of fat) of unfractionated and fractionated yellow meal worm oil (N = 2).

4. Concluding remarks

Insects represent a viable source of proteins and fats since they have a low environmental cost of production. However, its effective inclusion in food preparations or as a food ingredient relies on several factors including consumer's acceptance, rearing facilities and technological functionalities. A lot of work is needed regarding protein and fat extraction methods as well as on technological functionality of insect proteins and fats. Understanding the behaviour of insect proteins and fats during extraction and processing as well as better knowledge of the intrinsic properties of insect ingredients is required for the development of various food preparations. It is very likely that in the next few years, the scenario will be significantly different as there is an increase attention of the consumers and the food industry for the use of insects as food.

Author details

Daylan Amelia Tzompa Sosa* and Vincenzo Fogliano

*Address all correspondence to: daylancita@hotmail.com

Food Quality and Design Group, Wageningen University, Wageningen, the Netherlands

References

[1] Premalatha, M., T. Abbasi, and S.A. Abbasi, Energy-efficient food production to reduce global warming and ecodegradation: The use of edible insects. Renewable and Sustainable Energy Reviews, 2011. **15**(9): pp. 4357-4360. DOI: 10.1016/j.rser.2011.07.115.

[2] Van Huis, A., Potential of insects as food and feed in assuring food security. Annual Review of Entomology, 2013. pp. 563-583. DOI: 10.1146/annurev-ento-120811-153704

[3] Durst, P.B. and K. Shono, Edible forest insects: Exploring new horizons and traditional practices, in Forest insects as food: humans bite back, P.B. Durst, et al., Editors. 2010, FAO: Bangkok, Thailand.

[4] Yen, A.L., Edible insects: Traditional knowledge or western phobia? Entomological Research, 2009. **39**(5): pp. 289-298. 10.1111/j.1748-5967.2009.00239.x.

[5] Veldkamp, T., G. van Duinkerken, A. van Huis, C.M.M. Lakemond, E. Ottevanger, G. Bosch, and M.A.J.S. van Boekel, Insects as a sustainable feed ingredient in pig and poultry diets – a feasibility study, 2012, Wageningen UR Livestock Research: Wageningen, the Netherlands.

[6] Nasseri, A.T., S. Rasoul-Amini, M.H. Morowvat, and Y. Ghasemi, Single cell protein: Production and process. American Journal of Food Technology, 2011. **6**(2): pp. 103-116. DOI: 10.3923/ajft.2011.103.116.

[7] Wociór, A., D. Złotkowska, H. Kostyra, and E. Kostyra, Mycoprotein. Zywnosc. Nauka. Technologia. Jakosc/Food. Science Technology. Quality, 2010. **17**(5): pp. 5-16.

[8] Buono, S., A.L. Langellotti, A. Martello, F. Rinna, and V. Fogliano, Functional ingredients from microalgae. Food and Function, 2014. **5**(8): pp. 1669-1685. DOI: 10.1039/c4fo00125g.

[9] van Huis, A., J. van Itterbeeck, H. Klunder, E. Mertens, A. Halloran, G. Muir, and P. Vantomme, Edible insects: future prospects for food and feed security. FAO Forestry Paper; 1712013, Food and Agriculture Organization of the United Nations (FAO): Rome.

[10] Rumpold, B.A. and O.K. Schluter, Nutritional composition and safety aspects of edible insects. Molecular Nutrition & Food Research, 2013. **57**(5): pp. 802-823. DOI: 10.1002/mnfr.201200735.

[11] Zhao, X., J.L. Vázquez-Gutiérrez, D.P. Johansson, R. Landberg, and M. Langton, Yellow mealworm protein for food purposes – Extraction and functional properties. PLoS One, 2016. **11**(2). DOI: 10.1371/journal.pone.0147791.

[12] Yi, L.Y., C.M.M. Lakemond, L.M.C. Sagis, V. Eisner-Schadler, A. van Huis, and M.A.J.S. van Boekel, Extraction and characterisation of protein fractions from five insect species. Food Chemistry, 2013. **141**(4): pp. 3341-3348. DOI: 10.1016/j.foodchem.2013.05.115.

[13] Yi, L., M.A.J.S. van Boekel, S. Boeren, and C.M.M. Lakemond, Protein identification and in vitro digestion of fractions from Tenebrio molitor. European Food Research and Technology, 2016: pp. 1-13. DOI: 10.1007/s00217-015-2632-6.

[14] Rutten, M., T. J. Achterbosch, I. J. M. de Boer, J. C. Cuaresma, J. M. Geleijnse, P. Havlík, T. Heckelei, J. Ingram, A. Leip, S. Marette, H. van Meijl, L.-G. Soler, J. Swinnen, P. van't Veer, J. Vervoort, A. Zimmermann, K. L. Zimmermann, & M. Zurek, Metrics, models and foresight for European sustainable food and nutrition security: The vision of the

SUSFANS project. *Agricultural Systems, 2016. In press.* DOI: http://dx.doi.org/10.1016/j.agsy.2016.10.014

[15] Józefiak, D., A. Józefiak, B. Kierończyk, M. Rawski, S. Świątkiewicz, J. Długosz, and R.M. Engberg, Insects—A natural nutrient source for poultry—A review. Annals of Animal Science, 2016. **16**(2): pp. 297-313. DOI: 10.1515/aoas-2016-0010.

[16] Zielińska, E., B. Baraniak, M. Karaś, K. Rybczyńska, and A. Jakubczyk, Selected species of edible insects as a source of nutrient composition. Food Research International, 2015. **77**: pp. 460-466. DOI: 10.1016/j.foodres.2015.09.008.

[17] Beenakkers, A.M.T., D.J. Vanderhorst, and W.J.A. Vanmarrewijk, Insect lipids and lipoproteins, and their role in physiological processes. Progress in Lipid Research, 1985. **24**(1): pp. 19-67. DOI: 10.1016/0163-7827(85)90007-4.

[18] Fast, P.G., Insect lipids. Progress in the Chemistry of Fats and other Lipids, 1971. **11**: pp. 179-242. DOI: 10.1016/0079-6832(71)90006-1.

[19] Lease, H.M. and B.O. Wolf, Lipid content of terrestrial arthropods in relation to body size, phylogeny, ontogeny and sex. Physiological Entomology, 2011. **36**(1): pp. 29-38. DOI: 10.1111/j.1365-3032.2010.00767.x.

[20] Xiaoming, C., F. Ying, Z. Hong, and C. Zhiyong, Review of the nutritive value of edible insects, in Forest insects as food: humans bite back, P.B. Durst, et al., Editors. 2010, FAO: Bangkok, Thailand. pp. 85-92.

[21] Fontaneto, D., M. Tommaseo-Ponzetta, C. Galli, P. Rise, R.H. Glew, and M.G. Paoletti, Differences in fatty acid composition between aquatic and terrestrial insects used as food in human nutrition. Ecology of Food and Nutrition, 2011. **50**(4): pp. 351-367. DOI: 10.1080/03670244.2011.586316.

[22] Oonincx, D.G.A.B., S. van Broekhoven, A. van Huis, and J.J.A. van Loon, Feed conversion, survival and development, and composition of four insect species on diets composed of food by-products. PLoS One, 2015. **10**(12). ARTN e0144601. DOI: 10.1371/journal.pone.0144601.

[23] Downer, R.G.H. and J.R. Matthews, Patterns of lipid distribution and utilisation in insects. American Zoologist, 1976. **16**(4): pp. 733-745.

[24] Purschke, B., T. Stegmann, M. Schreiner, and H. Jäger, Pilot-scale supercritical CO_2 extraction of edible insect oil from *Tenebrio molitor* L. larvae—Influence of extraction conditions on kinetics, defatting performance and compositional properties. European Journal of Lipid Science and Technology, 2016. in press. DOI: 10.1002/ejlt.201600134.

[25] Tzompa-Sosa, D.A., L. Yi, H.J.F. van Valenberg, M.A.J.S. van Boekel, and C.M.M. Lakemond, Insect lipid profile: Aqueous versus organic solvent-based extraction methods. Food Research International, 2014. **62**: pp. 1087-1094. DOI: 10.1016/j.foodres.2014.05.052.

[26] Yang, L.F., S. Siriamornpun, and D. Li, Polyunsaturated fatty acid content of edible insects in Thailand. Journal of Food Lipids, 2006. **13**(3): pp. 277-285. DOI: 10.1111/j.1745-4522.2006.00051.x.

[27] Ushakova, N.A., E.S. Brodskii, A.A. Kovalenko, A.I. Bastrakov, A.A. Kozlova, and D.S. Pavlov, Characteristics of lipid fractions of larvae of the black soldier fly *Hermetia illucens*. Doklady Biochemistry and Biophysics, 2016. **468**(1): pp. 209-212. DOI: 10.1134/S1607672916030145.

[28] Dayrit, F.M., The properties of lauric acid and their Significance in coconut oil. Journal of the American Oil Chemists' Society, 2015. **92**(1): pp. 1-15. DOI: 10.1007/s11746-014-2562-7.

[29] Hinton Jr, A. and K.D. Ingram, Antimicrobial activity of potassium hydroxide and lauric acid against microorganisms associated with poultry processing. Journal of Food Protection, 2006. **69**(7): pp. 1611-1615.

[30] Ivanova-Petropulos, V., S. Mitrev, T. Stafilov, N. Markova, E. Leitner, E. Lankmayr, and B. Siegmund, Characterisation of traditional Macedonian edible oils by their fatty acid composition and their volatile compounds. Food Research International, 2015. **77**: pp. 506-514. DOI: 10.1016/j.foodres.2015.08.014.

[31] Gunstone, F.D., Vegetable oils, in Bailey's industrial oil and fat products—Edible oil and fat products: chemistry, properties, and health effects, F. Shahidi, Editor 2005, Wiley-Interscience USA: Hoboken, US.

[32] Ekpo, K.E., A.O. Onigbinde, and I.O. Asia, Pharmaceutical potentials of the oils of some popular insects consumed in southern Nigeria. African Journal of Pharmacy and Pharmacology, 2009. **3**(2): pp. 51-57.

[33] Tzompa-Sosa, D.A., J.M. Verbeek, and H.F.J. van Valenberg, Fractionation of insect oils: The case of yellow mealworm oil, in Inform 2016, American Oil Chemistry Society: Urbana, IL, USA. pp. 24-27.

[34] Ghotra, B.S., S.D. Dyal, and S.S. Narine, Lipid shortenings: A review. Food Research International, 2002. **35**(10): pp. 1015-1048. DOI: 10.1016/S0963-9969(02)00163-1.

[35] Narine, S.S. and A.G. Marangoni, Relating structure of fat crystal networks to mechanical properties: A review. Food Research International, 1999. **32**(4): pp. 227-248. DOI: 10.1016/S0963-9969(99)00078-2.

[36] Sato, K., & S. Ueno, (2014). Physical properties of fats in food. In Fats in Food Technology (Second edition. ed., pp. 1-38). in press. 2014. Chichester, West Sussex, UK: John Wiley & Sons, Ltd. DOI: 10.1002/9781118788745.ch1

[37] Patel, A.R. and K. Dewettinck, Edible oil structuring: An overview and recent updates. Food and Function, 2016. **7**(1): pp. 20-29. DOI: 10.1039/c5fo01006c.

Chapter 10

Entomophagy: Insects as Food

Tiencheu Bernard and Hilaire Macaire Womeni

Additional information is available at the end of the chapter

http://dx.doi.org/10.5772/67384

Abstract

Due to the increasing cost of animal proteins, food and feed insecurity, population growth, and increasing need for protein-rich food in the developed and less developed countries, alternative sources of protein-rich food are highly needed. Scientific research has shown that edible insects are a very rich source of proteins and other nutrients. Hence, insect consumption might help revolutionaries' food and feed insecurity and thus replace the conventional animal source. This work assesses the potential of insects as food for humans and feed for animals and gathers existing information and research on edible insects. The assessment is based on the most recent and complete data available from various sources and experts around the world, because lack of a complete data on edible insects reduces consumer confidence and limits integration of edible insect consumption with other food sources. Considering the nutritional, economic, and ecological advantages of edible insects over conventional livestock, much attention should therefore be given to their method of collection as this will help improve their availability. This could be achieved by improved conservation or by raising them as a minilivestock. Considering the economic, nutritional, and ecological advantages of this traditional food source, its promotion deserves more attention both from national governments and assistance programs.

Keywords: edible insects, entomophagy, minilivestock, food and feed security, conservation

1. Introduction

Entomophagy, the consumption of insects, is rooted in human evolutionary history [1]. Insects have played an important part in the history of human nutrition in Africa, Europe, Asia, and Latin America. Over 1900 species of insects are known worldwide to be part of human diets;

some important groups include grasshoppers, caterpillars, beetle grubs, wringed termites, bees, worms, ant brood, cicadas, and a variety of aquatic insects [2]. It is interesting to know that more than two billion people consume insects on a regular basis, and insect eating provides a significant proportion of the animal proteins consumed in some regions [3]. Because entomophagy is widely practiced, and because it compares favorably with nutrient and environmental aspects of conventional livestock rearing, it has the potential to contribute substantially to reducing undernutrition among an expanding global population [3].

2. Diversity of edible insect species in the world

Insects have been in existence for at least 400 million years, making them among the earliest land animals. They diverged as members of one of the largest subphyla in arthropods more than 390 million years ago experiencing a rapid evolution and radiation that is considered faster than any other group [4]. The class insects are the largest animal group on earth and constitute as much as 80% of the animal kingdom. Because of the nutritional importance of edible insects and their huge availability, this has attracted their consumption by more than two billion people on a daily basis [5].

Over 1900 species of edible insects in 300 ethnic groups in 113 countries worldwide have been recorded by various authors to be part of human diet (**Table 1**) [6]. According to Van Huis et al. [3], 246 species of edible insects have been reported in 27 countries in Africa. Another study

Order	Species	Location
Lepidoptera	*Anaphe panda* (Boisduval)	DRC, Zambia, Cameroon, Congo, CA Republic, Zimbabwe, Nigeria, Tanzania
	Cirina forda (Westwood)	DRC, Zambia, South Africa, Botswana, Burkina Faso, Nigeria, Mozambique, Namibia, Ghana, Togo, Chad
	Dactyloceras lucina (Drury)	DRC, Zambia, South Africa, Botswana, Burkina Faso, Nigeria, Mozambique, Namibia, Ghana, Togo, Chad
	Gynanisa ata Strand	DRC, Zambia, South Africa, Botswana, Burkina Faso
	Anaphe venata Butler	Zambia, Nigeria, Ivory Coast, Sierra Leone, Guinea, Liberia, Guinea Bissau
Orthoptera	*Acanthacris ruficornis* (Fabricius)	DRC, Zambia, South Africa, Botswana, Burkina Faso, Nigeria, Mozambique, Namibia, Ghana, Togo, Chad
	Ruspolia differens (Serville)	DRC, Zambia, South Africa, Cameroon, Congo, CA Republic, Zimbabwe, Burkina Faso, Malawi, Mali
	Zonocerus variegatus Linnaeus)	DRC, Zambia, South Africa, Cameroon, Zimbabwe, Kenya, Uganda, Tanzania, Malawi
Coleoptera	*Oryctes boas* (Fabricius)	DRC, Cameroon, Congo, CA Republic, Nigeria, Ivory Coast, Sao Tomé, Guinea, Ghana, Liberia
	Rhynchophorus phoenicis (Fabricius)	Thailand, Australia, Nigeria, Ivory Coast, Sierra Leone, Guinea, Liberia, Guinea Bissau DRC, Congo, Botswana

Order	Species	Location
Hymenoptera	*Apis mellifera* (Linnaeus)	Mexico, Cameroon, Congo, CA Republic, Nigeria, Angola, Ivory Coast, Niger, Sao Tomé, Guinea
	Carebara vidua (Smith)	DRC, Zambia, South Africa, Zimbabwe, Botswana, Malawi, Sudan, Kenya, South Sudan
	Carebara lignata Westwood	Zambia, South Africa, Zimbabwe, Botswana, Sudan, Mozambique, Namibia, South Sudan
Isoptera	*Macrotermes subhyalinus* (Rambur)	Zambia, Angola, Kenya, Togo, Burundi, Ivory Coast, Canada, the USA
	Macrotermes falciger (Gerstäcker)	Zambia, Zimbabwe, Burkina Faso, Burundi, Benin, Australia, the Netherlands
	Macrotermes natalensis (Haviland)	DRC, Cameroon, Congo, CA Republic, Nigeria, Burundi, South Africa, Zimbabwe, Nigeria, Malawi

Sources: Banjo et al. [8], Igwe et al. [9], Opara et al. [10], Kelemu et al. [11].

Table 1. Diversity and location of the most consumed edible insects in the world.

carried out 2 years later by Ramos-Elorduy noted that Africa is one of the most important hot spots of edible insect biodiversity in the world with 524 species recorded from 34 African countries [7]. These species are mostly of the orders Orthoptera, Lepidoptera, Coleoptera, Hymenoptera, and Isoptera. Below is a checklist of the most consumed edible insect species in the world and their orders and locations (**Figure 1**).

Figure 1. Recorded number of edible insect species, by country. Source: Centre for Geo Information, Wageningen University, based on data compiled by Jongema, 2012 [21]. Banjo et al, [8]; Igwe et al, [9]; Opera et al, [10]; Kelemu et al, [11].

3. Why eat insects?

Hunger and malnutrition is a serious problem in the ever-expanding human population. With the high rate at which the world population is growing, the world food supply should grow at the same rate, if not faster. Therefore, the search for new food sources including the identification and development of localized ethnic ones continues [3].

In most part of the world, particularly in Africa and Latin America, food resources are becoming increasingly scarce and the importing of foods is becoming more expensive. It is thus imperative to identify and develop indigenous food resources. To effectively respond not just to rapid population growth but also to other pressing challenges, researchers have turned their attention to insects not only because of their abundance, enormous biomass, and high-quality protein but also because of the time-honored practice among many culturally diverse peoples of Africa and Latin America of consuming live, roasted, and fried insects, providing them with a nutritious protein of good quality and high digestibility [12, 13]. The choice of insects as food is further strengthened by the fact that they also constituted rich sources of fat, vitamins, and minerals, especially iron and zinc [14].

4. Edible insects as food for humans and feed for animals

Insects are the most abundant and most diverse multicellular organisms on planet earth and are thought to account for about 80% of all species [15]. Numerous crops rely on them for pollination, and their importance extends into their other agricultural and human health benefit [16].

Over 1900 species of insects are known to be part of human diets, more than 2 billion people consume insects on a regular basis, and insect eating provides a significant proportion of the animal protein consumed in some regions [3, 17]. In fact, in many developing countries and among various cultures scattered throughout the world, insects remain a vital and preferred food and an essential source of protein, fat, minerals, and vitamins [18]. This is because some edible insects have been shown to have nutritional value that can be compared with meat and fish, while others have higher proportion of proteins, fat, and energy value [19]. This has become especially important as the need for alternative protein sources increases due to rapid urbanization in developing countries and the shifts in the composition of global food demand. Among the most important orders of insects consumed in the world are the Coleoptera, Hymenoptera, Isoptera, Lepidoptera, Odonata, and Orthoptera, and they are highly priced (**Figure 2**) [20]. Notable examples of these are the locusts, termites, worms, grasshoppers, caterpillars, palm weevils, and beetle grubs, among others.

Although insects were mainly recognized as pests affecting humans, plants, and animal health, insects play an essential role in minimizing food insecurity in addition to provide ecosystem services (such as pollination, waste degradation, and biological control). Insects also represent an important food source for a wide variety of animal species. Van Huis et al. outlined the important role of insects in assuring food and feed security. Below is a list of pictures showing the various edible insects consumed around the world (**Figure 3**) [3].

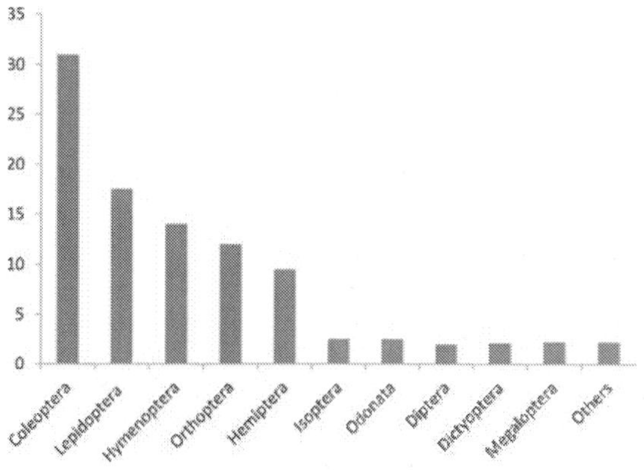

Figure 2. Number of insect species, by order, consumed worldwide. Source: Jongema [21].

Figure 3. Examples of edible insects.

5. Medicinal value of edible insects

Scientific validation of traditional wisdom in bioprospecting has assured greater significance. Edible insects have long been a significant dietary factor and remedy for illnesses in many regions of the world [22]. Traditional healers have used insects as medicine to treat various diseases in human beings and animals successfully. Some of these diseases include common fever, scabies, epilepsy, violent headaches, bronchitis, hemorrhage, and dog bite. Insects are also used to treat wound, to prevent gangrene, and to increase milk flow in lactating women, among others [23]. This treatment is finding modern usage in many hospitals. Also, chemicals produced by edible insects against self-defense have also been exploited by many researchers for the production of antibacterial and anticancer drugs. For instance, pierisin, a protein purified from pupa of cabbage butterfly, exhibits cytotoxic effects against human gastric cancer. Cecropin has also been reported to be cytotoxic against mammalian lymphoma and leukemia cells [22]. Despite the fact that edible insects have a high nutritional, economic, and medicinal value, they are often neglected. It is high time that scientists recognize this fact and begin to build on it given the benefits of these creatures to the environment and to human health.

6. Market value of edible insects

Many rural communities like those in Africa, Asia, and South America know that eating insects provides a valuable source of protein, minerals, and vitamins as well as a tasty snack and therefore must be in high demand. Crickets, grasshoppers, and locusts, for example, are a seasonal delicacy, while the giant water beetles are used in salads [22].

Considering the popularity of edible insects, it is not surprising that scores of species have been and are prominent items of commerce in the town and village markets of Africa and tropical and semitropical regions of the world [24]. In several areas in Africa, particularly in Zimbabwe, Nigeria, South Africa, Ivory Coast, and Zambia, many families make fairly good living from selling insects [25, 26]. These insects are mostly gathered from bushes and farmland by women and children (**Figure 4**), processed, and eaten or sold in school premises and

Figure 4. A picture showing women and children selling edible insects in local markets.

local and urban markets. Some of these insects are also processed and exported to shops and restaurants in cities in and out of the country [20, 27]. The commercialization of edible insects therefore provides significant income to many households in Africa. However, poorly understood and poorly organized market chains severely limit agribusiness in Africa. The market constraints encountered by farmers when attempting to diversify the production of edible insect business should therefore be documented.

7. Nutritional value of edible insects

The part played by insects in human nutrition cannot be underestimated [28]. Substantial evidence suggests that insects are a highly nutritious and healthy food source with high content of nutrients such as fats, proteins, amino acids, carbohydrates, vitamins, fibers, and minerals required by humans and animals [3]. However, the nutritional compositions of edible insects between and within species are highly variable depending upon the metamorphic stage, habitat, and diet of the insect as well as the preparation and processing methods applied before consumption [29]. Although the nutritional composition of some edible insects has previously been investigated in a number of countries, for example, India, the USA, Mexico, and Thailand, very few authors have reported a compiled nutritional composition of edible insects in the world. The lack of nutritional data on most edible insects may result in a reduction in consumer confidence and limits integration of insect consumption with other food sources. Hence, this work seeks to compile nutritional data of edible insects consumed in the world.

7.1. Dietary energy content of edible insects

With the growing world population, there are now more than 3.7 billion people suffering from malnutrition, mainly due to lack of protein and energy from food [30]. Also, new agricultural land is scarce to produce food for humans, and as such a greater proportion of people are eating resource-intensive animal protein than ever before. Livestock production is very expensive because it requires a large input of energy compared to the energy output [31]. These livestock also compete for nutrients and energy with humans.

Utilization of insects as a protein source could benefit insect conservation through habitat protection [32]. Insects are essential agents because they are able to exploit all organic sources in nature and are able to recycle organic waste and provide nutrients to farm animals and humans [33, 34]. Furthermore, edible insects have high quantities of polyunsaturated fat which provides majority of the energy for sustaining life. The energy contents of edible insects vary according to the species and region found and have also been found to be significantly higher than those of livestock and vegetables [19, 35]. Hence, edible insects may efficiently provide the necessary energy for the vital functions and survival of organisms.

Ramos Elorduy et al. [36] analyzed 78 insects from Oaxaca State, Mexico, and determined that caloric content was 293–762 kcal per 100 g of dry matter as shown in **Table 2**.

Location	Common name	Scientific name	Energy content kcal/100 g fresh weight
Australia	Australia plague locust, raw	*Chortoicetes terminifera*	499
Australia	Green (weaver) ant, raw	*Oecophylla smaragdina*	1272
Canada, Quebec	Red-legged grasshopper, whole, raw	*Melanoplus femurrubrum*	160
The USA, Illinois	Yellow mealworm, larva, raw	*Tenebrio molitor*	206
The USA, Illinois	Yellow mealworm, adult, raw	*Tenebrio molitor*	138
Ivory Coast	Termite, adult, dewinged, dried, flour	*Macrotermes subhyalinus*	535
Mexico Veracruz State	Leaf-cutter ant, adult, raw	*Atta mexicana*	404
Mexico Hidalgo State	Honey ant, adult, raw	*Myrmecocystus melliger*	116
Thailand	Field cricket, raw	*Gryllus bimaculatus*	120
Thailand	Giant water bug, raw	*Lethocerus indicus*	165
Thailand	Rice grasshopper, raw	*Oxya japonica*	149
Thailand	Grasshopper, raw	*Cyrtacanthacris tatarica*	89
Thailand	Domestic silkworm, pupa, raw	*Bombyx mori*	94
The Netherlands	Migratory locust, adult, raw	*Locusta migratoria*	179

Source: FAO [37].

Table 2. Examples of energy content of different processed insect species, by regions.

7.2. Protein and amino acid composition of edible insects

Proteins are organic compounds consisting amino acids which are the building blocks and could be either essential or nonessential. Protein is the basis of all organism activity and constitutes many important materials such as enzyme, hormones, and hemoglobin. Proteins are the most abundant biological macromolecules, occurring in all cells and all parts of cells. Proteins also occur in great variety; thousands of different kinds, ranging in size from relatively small peptides to huge polymers with molecular weights in the millions, may be found in a single cell. Moreover, proteins exhibit enormous diversity of biological function and are the most important final products of the information pathways. Proteins are the molecular instruments through which genetic information is expressed. Therefore, ensuring a steady source of protein is very important in providing energy for humans and animals.

Insects are potentially an important energy efficient source of protein in humans (**Table 3**), either through a direct consumption or as food supplements for stock [17]. Many local communities in the world, particularly in the Amazon region, attain about 8–70% of their dietary protein from insects [39]. The highest amount of crude protein is found in insects in the order Lepidoptera followed by Coleoptera, while the order Hymenoptera has the least. The high protein content is an indication that edible insects can be of value to man and animal ration and can eventually replace higher animal protein usually absent in the diet of rural dwellers in developing countries [8].

Insect order	Stage	Range (% protein)
Coleoptera	Adults and larvae	23–66
Lepidoptera	Pupae and larvae	14–68
Hemiptera	Adults and larvae	42–74
Homoptera	Adults, larvae, and eggs	45–57
Hymenoptera	Adults, pupae, larvae, and eggs	13–77
Odonata	Adults and naiad	46–65
Orthoptera	Adult and nymph	23–65

Source: Xiaoming et al. [40].

Table 3. Crude protein content, by insect order.

Xiaoming et al. [40] evaluated the protein content of 100 species from a number of insect orders. He showed that protein content was high and in the range of 13–77% of dry matter and that there was a large variation between and within the insect orders.

Edible insects have also been shown to have higher protein content, on a mass basis, than other animal and plant foods such as beef, chicken, fish, soybeans, and maize [41]. FAO compared protein content of insects, reptiles, cattle, and fish (**Table 4**) [42]. Results showed that the protein content of the selected insect species had higher protein contents than the fish and mammals. This strongly suggests that insects are an important source of protein and as such their consumption should be encouraged.

Many studies have also shown that edible insects are important source of amino acids (tryptophan and lysine). The inclusion of these insect species in diets could be of immense benefit in complementing lysine-poor staple cereals [43–45], and a host of others were able to demonstrate that an insect species *Tenebrio molitor* had a significant amount of essential and nonessential amino acids compared to beef (**Table 5**).

7.3. Fat and carbohydrate composition of edible insects

Fats and carbohydrates are important nutritive elements in the human body. They are the main energy source in the body, can reduce consumption of proteins, and help for detoxification [38].

Edible insects are also a good source of carbohydrates. In fact, the carbohydrate content of edible insects ranged from 6.71% in sting bug to 15.98% in cicada [47]. Carbohydrates in insects are formed mainly by chitin. Chitin is a macromolecular compound that has a high nutritional and health value. This chitin reduces serum cholesterol [48]. Also, recent report showed that considerable amounts of polysaccharide in edible insects might improve immune functioning of human body [49].

Animal group	Species and common name	Edible product	Protein content (g/100 g fresh weight)
Insect (raw)	Locusts and grasshoppers: *Locusta migratoria, Anacridium melanorhodon, Ruspolia differens*	Larva	14–18
	Locusts and grasshoppers: *Locusta migratoria, Anacridium melanorhodon, Ruspolia differens*	Adult	13–28
	Sphenarium purpurascens (chapulines, Mexico)	Adult	35–48
	Silkworm (*Bombyx mori*)	Caterpillar	10–17
	Palm worm beetles: *Rhynchophorus palmarum, R. phoenicis, Callipogon barbatus*	Larva	7–36
	Yellow mealworm (*Tenebrio molitor*)	Larva	14–25
	Crickets	Adult	8–25
	Termites	Adult	13–28
Cattles	Turtles: *Chelodina rugosa, Chelonia depressa*	Flesh	25–27
		Intestine	18
		Liver	11
		Heart	17–23
		Liver	12–27
Fish (raw)	Finfish	Tilapia	16–19
		Mackerel	16–28
		Catfish	17–28
	Crustaceans	Lobster	17–19
		Prawn (Malaysia)	16–19
		Shrimp	13–27
	Mollusks	Cuttlefish	15–18

Sources: FAO [37, 42].

Table 4. Comparison of average protein content among insects, fish, reptiles, and cattle.

Fat is the most energy-dense macronutrient in food. It consists of triglycerides and three fatty acid molecules which are the backbone. Fatty acid can either be saturated, unsaturated, or essential. Dietary intervention and epidemiological studies showed that fatty acid intake

Amino acid	*T. molitor* g/kg dry matter	Beef g/kg dry matter
Essential		
Isoleucine	24.7	16
Leucine	52.2	42
Lysine	26.8	45
Methionine	6.3	16
Phenylalanine	17.3	24
Threonine	20.2	25
Tryptophan	3.9	-
Valine	28.9	20
Semi-essential		
Arginine	25.5	33
Histidine	15.5	20
Methionine + cysteine	10.5	22
Tyrosine	36.0	22
Nonessential		
Alanine	40.4	30
Aspartic acid	40.0	52
Cysteine	4.2	5.9
Glycine	27.3	24
Glutamic acid	55.4	90
Proline	34.1	28
Serine	25.2	27
Taurine	210	-

Sources: Finke [45], Oonincx [46].

Table 5. Average amino acid content in *Tenebrio molitor* and beef in g/kg dry matter.

played a key role in human health. Reduction in dietary saturated fatty acids can decrease factor VII coagulant activity, which was implicated as a risk factor of cardiovascular disease [50]. Scientific literature has shown that consumption of PUFAs has several health benefits to humans such as reduction of glucose tolerance, thus reducing risk of diabetes and blood pressure [51], and prevention of insulin resistance [52], decreasing thrombotic tendency by inhibition of thromboxane A_2 formation [53] and lowering blood pressure [51, 54]. Womeni et al. [55] investigated the content and composition of fat extracted from several insects (**Table 6**). They showed that their oils are rich in polyunsaturated fatty acids and frequently contain the essential linoleic and α-linolenic acids. The nutritional importance of these two essential fatty acids is well recognized, mainly for the healthy development of children and infants [56].

Many recent studies have shown that edible insects are a rich source of fatty acids, particularly polyunsaturated fat [19, 57]. These fatty acids differ and depend on many factors such as species reproduction stages [58, 59] and season or life stage [59].

Edible insect species	Fat content (% of dry matter)	Composition of main fatty acids (% of oil content)	Type of fatty acid
African palm weevil (*Rhynchophorus phoenicis*)	54%	Palmitoleic acid 38%	MUFA
		Linoleic acid 45%	PUFA
Edible grasshopper (*Ruspolia differens*)	67%	Palmitoleic acid 28%	MUFA
		Linoleic acid 46%	PUFA
		a-Linolenic acid	PUFA
Termites (*Macrotermes* sp.)	49%	Palmitic acid	SFA
		Oleic acid	MUFA
		Stearic acid 9%	SFA
Saturniid caterpillar (*Imbrasia* sp.)	24%	Palmitic acid 8%	SFA
		Oleic acid 9%	MUFA
		Linoleic acid 7%	PUFA
		a-Linolenic acid 38%	PUFA
Variegates grasshopper (*Zonocerus variegates*)	9%	Palmitoleic acid 24%	MUFA
		Oleic acid 11%	MUFA
		Linoleic acid 21%	PUFA
		a-Linolenic acid 15%	PUFA
		g-Linolenic acid 23%	PUFA

Source: Womeni et al. [55]. PUFA= Polyunsaturated Fatty Acids; MUFA= Monounsaturated Fatty Acids; SFA= Saturated Fatty Acids.

Table 6. Fat content and randomly selected fatty acids of several edible insect species consumed in Cameroon.

7.4. Mineral composition of edible insects

Minerals are known to play important metabolic and physiologic roles in the living system. Iron, zinc, copper, and manganese strengthen the immune system as antioxidant and cofactors of enzyme [60]. Similarly, magnesium, zinc, and selenium prevent cardiomyopathy, muscle degeneration, growth retardation, impaired spermatogenesis, immunologic dysfunction, and bleeding disorder [61]. Iron deficiency is a major problem in women's diets in the developing world, particularly among pregnant women, and especially in Africa [62]. Magnesium is needed for more than 300 biochemical reactions in the body. It helps to maintain normal muscle and nerve function, keeps the heart rhythm steady, and regulates blood sugar levels [63].

Analysis of normal elements showed that edible insects are rich in nutritious elements such as potassium and sodium (cricket nymph), calcium (adult cricket), copper (mealworm), iron (axayacatl), zinc (cricket), manganese (cricket), and phosphorus [3]. Therefore, edible insects can supply the necessary nutritive elements for human body functions and could be consumed along with other food and animals rich in other essential minerals to further complement the diet of these insects. The mineral contents of some edible insects in Nigeria are shown in **Table 7**.

Insect species	Ca	P	Fe	Mg	Ash
Megachile nigeriensis	1.00*	14.90*	9.56*	60.96*	7.60*
Macrotermes bellicosus	21	136	27	0.15	2.90
M. natalensis	18	114	29	0.26	1.90
Brachytrupes spp.	9.21	126.9	0.68	0.13	1.82
Circus aeruginosus	4.40	100.2	0.35	0.09	2.10
Z. variegatus	42.40	131.2	1.96	8.21	1.20
Argiope trifasciata	61.28	136.4	18.2	6.14	4.21
Anaphe infracta	8.56	111.3	1.78	1.01	1.60
Annona reticulata	10.52	102.4	2.24	2.56	2.50
Lepidoptera litoralia	12.00	9.00	19.50	0.50	4.30
A. venata	8.57	100.5	2.01	1.56	3.20
C. forda	8.24	111.0	1.79	1.87	1.50
A. mellifera	15.4	125.5	25.2	5.23	2.20
O. boas	45.68	130.2	2.31	6.62	1.50
Oryctes monoceros	NA	NA	85.00	175.00	10.50
Gymnelia lucens	NA	NA	NA	NA	6.40
R. phoenicis	54.1	685.0	30.80	131.8	2.70
Aphodius rufipes	42.16	131.2	30.82	11.72	2.74

*mg/kg body weight; NA, not available.

Sources: Banjo et al. [8], Ifie and Emeruwa [64], Finke [45]; Paiko et al. [65]; Solomon and Pisca [66], Johnson [67].

Table 7. Mineral and ash (mg/100 g) contents in some edible insect species.

7.5. Vitamin composition of edible insects

Vitamins are a group of organic compounds that are necessary for metabolism in human bodies. Vitamins cannot be synthesized in the human body; they must be supplied constantly by food [38]. Vitamin C, also called ascorbic acid, serves as a reducing agent (an antioxidant), while vitamin B comprises of components of coenzymes. Vitamins K and A are required for normal blood clotting and proper vision, respectively. Many studies have shown that edible insects contain appreciable amounts of vitamins [9, 44]. The high vitamin content of edible insects presents them as a highly potentially good source of food supplement for malnourished people and animals. The vitamin contents of some insect orders are shown in **Table 8**. Each of the orders contains appreciably high amounts of vitamins A, B2, and C.

Order	N	Vitamin A		Vitamin B2		Vitamin C	
		Range	Mean	Range	Mean	Range	Mean
Isoptera	3	0.026–0.05	0.14	1.54–1.98	1.69	3.01–17.76	8.06
Coleoptera	3	0.086–0.125	0.11	0.08–2.62	1.64	4.25–7.59	5.75
Orthoptera	3	0.0–0.068	0.03	0.03–0.08	0.06	0.0–8.64	3.21
Lepidoptera	5	0.028–0.034	0.09	0.09–2.21	1.50	1.95–4.52	2.83
Hymenoptera	1				3.24		10.25

N, number of insect species in each order. Sources: Banjo et al. [8], Ifie and Emeruwa [64], Igwe et al. [9].

Table 8. Vitamin (mg/100 g) contents in some edible insects in Nigeria.

8. Edible insects as an engine for improving/replacing livestock rearing

Land, water, and energy resources are declining, so these resources need to be conserved and managed to produce more food [68]. Also, animal husbandry competes for these vital resources, as the land is occupied by the production of feed and cannot be used to produce more food for humans [32]. It is very expensive to carry out livestock farming. This is because they consume large amounts of energy than they produce [68]. For example, livestock consume 77 million tons of protein in feedstuff that is potential for human nutrition to produce 58 million tons of protein [69]. Insect culture, on the other hand, requires little areas [70]. Also, many of the edible insect species do not compete with human beings for food resources. Equally, insect farming requires little water, which is significant because water shortages already exist throughout the world and are likely to increase. Hence, insects are nick named "minilivestock" [71].

9. Environmental (ecological) opportunities of insect rearing

Insects deliver a host of ecological services fundamental to the survival of humankind. Utilization of insects as a protein source could benefit insect conservation through habitat protection [32]. Insects are essential agents feeding on organic matter in nature, and they efficiently exploit all organic sources. Insects also recycle organic waste and provide nutrients for farm animals [33, 34]. Many insect species are absolutely necessary to improve soil fertility. This is because insects play an important role in breaking down waste products until it is fit to be consumed by fungi and bacteria, thus releasing minerals and nutrients which become readily available in the soil for plant uptake, hence improving soil fertility. Animal carcasses, for example, are consumed by fly maggots and beetle larvae. Dung beetles—of which there are about 400 known species—also play a significant role in decomposing manure. Hence, insects could be used as efficient bio-transformers to convert abundant, low-cost organic wastes into animal biomass rich in proteins and suitable for use in animal nutrition [72].

10. Beneficial roles of insects for humans

Besides serving as sources of food, edible insects provide humans with a variety of other valuable products. A huge variety of insect species are known to have remarkable commercial and pharmaceutical values. For example, bees and silkworm have been shown to produce massive tons of honey and silk, respectively. These products can be sold in the local as well as in the international markets [72], while silkworms produce more than 90,000 tons of silk [73]. Also carmine, a red dye produced by scale insects of the order Hemiptera, is used to color foods, textiles, and pharmaceuticals. Resilin, a rubberlike protein that enables insects to jump, has been used in medicine to repair arteries because of its elastic properties [74]. In addition to this, other products produced by edible insects such as honey, propolis, royal jelly, and venom have been used in treating traumatic and infected wounds and burns [75]. Furthermore, insect products have also been used in engineering methods in the production of biomaterials [76].

11. Conclusion

Sustainably meeting global food demands is one of humanity's greatest challenges and has attracted considerable attention in the past few years [77]. There is general consensus on agriculture's positive contribution to food security through its role in increasing availability of affordable food and the incomes of the poor. Within the context of sustainable diet, the use of insects as food and feed has a significant role to play in assuring food security and improving livelihood of the African people. Edible insects are rich in protein and amino acids, especially essential amino acids which are necessary for the human body. They can also supply unsaturated fatty acids, minerals, vitamins, and carbohydrates, which have an excellent nutritive value. They are also of valuable importance medically, commercially, and ecologically. These edible insects should therefore be taken into consideration for a world in which human nutrition has been a huge problem.

12. The way forward

Recent studies have identified four important challenges that must be addressed in order to tap the huge potentials that edible insects offer for enhancing food and feed security.

To begin with, more work on the nutritional values of edible insects is needed in order to establish insects as food. Also, insect farming should be compared with livestock farming in order to determine which one of them is more environmentally damaging or environmentally friendly. Furthermore, there should be a further clarification on the socioeconomic importance of edible insects in enhancing food security. Lastly, clear comprehensive legal framework at (inter)national levels is needed to pave the way for more investment, leading to the full development of production and international trade in insect products as food and feed sources.

Author details

Tiencheu Bernard[1,2]* and Hilaire Macaire Womeni[2]

*Address all correspondence to: tiencheu.bernard@ubuea.cm

1 Department of Biochemistry, Faculty of Science, University of Buea, Buea, Cameroon

2 Department of Biochemistry, Faculty of Science, University of Dschang, Dschang, Cameroon

References

[1] Fontaneto D, Tommaseo-Ponzetta M, Galli C, Risé P, Glew RH, Paoletti MG. Differences in fatty acid composition between aquatic and terrestrial insects used as food in human nutrition. Ecol. Food Nutr. 2011; 50:351–367. doi: 10.1080/03670244.2011.586316.

[2] Bodenheimer FS. Insects as Human Food: A Chapter of the Ecology of Man. 1951. The Hague: Dr. W. Junk Publishers. doi: http://dx.doi.org/10.1007/978-94-017-6159-8.

[3] Van Huis A et al. Edible Insects: Future Prospects for Food and Feed Security. 2013. Roma: FAO: 1–201.

[4] Gaunt MW, Miles MA. An insect molecular clock dates the origin of the insects and accords with palaeontological and biogeographic landmarks. Mol. Biol. Evol. 2002; 19:748–761.

[5] Premalatha M, Abbasi T, Abbasi T, Abbasi SA. Energy efficient food production to reduce global warming and ecodegradation: the use of edible insects. Renew. Sustain. Energ. Rev. 2011; 15:4357–4360.

[6] MacEvilly C, Bugs in the system. Nutr. Bull. 2000; 25:267–268.

[7] Ramos-Elorduy J. Insects a hopeful food. In: Paoletti M, editor. Ecological Implications of Minilivestock for Sustainable Development. USA: Science Publishers Inc:2005. pp. 263–291.

[8] Banjo AD, Lawal OA, Songonuga EA. The nutritional value of fourteen species of edible insects in Southwestern Nigeria. Afr. J. Biotechnol. 2006; 5(3):298–301.

[9] Igwe CU, Ujowundu CO, Nwaogu LA, Okwu GN. Chemical analysis of an edible African termite, *Macrotermes nigeriensis;* a potential antidote to food security problem. Biochem. Anal. Biochem. 2011; 1:105. doi: 10.4172/2161-1009.1000105.

[10] Opara MN, Sanyigha FT, Ogbuewu IP, Okoli IC. Studies on the production trend and quality characteristics of palm grubs in the tropical rainforest zone of Nigeria. J. Agric. Technol. 2012; 8(3):851–860.

[11] Kelemu S, Niassy B, Torto B, Fiaboe K, Affognon H, Tonnang H, Maniania NK, Ekesi S. African edible insects for food and feed: inventory, diversity, commonalities and

contribution to food security. J. Insects Food Feed. 2015; 1(2):103–119. doi: http://dx.doi.org/10.3920/JIFF2014.0016.

[12] Conconi JRE. Insects as a source of proteins in the future (prospective). Reg. Sec. Educ. Pub. 1974: 1639/74.

[13] Wang, D, Bai YT, Li JH, Zhang CX. Nutritional value of the field cricket (Gryllus testaceus Walker). J. Entomol. Sin. 2004; 11:275–283. doi: 10.1111/j.1744-7917.2004.tb00424.x

[14] Oliveira JFS, de Carvalho JP, de Sousa RFXB, Simao MM. The nutritional value of four species of insects consumed in Angola. Ecol. Food Nutr. 1976; 5:91–97.

[15] Scaraffia PY, Miesfeld RL. Insect Biochemistry/Hormones. BCH2 00093. 2012. Tuczon, AZ, USA: University of Arizona.

[16] Dzerefos CM, Witkowski ETF, Toms R. Comparative ethnoentomology of edible stinkbugs in southern Africa and sustainable management considerations. J. Ethnobiol. Ethnomed. 2013; 9:20. doi: 10.1186/1746-4269-9-20.

[17] De Foliart GR. The human used of insects as food and feed. Bull. Entomol. Soc. Am. 1989; 35:22–35.

[18] Durst PB, Shono K. Edible forest insects: exploring new horizons and traditional practices. In: Durst PB, Johnson DV, Leslie RL, Shono K. Forest Insects as Food: Humans Bite Back, Proceedings of a Workshop on Asia-Pacific Resources and Their Potential for Development. 2010: pp. 1–4. Bangkok: FAO Regional Office for Asia and the Pacific.

[19] De Foliart GR. Insects as human food. Crop Prot. 1992; 11:395–399.

[20] Fasoranti JO, Ajiboye DO. Some edible insects of K Wara State, Nigeria. Am. Entomol. 1993; 39(2):116.

[21] Jongema Y. List of Edible Insect Species of the World. 2012. Wageningen: Laboratory of Entomology, Wageningen University. (available at www.ent.wur.nl /UK/Edible +insects/Worldwide+species+list/).

[22] Srivastava JK, Gupta S. Health promoting benefits of chamomile in the elderly population. In: Watson Ronald R, editor. Complementary and Alternative Therapies in the Aging Population. 2009. Elsevier Inc, Academic Press.

[23] Tango M, Insect as human food, Published in 1981, reviewed in Food Insects Newslett, (1994); 7(3):3–4.

[24] DeFoliart GR. The Human Use of Insects as Food Resource: A Bibliographic Account in Progress. 2002. Wisconsin, USA: Department of Entomology, University of Wisconsin-Madison. (also available at: www.foodinsects.com/book7_31/The%20 Human %20 Use%20of%20Insects%20as%20a%20Food%20Resource.htm).

[25] Balinga MP, Mapunzu PM, Moussa JB. And N'gasse G, Contribution of forest insects to food security - The example of Central African caterpillars. Food and Agriculture Organization of the United Nations, Rome, Italy, 2004. 117 pp. Available at: http://www.fao.org/docrep/019/j3463e/j3463e.pdf

[26] Agbidye FS, Ofuya TI, Akindele SO. Marketability and nutritional qualities of some edible forest insects in Benue State. Nigeria. Pak. J. Nutr. 2009; 8:917–922. doi: 10.3923/pjn.2009.917.922.

[27] Chavunduka DM. Insects as source of food the African. Rhod. Sci. News. 1975; 9:217–220.

[28] Sirithon S, Pornpisanu T. Insects as a delicacy and a nutritious food in Thailand. Chapter 16 from Using Food Science and Technology to Improve Nutrition and Promote National Development. In: Robertson, GL, Lupien, JR (Eds). © International Union of Food Science & Technology. 2008.

[29] Rumpold BA, Schlüter OK. Mol nutritional composition and safety aspects of edible insects. Nutr. Food Res. 2013; 57(5):802–823. doi: 10.1002/mnfr.201200735.

[30] Olaf M, Michael K. Malnutrition and health in developing countries. CMAJ. 2005; 173(3):279–286.

[31] Mlcek J, Rop O, Borkovcova M, Bednarova M. A comprehensive look at the possibilities of edible insects as food in Europe – a review. Pol. J. Food Nutr. Sci. 2014; Vol. 64(3. pp. 147–157.

[32] Steinfeld H, Gerber P, Wassenaar T, Castel V, Rosales M, deHaan C. Livestock's Long Shadow, Environmental Issues and Options. 2006. 1st ed. Rome, Italy, cap: FAO. Livestock as a major player in global environmental issues, 24 p.

[33] Myers HM, Tomberlin JK, Lambert BD, Kattes D. Development of black soldier fl y (*Diptera: Stratiomyidae*) larvae fed dairy manure. Evniron. Entomol. 2008; 37:11–15.

[34] Diener S, Zurbrügg C, Tockner K. Conversion of organic material by black soldier fl y larvae: establishing optimal feeding rates. Waste Manag. Res. 2009; 27:603–610.

[35] Krause M, Mahan LM. Food, Nutrition and Diet Therapy. 2003. 11th ed. St. Louis, CA, USA: W. B. Saunders Co..

[36] Ramos-Elorduy J, Pino JM, Prado EE, Perez MA, Otero JL, deGuevara OL. Nutritional value of edible insects from the state of Oaxaca, Mexico. J. Food Compos. Anal. 1997; 10(2):142–57.

[37] FAO, 2012. FAOSTAT. Food and Agriculture Organization of the United Nations FAO. 2009. The State of Food and Agriculture: Livestock in the Balance. Rome: Food and Agriculture Organization of the United Nations (FAO).

[38] Alamu OT, Amao AO, Nwokedi CI, Oke OA, Lawa IO. Diversity and nutritional status of edible insects in Nigeria: a review. Int. J. Biodivers. Conserv. 2013; 5(4):215–222.

[39] Paoletti MG, Dufour DL. Edible invertebrates among Amazonian Indians: a critical review of disappearing knowledge. 2005. In: Paoletti MG, ed. Ecological Implications of Minilivestock. Enfield, NH, USA: Science Pub. 648 p.

[40] Xiaoming C, Ying F, Hong Z, Zhiyong C. Review of the nutritive value of edible insects. In Durst PB, Johnson DV, Leslie RL, Shono K, eds. Forest Insects as Food: Humans Bite

Back, Proceedings of a Workshop on Asia-Pacific Resources and Their Potential for Development. 2010. Bangkok: FAO Regional Office for Asia and the Pacific.

[41] Teffo LS, Toms RB, Eloff JN. Preliminary data on the nutritional composition of the edible stink-bug, Encosternum delegorguei Spinola, consumed in Limpopo province, South Africa. South Afr. J. Sci. 2007; 103:434–436.

[42] FAO. 2004. Contribution of forest insects to food security. The example of the Central African caterpillars (available at: www.fao.org/docrep/007/j3463e/j346 3f00.htm)

[43] Elemo BO, Elemo GN, Makinde MA, Erukainure OL. Chemical evaluation of African palm weevil, *Rhynchophorus phoenicis,* larvae as a food source. J. Insect Sci. 2011; 11:146. available online: insectscience.org/11.146.

[44] Ekpo KE, Onigbinde AO. Nutritional Potentials of the larva of *Rhynchophorus phoenicis* (F). Pak. J. Nutr. 2005; 4(5):287–290.

[45] Finke MD, Complete nutrient composition of commercially raised invertebrates used as food for insectivores. Zoo Biol. 2002; 21:269–285.

[46] Oonincx DGAB, Dierenfeld ES, An investigation into the chemical composition of alternative invertebrate prey. Zoo Biol. 2012; 31:40–54.

[47] Raksakantong P, Meeso N, Kubola J, Siriamornpun S. Fatty acids and proximate composition of eight Thai edible terricolous insects. Food Res. Int. 2010; 43:350–355.

[48] Burton OT, Zaccone P. The potential role of chitin in allergic reactions. Trends Immunol. 2007; 28:419–422.

[49] Sun L, Feng Y, He Z, Ma T, Zhang X. Studies on alkaline solution extraction of polysaccharide from silkworm pupa and its immunomodulating activities. Forest Res. 2007; 20:782–786.

[50] Lefevre M, Krisetherton PM, Zhao G, Tracy RP. Dietary fatty acids, hemostasis, and cardiovascular disease risk. J. Am. Diet Assoc. 2004; 104:410–419.

[51] Sirtori CR, Galli C. N-3 fatty acid and diabetes. Biomed. Phamacother. 2002; 56:397–406.

[52] Mori Y, Murakawa Y, Katoh S, et al. Influence of highly purified eicosapentaenoic acid ethyl ester on insulin resistance in the Otsuka Long-Evans Tokushima Fatty rat, a model of spontaneous non-insulin-dependent diabetes mellitus. Metabolism. 1997; 46:1458–1464.

[53] Nieuwenhuys CM, Hornstra G. The effects of purified eicosapentaenoic and docosahexaenoic acids on arterial thrombosis tendency and platelet function in rats. Biochim. Biophys. Acta. 1998; 1390:313–322.

[54] Asaf AQ, Saeed AS, Farooq AK. Effects of stabilized Rice bran, its soluble and fiber fractions on blood glucose levels and serum lipid parameters in humans with diabetes mellitus types I and II. J. Nutr. Biochem. 2002; 13:175–187.

[55] Womeni HM, Linder M, Tiencheu B, Mbiapo, FT, Villeneuve P, Fanni J, Parmentier, M. Oils of insects and larvae consumed in Africa: potential sources of polyunsaturated fatty acids. OCL – Oléagineux, Corps Gras, Lipides. 2009; 16(4):230–235.

[56] Michaelsen KF, Lauritzen L, Mortensen EL. Effects of breast-feeding on cognitive function, in Breast-feeding: early influences on later health. 2009. In: Goldberg G, Prentice A, Prentice A, Filteau S, Simondon K, eds. (Netherlands: Springer):199–215. 10.1007/978-1-4020-8749-3_15.

[57] Feng Y, Chen XM, Wang SY, Ye SD, Chen Y. The nutritive elements analysis of bamboo insect and review on its development and utilization value. Forest Res. 2000; 13:188–191.

[58] Stanley-Samuelson DW, Jurenka RA, Cripps C, Blomquist JG, De Renobales M, Loher W, Blomquist JG. Fatty acids in insects: composition, metabolism, and biological significance. J. Insect Biochem. Physiol. 1988; 9:1–33.

[59] Pennino M, Dierenfeld ES, Behler JL. Retinol, alpha-tocopherol and proximate nutrient composition of invertebrates used as feed. Int. Zoo Yearb. 1991; 30:143–149.

[60] Talwar GP, Srivastava LM, Mudgil KD. Textbook of Biochemistry and Human Biology. 1989. India: Prentice Hall of India Private Limited.

[61] Chaturvedi VC, Shrivastava R, Upreti RK. Viral infections and trace elements: a complex interaction. Curr. Sci. 2004; 87:1536–1554.

[62] Orr B. (Improvement of women's health linked to reducing widespread anemia. Int. Health News. 1986; 7:3.

[63] Saris NE, Mervaala E, Karppanen H, Khawaja JA, Lewenstam A. Magnesium: an update on physiological, clinical, and analytical aspects. Clin. Chim. Acta. 2000; 294:1–26.

[64] Ifie I, Emeruwa CH. Nutritional and anti-nutritional characteristics of the larva of *Oryctes monoceros*. Agric. Biol. J. N. Am. 2011; 2(1):42–46

[65] Paiko YB, Dauda BEN, Salau RB, Jacob JO. Preliminary data on the nutritional potentials of the larvae of edible dung beetle consumed in Paikoro Local Government Area of Niger State. Nigeria. Cont. J. Food Sci. Technol. 2012; 6(2): 38–42.

[66] Solomon M, Prisca N. Nutritive value of Lepidoptara litoralia (edible caterpillar) found in Jos Nigeria: implication for food security and poverty alleviation. Afr. J. Food Agric. Nutri. Develop. 2012; 12(6):6737–6747.

[67] Johnson DV. The contribution of edible forest insects to human nutrition and to forest management. Proceedings of a Workshop on Asia-Pacific Resources and their Potential for Development, 19–21 February 2008, Chiang Mai, Thailand. Bangkok. 2010. United Nations: Food Agriculture Organisation:5–22.

[68] Pimentel D, Livestock production and energy use. ENC Energ. 2004; 3:671–676.

[69] Ramos-Elorduy J. Energy supplied by edible insects from Mexico and their nutritional and ecological importance. Ecol. Food Nutr. 2008; 47:280–297.

[70] Pimentel D, Energy and land constraints in food production. Science. 1980; 190:754–761.

[71] Paoletti MG, Dreon AL. Minilivestock, environment, sustainability, and local knowledge disappearance. In: Paoletti MG, ed. Ecological Implications of Minilivestock. 2005. Enfield, NH, USA: Science Pub.: pp. 1–18. doi: 10.1038/ncomms5151.

[72] FAO. The State of Food and Agriculture: Livestock in the Balance. 2009. Rome: Food and Agriculture Organization of the United Nations (FAO).

[73] Yong-woo L. Silk Reeling and Testing Manual. 1999. Rome: Publishers, FAO Agric Services. No. (136). pp. 5–35. Iron, zinc, copper and manganese strengthen the immune system as antioxidant and cofactors of enzyme.

[74] Elvin CM, Carr AG, Huson MG, Maxwell JM, Pearson RD, Vuocolo T, Liyou NE, Wong DCC, Meritt DJ, Dixon NE. Synthesis and properties of crosslinked recombinant pro-resilin. Nature. 2005; 437:999–1002. doi: :10.1038/nature04085.

[75] Van Huis A, Insects as food in Sub-Saharan Africa. Insect science and its application. 2003; 23:163–185. doi: 0191-9040/03.

[76] Lewis SE. Insects of the Klondike Mountain Formation. 1992. Republic Washington: Washington geo; 20(3):15–19.

[77] West PC, Gerber JS, Engstrom PM, Mueller ND, Brauman KA, Carlson KM, Cassidy ES, Johnston M, MacDonald GK, Ray DK, Siebert S. Leverage points for improving global food security and the environment. Science. 2014; 345:325–328. DOI: 10.1126/science.1246067.